D1486198

Grundlehren der mathematischen Wissenschaften 244

A Series of Comprehensive Studies in Mathematics

Daniel S. Kubert
Serge Lang

Modular Units

Springer-Verlag
New York Heidelberg Berlin

AMS Subject Classifications: 10D99, 12A45

Library of Congress Cataloging in Publication Data

Kubert, Daniel S.
Modular units.
(Grundlehren der mathematischen Wissenschaften; 244)
Bibliography: p.
Includes index.
1. Algebraic number theory. 2. Class field theory. 3. Modules (Algebra) I. Lang, Serge, 1927– II. Title.
QA247.K83 512′.74 81-824 AACR2

Printed in the United States of America.

9 8 7 6 5 4 3 2 1

ISBN 0-387-90517-0 Springer-Verlag New York Heidelberg Berlin
ISBN 3-540-90517-0 Springer-Verlag Berlin Heidelberg New York

Contents

Introduction

In the present book, we have put together the basic theory of the units and cuspidal divisor class group in the modular function fields, developed over the past few years.

Let $\mathfrak{H}$ be the upper half plane, and N a positive integer. Let $\Gamma(N)$ be the subgroup of $SL_2(\mathbf{Z})$ consisting of those matrices $\equiv 1 \bmod N$. Then $\Gamma(N)\backslash\mathfrak{H}$ is complex analytic isomorphic to an affine curve $Y(N)$, whose compactification is called the modular curve $X(N)$. The affine ring of regular functions on $Y(N)$ over $\mathbf{C}$ is the integral closure of $\mathbf{C}[j]$ in the function field of $X(N)$ over $\mathbf{C}$. Here j is the classical modular function. However, for arithmetic applications, one considers the curve as defined over the cyclotomic field $\mathbf{Q}(\boldsymbol{\mu}_N)$ of N-th roots of unity, and one takes the integral closure either of $\mathbf{Q}[j]$ or $\mathbf{Z}[j]$, depending on how much arithmetic one wants to throw in.

The units in these rings consist of those modular functions which have no zeros or poles in the upper half plane. The points of $X(N)$ which lie at infinity, that is which do not correspond to points on the above affine set, are called the cusps, because of the way they look in a fundamental domain in the upper half plane. They generate a subgroup of the divisor class group, which turns out to be finite, and is called the cuspidal divisor class group. The trivial elements are represented precisely by the divisors of units as above, called modular units.

Investigation of units and divisor, or ideal, class groups is a classical activity. In the cyclotomic case, some basic problems remain open (Iwasawa-Leopoldt conjecture and Kummer-Vandiver conjecture—cf. *Cyclotomic Fields*). Both in this case, and the case of the cuspidal divisor class group, the class group is a module over a group ring $\mathbf{Z}[G]$, where $G \approx (\mathbf{Z}/N\mathbf{Z})^*$ in the cyclotomic case, and over a suitable Cartan group in the modular case. Contrary to the cyclotomic ideal class group, it is now possible to exhibit the cuspidal divisor class group as a cyclic module, since one can immediately identify the cusps with elements of the Cartan group. The kernel is then the analogue of the Stickelberger ideal, and corresponds to the divisors of units, mentioned above, which can be completely described. These units are the analogues of cyclotomic units.

The classical cyclotomic numbers $e^{2\pi i/N} - 1$ satisfy certain relations, arising from the identity

$$\prod_{\zeta^N = 1} (\zeta T - 1) = T^N - 1.$$

These relations can be axiomatized, and are called distribution relations. In Chapter 1, we have summarized this algebraic theory, independent of all applications. We emphasize the universal properties and the general algebraic structure, as distinguished from the p-adic properties which were emphasized by Iwasawa and Mazur. The general theory has applications going beyond the present book, and to other fields.

It is worthwhile here to review some facts from the theory of cyclotomic units in light of our present approach. For $x \in \mathbf{Q}/\mathbf{Z}$, $x \neq 0$, we define

$$g(x) = e^{2\pi i x} - 1.$$

We view the map

$$x \mapsto g(x)$$

as a map $\mathbf{Q}/\mathbf{Z} - \{0\} \to \mathbf{C}^*$, that is into the multiplicative group of complex numbers. We call the numbers $g(x)$ cyclotomic numbers. They have the following properties.

(1) If the denominator of x is composite, then $g(x)$ is a unit, in the sense of Dirichlet. If the denominator of x is a prime power p^n, then $g(x)$ is a p-unit.

(2) The map satisfies the distribution relation, that is for any positive integer N, we have

$$\prod_{Ny = x} g(y) = g(x).$$

Furthermore, the distribution g is essentially even, that is

$$g(-x) = g(x)\zeta$$

for some root of unity ζ. If we view its values as lying in $\mathbf{C}^*/\boldsymbol{\mu}$ (where $\boldsymbol{\mu}$ is the group of all roots of unity) then it is an even distribution.

(3) Up to 2-torsion, g is the universal even distribution. (Cf. [Bass].)

(4) The group $\mathbf{Z}(N)^* = (\mathbf{Z}/N\mathbf{Z})^*$ operates by multiplication on $(1/N)\mathbf{Z}/\mathbf{Z}$. For $a \in \mathbf{Z}(N)^*$, let σ_a be the corresponding automorphism of $\mathbf{Q}(\boldsymbol{\mu}_N)/\mathbf{Q}$, and let $G(N) = \text{Gal}(\mathbf{Q}(\boldsymbol{\mu}_N)/\mathbf{Q})$. Then for $x \in (1/N)\mathbf{Z}/\mathbf{Z}$ we have

$$g(x)^{\sigma_a} = g(ax).$$

(5) The index of the cyclotomic units in the group of all units is essentially the class number of the real subfield. It is equal to that class number in the case of prime power level.

The study of the universality of the constructed units involves both determining the rank and the torsion. The rank is obtained by establishing an isomorphism between the given distribution (tensored with $\mathbf{Q}$) and a "Stickelberger distribution" using regulator maps. We decompose the image of the distribution into eigenspaces for the group $G(N)$ mentioned above, and find that each eigenspace is generated by a certain character sum. The contribution of this eigenspace to the rank then depends on whether this character sum is zero or not.

We shall find that the above pattern is repeated in several contexts. In this book, we study it in the context of modular units in the modular function field, and their specializations to the complex multiplication case.

Let F_N denote the function field of $X(N)$ over $\mathbf{Q}(\boldsymbol{\mu}_N)$ and let F_∞ be the union of all F_N. Elements of F_N are said to be of level N. In Chapter 2, we exhibit a group of units in F_∞ which fits the above pattern. After describing their formal properties and their universality in the light of Chapter 1, we show that they generate all units, and determine precisely which ones have a given level. This allows us to describe precisely the group of cuspidal divisor classes as a group ring modulo a certain ideal, and to analyze precisely the order of this group by computing the index of that ideal. In the present book, we limit the computation of the order to the prime power level. The general case has to use more complicated combinatorial methods due to Sinnott in the cyclotomic case, and Kubert for the general Cartan group. Cf. Yu's thesis.

The situation here is not merely analogous to that of the cyclotomic case, but they influence each other in ways which are now only beginning to be investigated. For instance, Wiles has introduced Mazur's Eisenstein ideal into the picture relating ideal class groups in the cyclotomic case with cuspidal divisor class groups, cf. also Mazur–Wiles. In this book, no direct connection will be made with the cyclotomic case.

After the study of the generic case, we are then concerned with specializations, which can occur essentially in three contexts: non-integral j-invariants, j-invariants with complex multiplication, integral j-invariants without complex multiplication. The case of non-integral j-invariants can be studied in a manner fairly close to that of the generic case by use the Tate parametrization. We give results in this direction in Chapter 7.

We continue in Chapter 8 with some applications to Diophantine analysis. The fact that the cusps of $X(N)$ are of finite order in the Jacobian had originally been proved by Manin and Drinfeld. (However, their proof, using Hecke operators, did not give an explicit bound for the order of the cusps. This came only from the specific nature of the modular units.) Such a phenomenon is always remarkable, because the Manin-Mumford conjecture asserts that on a curve of genus $\geqq 2$, there is only a finite number of points which are torsion points in the Jacobian. This conjecture is known to be true in the case of complex multiplication, cf. Chapter 8, §3.

We also give some applications as in Demjanenko and Kubert, concerning the uniform boundedness of the order of torsion points, and the application to a direct proof for the finiteness of integral points on modular curves, following the Gelfond idea, complemented by Baker's inequalities on linear combinations of logarithms of algebraic numbers.

We then deal with the application of the generic theory to the construction of units in complex multiplication fields, in the manner of Siegel, Ramachandra, and especially Robert. In particular, we recover the results of Robert, giving the index of a special group of units, used for instance by Coates-Wiles in their work on the Birch-Swinnerton-Dyer conjecture. However, since we have a more flexible and complete theory of the generic units, especially involving the Klein forms, it is possible to define a group bigger than Robert's group. This allows for a more refined index computation due to Kersey, some of whose results are included here for the first time. He collaborated on Chapter 12, and Chapter 13 is entirely due to him. In particular, he succeeded in eliminating all extraneous factors from the index of the modular units in the group of all units.

We conclude the book with an appendix giving the periods of differentials of third kind associated with the Siegel functions, and the associated Dedekind-type sums. This appears isolated in the present context, but is used elsewhere to determine more precisely the 2-torsion in the cuspidal divisor class group, and various integrality properties in several contexts. The section is included for the convenience of the reader, for easy reference, and as an introduction to other applications.

We are indebted to G. Ligozat and G. Robert for a careful reading of the manuscript (the function field part and the complex multiplication part respectively), and for a number of useful suggestions and corrections.

D. Kubert
S. Lang

Notation

If R is a ring and $\mathfrak{a}$ an ideal we let $R(\mathfrak{a}) = R/\mathfrak{a}$. Thus $\mathbf{Z}(N) = \mathbf{Z}/N\mathbf{Z}$ and $\mathbf{Z}(N)^*$ is the group of invertible elements in $\mathbf{Z}/N\mathbf{Z}$. This notation is compatible with viewing elements of R as functions on the set of ideals, with values in the residue class rings. For instance, for a prime ideal $\mathfrak{p}$, and an element $x \in R$, $x(\mathfrak{p})$ is the value of x at $\mathfrak{p}$, viewing $\mathfrak{p}$ as a point in spec R.

$\boldsymbol{\mu}$ denotes the group of all roots of unity.
$\boldsymbol{\mu}_N$ = subgroup of elements of order N in $\boldsymbol{\mu}$.
$\boldsymbol{\mu}_F$ = subgroup of elements of $\boldsymbol{\mu}$ lying in a field F.
F^{ab} = maximal abelian extension of a field F.
F^{a} = algebraic closure of F.
$|X|$ = cardinality of a set X.

If A is an abelian group then $A^{(p)}$ denotes its p-primary part (subgroups of elements annihilated by a power of p).

CHAPTER 1

Distributions on Toroidal Groups

In recent years, it has become clear that the notion of distribution is playing a central role in certain aspects of number theory. "Distribution" is here taken in the following sense: in the simplest case, it is a function

$$\varphi : \mathbf{Q}/\mathbf{Z} \to \text{abelian group},$$

satisfying the relations

$$\sum_{\nu=0}^{N-1} \varphi\left(x + \frac{\nu}{N}\right) = \varphi(Nx)$$

for all $x \in \mathbf{Q}/\mathbf{Z}$ and all positive integers N. Historically, such relations occur in classical analysis, without receiving a name, or without their general structure being analyzed independently. A version occurs more recently in the context of p-adic integration theory as offshoot of Iwasawa theory, as defined by Mazur on projective systems [M-SwD], p. 36. Our point of view is quite different. First distributions occur on injective systems as above; and second they are studied from the point of view of their universal properties rather than their p-adic congruence properties (cf. [KL 3] and [KL 5], [Ku 3] and [Ku 4]). For distributions in the theory of cyclotomic fields, cf. for instance [L 8], Chapter 2, and the bibliography given there.

We carry out the basic theory on $\mathbf{Q}^k/\mathbf{Z}^k$ and also introduce the Cartan group which makes the theory look entirely analogous to that on $\mathbf{Q}/\mathbf{Z}$, with groups of automorphisms analogous to the groups $\mathbf{Z}(N)^*$, familiar from cyclotomic theory. For the applications to modular functions, we shall deal principally with the case $k = 2$.

§1. The Cartan Group

Let k be a positive integer. In the applications to modular units, for the most part $k = 2$. Let p be a prime number, and let $\mathfrak{o}_p$ be the ring of integers in the unramified extension of $\mathbf{Q}_p$ of degree k. The units $\mathfrak{o}_p^*$ operate on $\mathfrak{o}_p$ by multiplication, and if we select a basis of $\mathfrak{o}_p$ over $\mathbf{Z}_p$, then we obtain a natural embedding

$$\mathfrak{o}_p^* \to \mathrm{GL}_k(\mathbf{Z}_p).$$

We call either $\mathfrak{o}_p^*$ or its image in $\mathrm{GL}_k(\mathbf{Z}_p)$ the **Cartan group** at the prime p, and sometimes denote it by C_p.

The elements of $\mathfrak{o}_p^*$ in $\mathfrak{o}_p$ are characterized by the fact that when we write them in terms of a basis over $\mathbf{Z}_p$, then at least one coefficient is a unit.

For each positive integer n we let

$$\mathfrak{o}(p^n) = \mathfrak{o}_p(p^n) = \mathfrak{o}_p/p^n\mathfrak{o}_p.$$

Then $\mathfrak{o}_p(p^n)^* \approx C(p^n)$, where $C(p^n)$ is the reduction of C_p mod p^n.

We have an isomorphism

$$\frac{1}{p^n}\mathfrak{o}_p/\mathfrak{o}_p \approx \mathfrak{o}(p^n)$$

given by multiplication with p^n. Via this isomorphism, the Cartan group $C(p^n)$ operates on the $\mathbf{Z}_p$-module

$$Z_{p^n} = \frac{1}{p^n}\mathfrak{o}_p/\mathfrak{o}_p,$$

and operates simply transitively on the set of primitive elements

$$Z_{p^n}^* = \left(\frac{1}{p^n}\mathfrak{o}_p/\mathfrak{o}_p\right)^*,$$

i.e., those elements which have precise order p^n in the additive group Z_{p^n}.

Now let N be an integer > 1 and let

$$N = \prod p^{n(p)}$$

be its prime factorization. We let

$$\mathfrak{o}_N = \prod_{p|N} \mathfrak{o}_p \quad \text{and} \quad \mathfrak{o}_N^* = \prod_{p|N} \mathfrak{o}_p^*.$$

We may then form

$$\mathfrak{o}_N(N) = \mathfrak{o}_N/N\mathfrak{o}_N = \prod \mathfrak{o}_p(p^{n(p)}),$$

and similarly for the units $\mathfrak{o}_N(N)^*$. We have an isomorphism

$$Z_N = \frac{1}{N}\mathfrak{o}_N/\mathfrak{o}_N \approx \mathfrak{o}_N(N) = \mathfrak{o}_N/N\mathfrak{o}_N.$$

The composite Cartan group is then

$$C_N = \mathfrak{o}_N^*,$$

and admits an embedding in $\mathrm{GL}_k(\mathbf{Z}_N)$,

$$C_N \to \mathrm{GL}_k(\mathbf{Z}_N) \quad \text{where } \mathbf{Z}_N = \prod_{p|N} \mathbf{Z}_p,$$

after a basis of $\mathfrak{o}_N$ over $\mathbf{Z}_N$ has been chosen. The Cartan group C_N may be reduced mod N, to obtain

$$C(N) = \mathfrak{o}_N(N)^* = \prod_{p|N} C(p^{n(p)}).$$

The group $C(N)$ operates simply transitively on the set of **primitive elements**

$$Z_N^* = \left(\frac{1}{N}\mathfrak{o}_N/\mathfrak{o}_N\right)^*.$$

These are the elements which have period exactly N in Z_N.

For each p let $\mathbf{K}_p$ be the quotient field of $\mathfrak{o}_p$. Define

$$\mathbf{K} = \bigoplus_p \mathbf{K}_p, \qquad \mathfrak{o} = \bigoplus_p \mathfrak{o}_p.$$

Then we can also use the notation

$$Z_N = (\mathbf{K}/\mathfrak{o})_N.$$

This is the group of elements of period dividing N in $\mathbf{K}/\mathfrak{o}$.

Suppose $M|N$. Then we have a commutative diagram

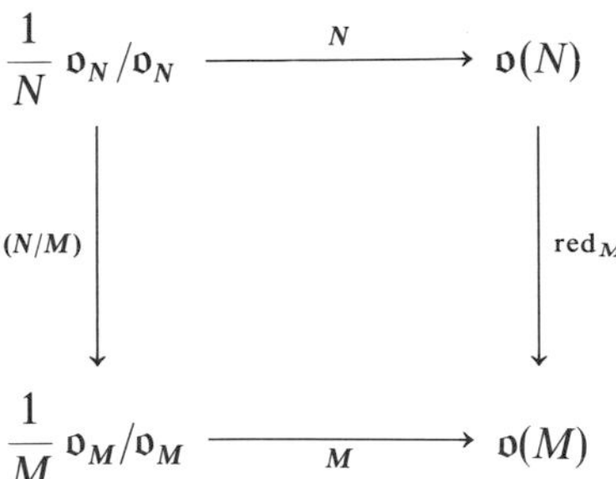

The horizontal maps are given by multiplication with N and M respectively. The left vertical map is multiplication by N/M, and the right vertical map is reduction mod M.

Under the vertical maps, we have two projective systems which are isomorphic to each other. They are the basic projective systems which will arise constantly in this book.

Remark. Only the unramified Cartan group will have applications in this book. Of course, one can also consider the general algebraic theory (independently of modular forms) for the other Cartan groups. We take for $\mathfrak{o}_p$ the ring of integers in an arbitrary extension of $\mathbf{Q}_p$ of degree k. Relative to a basis of $\mathfrak{o}_p$ over $\mathbf{Z}_p$, we get an embedding of $\mathfrak{o}_p^*$ into $\mathrm{GL}_k(\mathbf{Z}_p)$. Let $e_1 = {}^t(1, 0, \ldots, 0)$ be the first vertical unit vector, and let $G_{p,\infty}$ be the isotropy group. Then there is product decomposition

$$\mathrm{GL}_k(\mathbf{Z}_p) = C_p \cdot G_{p,\infty}$$

valid for any Cartan, and which will be interpreted in terms of modular functions for $k = 2$ and the unramified Cartan in Chapter 2, §3.

§2. Distributions

Instead of forming the projective limit, we may also form the injective limit

$$\frac{1}{N^\infty}\mathfrak{o}_N/\mathfrak{o}_N = \bigoplus_{p|N} \frac{1}{p^\infty}\mathfrak{o}_p/\mathfrak{o}_p.$$

If we select a basis for $\mathfrak{o}_N$ over $\mathbf{Z}_N$, then we obtain an isomorphism

$$\frac{1}{N^\infty}\mathfrak{o}_N/\mathfrak{o}_N \approx (\mathbf{Q}/\mathbf{Z})^k[N^\infty].$$

The right hand side is by definition the subgroup of elements in $\mathbf{Q}^k/\mathbf{Z}^k$ annihilated by a power of N. Although it is often convenient to consider this injective limit, it is necessary to give the theory also at the finite levels N. At a given level N, we have an isomorphism

$$\frac{1}{N}\mathfrak{o}_N/\mathfrak{o}_N \approx \frac{1}{N}\mathbf{Z}^k/\mathbf{Z}^k.$$

In the applications, a basis has been fixed, and we identify these two groups.

We first fix N and let $\mathfrak{o} = \mathfrak{o}_N$. Let w be a homomorphism from the multiplicative semi group generated by the primes dividing N into a semi group operating on an abelian group A. Let

$$h: \frac{1}{N}\mathfrak{o}/\mathfrak{o} \to A$$

be a function into A. We say that h is a **distribution of weight** w if for each divisor M of N and each element $x \in Z_M = (1/M)\mathfrak{o}/\mathfrak{o}$ we have

$$\boxed{w(N/M) \sum_{(N/M)y = x} h(y) = h(x).}$$

The sum on the left is taken over all elements $y \in Z_N$ such that $(N/M)y = x$. If y_0 is one such element, then all the others are of the form

$$y_0 + \frac{v}{D} \quad \text{with } v \in \mathfrak{o}(D),\ D = N/M.$$

If we **define $\mathbf{h_M}$** by

$$h_M(b) = w(M)h\left(\frac{b}{M}\right) \quad \text{for } b \in \mathfrak{o}(M),$$

then we may rewrite the distribution relation as

$$\boxed{\sum_{a \equiv b(M)} h_N(a) = h_M(b)}$$

for $a \in \mathfrak{o}(N)$ and $b \in \mathfrak{o}(M)$. This is the corresponding relation on the projective system of groups $\mathfrak{o}(M)$ for $M|N$, whereas the other relation was on the injective system of groups Z_M for $M|N$. Observe that on the projective system, the relation occurs without weight.

Let f be a function on $\mathfrak{o}(N)$. We define

$$S_N(f_N, h_N) = \sum_{a \in \mathfrak{o}(N)} f(a)h_N(a).$$

In practice, the values of f and h lie in a ring so the product $f(a)h_N(a)$ is defined. [It is suggestive to write the above sum in the form of an integral

$$\int f\, dh,$$

but this aspect of the question will not enter into our considerations.] What is important for us is that the value of the sum is independent of the level in the following sense. Suppose that $f = f_N$ factors through the reduction mod M, i.e. factors through a function f_M on $\mathfrak{o}(M)$. Then we have

$$S_N(f_N, h_N) = S_M(f_M, h_M).$$

This is immediate from the distribution relation.

If χ is a character on $C(N)$, we define as usual the function χ_N to be the same as χ on $C(N)$ but 0 on elements of $\mathfrak{o}(N)$ which are not prime to N. We suppose now that h takes its values in a module over the ring generated by the values of χ over the integers. We also assume that multiplication by elements of this ring commutes with multiplication by elements $w(p)$ for $p|N$. All of this is trivially satisfied in practice, when the distribution has values in a field and all the operators come from elements of the multiplicative group. The reader may think of this case all the way through, but the computations are purely formal, and we might as well lay the foundations in some generality.

Theorem 2.1. *Let χ be a character on $C(N)$ with conductor m, so χ factors through χ_m on $C(m)$. Then*

$$S_N(\chi_N, h_N) = \prod_{\substack{p|N \\ p \nmid m}} (1 - \chi_m(p)w(p))S_m(\chi_m, h_m).$$

Proof. Let $M|N$ and suppose first M, N have the same prime factors. Then the equality

$$S_N(\chi_N, h_N) = S_M(\chi_M, h_M)$$

is simply a special case of the distribution relation, and of the formula which we have already observed for an arbitrary function on $\mathfrak{o}(N)$, factoring through $\mathfrak{o}(M)$.

Suppose now that $N = pM$ and $p \nmid M$. It will suffice to prove that

$$S_N(\chi_N, h_N) = (1 - \chi(p)w(p))S_M(\chi_M, h_M).$$

We have

$$\begin{aligned} S_N(\chi_N, h_N) &= \sum_{a \in \mathfrak{o}(N)} \chi_N(a)h_N(a) \\ &= \sum_{b \in \mathfrak{o}(M)} \sum_{\substack{x \in \mathfrak{o}(N) \\ x \equiv b(M)}} \chi_N(x)h_N(x). \end{aligned}$$

Now $\chi_N(x) = \chi_M(b)$ for $x \equiv b \bmod M$ and $(x, p) = 1$, whereas $\chi_N(x) = 0$ if $x = pc$ for some c. Hence

$$S_N(\chi_N, h_N) = \sum_{b \in \mathfrak{o}(M)} \chi_M(b) \sum_{\substack{x \in \mathfrak{o}(N) \\ x \equiv b(M)}} h_N(x) - \sum_{\substack{a \in \mathfrak{o}(N) \\ a = pc}} \chi_M(a)h_N(a).$$

In the last sum we make the change of variables $a = pc$ where $c \in \mathfrak{o}(M)$, to get

$$\begin{aligned} \sum_{c \in \mathfrak{o}(M)} \chi_M(pc)h_N(pc) &= \chi_M(p)w(p) \sum_{c \in \mathfrak{o}(M)} \chi_M(c)w(M)h\left(\frac{c}{M}\right) \\ &= \chi_M(p)w(p)S_M(\chi_M, h_M). \end{aligned}$$

This proves the theorem.

Remark 1. In applications, one is interested in knowing whether $S_N(\chi_N, h_N) = 0$. If the weight is a number of absolute value 1, and in particular, if the weight is 1, then the factor in front on the right hand side may be 0. If the weight is 1, then this factor is 0 if and only if $\chi_m(p) = 1$. On the other hand, if the weight has absolute value $\neq 1$, then this factor is $\neq 0$, and the nonvanishing is reduced to the absolute sum $S_m(\chi_m, h_m)$, taken at the level of the character. This sum will be denoted without indices, that is

$$S(\chi, h) = S_m(\chi_m, h_m).$$

Remark 2. Throughout we have worked invariantly using distributions defined on $(1/N)\mathfrak{o}/\mathfrak{o}$. Since we have an isomorphism

$$\frac{1}{N}\mathfrak{o}/\mathfrak{o} \approx \frac{1}{N}\mathbf{Z}^k/\mathbf{Z}^k$$

everything we have said may be transposed to distributions defined on $(1/N)\mathbf{Z}^k/\mathbf{Z}^k$ by the same formula that we have used. In the application to modular units, this will be done with $k = 2$.

§3. Stickelberger Distributions

We wish to take the group ring of $C(N)$. For this it is convenient to use a group, denoted by $G(N)$, isomorphic to $C(N)$ under a map

$$a \mapsto \sigma_a.$$

One may think of $G(N)$ as a Galois group, analogous to the Galois group of N-th roots of unity over the rationals. In applications, such a group does indeed arise as a Galois group of some covering or other. If $k = 1$, then $C(N) = \mathbf{Z}(N)^*$.

Let h be a distribution on $Z_N = (1/N)\mathfrak{o}/\mathfrak{o}$, with values in an algebraically closed field F of characteristic 0. We define the **Stickelberger distribution associated with** h by the formula

$$\mathrm{St}_h(x) = \sum_{a \in C(N)} h(ax)\sigma_a^{-1}.$$

Since the mapping $x \mapsto ax$ is an automorphism of Z_N, each coordinate function $h(ax)$ is a distribution, so St_h is a distribution. Note that St_h has values in the group algebra $F[G(N)]$. By the **F-rank** of St_h we mean the dimension of the F-subspace generated by the values $\mathrm{St}_h(x)$ for $x \in Z_N$. We shall analyze this rank in terms of eigenspaces for the action of $G(N)$.

Let $M | N$. We have a natural embedding of the group spaces (as vector spaces)

$$F[G(M)] \to F[G(N)]$$

obtained by defining

$$\sigma_b \mapsto \sum_{a \equiv b(M)} \sigma_a,$$

and extending to the group space by linearity. Thus each element σ_b with $b \in C(M)$ is mapped on the formal sum of elements in $G(N)$ which map on σ_b under the reduction mod M.

We observe that if $x \in Z_M$ then the element

$$\mathrm{St}_{M,h}(x) = \sum_{b \in C(M)} h(bx)\sigma_b^{-1}$$

when injected in $F[G(N)]$ is precisely $\mathrm{St}_h(x)$ as defined previously. Thus the distribution St_h, which a priori should be indexed by N, actually need not be if we view its values as lying in the injective limit

$$\lim_{\substack{\rightarrow \\ N}} F[G(N)].$$

If h is a distribution defined on all of $\mathbf{K}/\mathfrak{o}$, then St_h is also defined on $\mathbf{K}/\mathfrak{o}$ with values in this injective limit.

For each character χ of $C(N) \approx G(N)$ we have the usual idempotent

$$e_\chi = \frac{1}{|C(N)|} \sum_{a \in C(N)} \bar{\chi}(a)\sigma_a.$$

We then have the following properties. We write $\mathrm{St}_{N,h} = \mathrm{St}_h$, and we let M be a divisor of N. We let cond χ be the conductor of χ.

ST 1. *If* cond χ *does not divide* M, *then*

$$\mathrm{St}_h\left(\frac{1}{M}\right)e_\chi = 0.$$

ST 2. *If* cond χ *divides* M *and has the same prime factors as* M, *then*

$$\mathrm{St}_h\left(\frac{1}{M}\right)e_\chi = S(\bar{\chi}, h)w(M)^{-1}(C(N) : C(M))e_\chi.$$

ST 3. *If* cond $\chi = m$, *and* $m \mid M$, *then*

$$\mathrm{St}_h\left(\frac{1}{M}\right)e_\chi = S(\bar{\chi}, h) \prod_{\substack{p \mid M \\ p \nmid m}} (1 - \chi_m(p)w(p))w(M)^{-1}(C(N) : C(M))e_\chi.$$

Proofs. We have

$$\begin{aligned}\mathrm{St}_h\left(\frac{1}{M}\right)e_\chi &= \sum_{a \in C(N)} h\left(\frac{a}{M}\right)\bar{\chi}(a)e_\chi \\ &= \sum_{b \in C(M)} h\left(\frac{b}{M}\right) \sum_{a \equiv b(M)} \bar{\chi}(a)e_\chi.\end{aligned}$$

If the conductor of χ does not divide M, then χ is non-trivial on the kernel of the reduction map $C(N) \to C(M)$, and consequently the sum on the right is 0, thus proving **ST 1**.

Suppose that the conductor of χ divides M. Then

$$\bar{\chi}(a) = \bar{\chi}(b)$$

on the right, so

$$\begin{aligned}\mathrm{St}_h\left(\frac{1}{M}\right)e_\chi &= \sum_{b\in C(M)} h\left(\frac{b}{M}\right)\bar{\chi}(b)(C(N):C(M))e_\chi \\ &= S_M(\bar{\chi}_M, h_M)w(M)^{-1}(C(N):C(M))e_\chi.\end{aligned}$$

The last assertions follow from Theorem 2.1.

The values at elements other than $1/M$ are immediately derived from the previous ones as follows.

ST 4.
$$\mathrm{St}_h\left(\frac{c}{M}\right)e_\chi = \chi(c)\mathrm{St}_h\left(\frac{1}{M}\right)e_\chi.$$

This comes from changing variables $a \mapsto c^{-1}a$ in the definition of $\mathrm{St}_h(c/M)$. It shows that to determine the image of the Stickelberger distribution as a vector space, the essential features are contained in the values $\mathrm{St}_h(1/M)$. Precisely:

Theorem 3.1. *Let A_N be the vector space generated over F by the image of St_h. Then the χ-eigenspace $A_N(\chi)$ is generated over F by the single element*

$$S(\bar{\chi}, h),$$

and consequently has dimension 0 or 1 according as this element is $= 0$ or $\neq 0$. The F-rank of the Stickelberger distribution is equal to the number of characters χ such that $S(\bar{\chi}, h) \neq 0$.

Since distributions are defined on an abelian group, it makes sense to say that they are even or odd, in other words

$$h(x) = h(-x) \quad \text{for } h \text{ even} \quad \text{and} \quad h(-x) = -h(x) \quad \text{for} \quad h \text{ odd}.$$

If h is even, then its associated Stickelberger distribution is even, and similarly for the odd case. It is also clear from the definition of the sum that unless h

and χ have the same parity (i.e. both even or both odd), then

$$S(\chi, h) = 0.$$

Hence dealing with an even (resp. odd) distribution immediately eliminates half the possibilities for the rank, unless $N = 2$, $k = 1$, in which case every distribution is even.

§4. Lifting Distributions from Q/Z

If the weight of a distribution is of the form $w(N) = N^s$ for some number s, then we say that the distribution is of **degree** s. These are the weights which will arise in the applications. Specifically, we shall deal with distributions on $\mathbf{Q}/\mathbf{Z}$ of degree $k - 1$ and lift these to distributions of degree 0 on $\mathbf{K}/\mathfrak{o}$. This is done by means of an auxiliary mapping as follows.

Let

$$T: \prod_p \mathfrak{o}_p \to \prod_p \mathbf{Z}_p$$

be a surjective homomorphism. We assume that T maps each factor $\mathfrak{o}_p$ onto $\mathbf{Z}_p$, and that each local homomorphism is $\mathbf{Z}_p$-linear. For instance, the trace is natural, but in special cases, when a basis is selected, others also arise naturally.

For each N we obtain a corresponding homomorphism

$$\mathfrak{o}_N \to \mathbf{Z}_N$$

and also

$$T_N : \mathfrak{o}(N) \to \mathbf{Z}(N).$$

If $x \in (K/\mathfrak{o})_N$ then $Nx \in \mathfrak{o}(N)$ and $T(Nx) \in \mathbf{Z}(N)$. Then

$$\frac{T(Nx)}{N} \in \frac{1}{N}\,\mathfrak{o}/\mathfrak{o}.$$

By abuse of notation, we shall write

$$T(x) = \frac{T(Nx)}{N}.$$

We may then form the composite function $h \circ T$, defined on $(1/N)\mathfrak{o}/\mathfrak{o}$.

Theorem 4.1. *Let h be a distribution of weight w on $(\mathbf{Q}/\mathbf{Z})_N$. Then the function*

$$h \circ T(x) = h\left(\frac{T(Nx)}{N}\right) \quad \textit{for } x \in (\mathbf{K}/\mathfrak{o})_N$$

is a distribution of weight $w(N)N^{1-k}$. In particular, if $w(N) = N^{k-1}$ then $h \circ T$ is a distribution of weight 1.

Proof. It is convenient to write the function $h \circ T$ also on the projective system in the form

$$h\left(\frac{Ta}{N}\right) \quad \text{with } a \in \mathfrak{o}(N).$$

The formula to be proved then reads

$$(N/M)^{1-k} \sum_{\substack{a \in \mathfrak{o}(N) \\ a \equiv b(M)}} h_N \circ T_N(a) = h_M T_M(b)$$

with b fixed in $\mathfrak{o}(M)$. As in §2, $h_N(r) = w(N)h(r/N)$. But we have

$$w(N) \sum_{\substack{a \in \mathfrak{o}(N) \\ a \equiv b(M)}} h\left(\frac{Ta}{N}\right) = w(N) \sum_{\substack{r \in \mathbf{Z}(N) \\ r \equiv T_M b(M)}} h\left(\frac{r}{N}\right) t_N(r),$$

where $t_N(r)$ is the number of elements $a \in \mathfrak{o}(N)$ such that

$$T_N a = r \quad \text{and} \quad a \equiv b \bmod M.$$

A simple argument using elementary divisors shows that

$$t_N(r) = (N/M)^{k-1}.$$

If we now use the distribution relation on h on the right hand side, we find precisely the desired expression to prove the theorem.

The distribution $h \circ T$ will be called the **lifted distribution** of h by T.

§5. Bernoulli-Cartan Numbers

For each real number t, we let $\{t\}$ be the representative of t mod $\mathbf{Z}$ such that

$$0 < \{t\} \leqq 1$$

and we let $\langle t \rangle$ be the representative such that

$$0 \leqq \langle t \rangle < 1.$$

If t is not an integer, then the two values coincide. The first value is adjusted to fit the formalism of the **Hurwitz zeta function**, which is defined for a real number u with $0 < u \leqq 1$ by

$$\zeta(s, u) = \sum_{n=0}^{\infty} \frac{1}{(n+u)^s}.$$

It is verified at once that the function

$$t \mapsto \zeta(s, \{t\})$$

is a distribution of degree $-s$ on $\mathbf{Q}/\mathbf{Z}$ (even on $\mathbf{R}/\mathbf{Z}$). Making the notation explicit, this means that for every positive integer D and $r \in \mathbf{R}/\mathbf{Z}$ we have

$$\text{(1)} \qquad D^{-s} \sum_{Dt=r} \zeta(s, \{t\}) = \zeta(s, \{r\}).$$

If T is a given mapping as in the preceding section, then we have the corresponding distribution on $Z_N = (\mathbf{K}/\mathfrak{o})_N$. Making Theorem 4.1 explicit in the present case yields the distribution relation for $M|N$:

$$\text{(2)} \qquad N^{-s-(k-1)} \sum_{\substack{y \in \mathfrak{o}(N) \\ y \equiv x(M)}} \zeta\left(s, \left\{\frac{Ty}{N}\right\}\right) = M^{-s-(k-1)} \zeta\left(s, \left\{\frac{Tx}{M}\right\}\right).$$

We define the **L-series** for a character $\chi = \chi_N$ on $C(N)$:

$$L_N(s, \chi_N, T_N) = N^{-s-(k-1)} \sum_{\alpha \in C(N)} \chi(\alpha) \zeta\left(s, \left\{\frac{T\alpha}{N}\right\}\right).$$

Proposition 5.1. *Let $M|N$ and suppose χ factors through $C(M)$. Then*

$$L_N(s, \chi_N, T_N) = \prod_{\substack{p|N \\ p \nmid M}} \left(1 - \frac{\chi_M(p)}{p^{s+k-1}}\right) L_M(s, \chi_M, T_M).$$

Proof. Since the L-series is merely the sum defined abstractly in §2 for any distribution, the proposition follows from Theorem 2.1.

Let χ be a character on $C(N)$ and let M be its conductor. We define (without subscript)

$$L(s, \chi, T) = L_M(s, \chi_M, T_M).$$

Proposition 5.1 gives us the reduction of the value of the L-series to the case when χ is primitive.

Let m be an integer $\geqq 1$. We have

$$L_N(1 - m, \chi_N, T_N) = N^{m-k} \sum_{\alpha \in C(N)} \chi(\alpha)\zeta\left(1 - m, \left\{\frac{T\alpha}{N}\right\}\right),$$

and by a classical theorem of Hurwitz, cf. for instance [L 7], Chapter XIV, §1, we have

$$\zeta(1 - m, u) = -\frac{1}{m}\mathbf{B}_m(u),$$

where $\mathbf{B}_m$ is the m-th Bernoulli polynomial. Hence the value of $L(1 - m, \chi, T)$ may be viewed as the appropriate **Bernoulli number** with respect to the Cartan group under consideration, i.e.

$$\boxed{L(1 - m, \chi, T) = -\frac{1}{m}B_{m,\chi},}$$

where, assuming χ primitive, we define

$$\boxed{B_{m,\chi} = B_{m,\chi,k,T} = N^{m-k} \sum_{\alpha \in C(N)} \chi(\alpha)\mathbf{B}_m\left(\left\{\frac{T\alpha}{N}\right\}\right).}$$

The dependence on T is almost nil. As $\mathbf{Z}(N)$-module, it is standard that $\mathfrak{o}(N)$ is self dual, so if T' is another surjective $\mathbf{Z}(N)$-linear map, then there exists $\beta \in \mathfrak{o}(N)^*$ such that

$$T'(x) = T(\beta x).$$

Making a change of variables in the sum, we see that the Bernoulli numbers formed with T' instead of T differ by a factor $\bar{\chi}(\beta)$, which is a root of unity, of no consequence for the applications which deal either with the non-vanishing, or with the absolute value (ordinary or p-adic).

Remark. In some applications, it is useful to distinguish cases depending on parity, and to work with $C(N)/\pm 1$ whenever k is even. The reader should check each time when using the general Bernoulli numbers which normalization has been taken.

Assume that χ is primitive, so has conductor N. Write

$$\begin{aligned} L(s, \chi, T) &= N^{-s-(k-1)} \sum_{x=1}^{N} f_\chi(x)\zeta\left(s, \frac{x}{N}\right) \\ &= N^{1-k}\zeta(s, f_\chi), \end{aligned}$$

where f_χ is the function on $\mathbf{Z}(N)$ given by

$$f_\chi(x) = \sum_{T\alpha \equiv x(N)} \chi(\alpha),$$

and the sum is taken over all α in $C(N)$ such that $T\alpha \equiv x \bmod N$.

A well known expression for the partial zeta function formed with an arbitrary function f on $\mathbf{Z}(N)$ is

$$\zeta(s, f) = \frac{1}{2\pi i}\left(\frac{2\pi}{N}\right)^s \Gamma(1-s)[\zeta(1-s, \hat{f}^-)e^{\pi i s/2} - \zeta(1-s, \hat{f})e^{-\pi i s/2}]$$

where f^- is the function such that $f^-(x) = f(-x)$, and the Fourier transform $\hat{f}$ is given by the formula

$$\hat{f}(n) = \sum_{x=1}^{N} f(x)e^{-2\pi i x n/N}.$$

See for instance [L 7], Chapter XIV, Theorem 2.1. Then

$$\begin{aligned} \hat{f}_\chi(n) &= \sum_{x=1}^{N} \sum_{T\alpha = a} \chi(\alpha)e^{-2\pi i x n/N} \\ &= \sum_{\alpha \in C(N)} \chi(\alpha)e^{-2\pi i n T\alpha/N} \\ &= S(\chi, T)\bar{\chi}(-n), \end{aligned}$$

where $S(\chi, T)$ is the Gauss sum on the Cartan group, formed with the additive character arising from the surjective map

$$T_N : \mathfrak{o}(N) \to \mathbf{Z}(N).$$

Thus

$$S(\chi, T) = \sum_{\alpha} \chi(\alpha)e^{2\pi i T\alpha/N}.$$

Note that $\hat{f}_\chi(n) = 0$ if $(n, N) \neq 1$. So we find

$$\hat{f}_\chi^-(n) = S(\chi, T)\bar{\chi}(n),$$

whence we obtain an expression for the L-function on the Cartan group in terms of an ordinary L-function on $\mathbf{Z}(N)$, namely

Theorem 5.2. $L(s, \chi, T) =$

$$\frac{N^{1-k}}{2\pi i}\left(\frac{2\pi}{N}\right)^s \Gamma(1-s)S(\chi, T)[e^{\pi is/2} - \chi(-1)e^{-\pi is/2}]L_N(1-s, \bar{\chi}_{\mathbf{Z}})$$

where $\chi_{\mathbf{Z}}$ *is the restriction of* χ *to* $\mathbf{Z}(N)$.

We assumed that χ is primitive, but of course $\chi_{\mathbf{Z}}$ need not be primitive. It is easy to give the value of the primitive L-series in terms of the non-primitive one by multiplying with the appropriate factors involving the primes dividing N. We also note that the factors in the analytic expression for the L-function on the Cartan group are essentially the same as the corresponding factors for the ordinary L-function. Consequently, after some cancellations, we find:

Theorem 5.3.

$$L(s, \chi, T) = N^{1-k}\frac{S(\chi, T)}{S_{\mathbf{Z}}(\chi_{\mathbf{Z}})}\prod_{\substack{p \mid N \\ p \nmid c}}\left(1 - \frac{\bar{\chi}(p)}{p^{1-s}}\right)L(s, \chi_{\mathbf{Z}})$$

where $c = c(\chi_{\mathbf{Z}})$ *is the conductor of* $\chi_{\mathbf{Z}}$, *and* $S_{\mathbf{Z}}(\chi_{\mathbf{Z}})$ *is the standard Gauss sum formed with the standard additive character* $x \mapsto e^{2\pi ix}$ *for* $x \in \mathbf{Z}(N)$.

Evaluating the L-function in Theorem 5.3 at negative integers yields the desired values in terms of ordinary Bernoulli numbers.

From the analytic expression of Theorem 5.2, we know at which negative integers does the L-function vanish, and we summarize this as follows:

Corollary 1.

(i) $L(0, \chi, T) = 0$ *if* χ *is even,* $\chi_{\mathbf{Z}}$ *non-trivial.*
$L(0, \chi, T) \neq 0$ *if* χ *is even,* $\chi_{\mathbf{Z}}$ *is trivial, or if* χ *is odd.*

(ii) *Let* m *be an integer* $\geqq 2$. *Then* $L(1-m, \chi, T) \neq 0$ *if and only if* χ *and* m *have the same parity.*

Proof. The vanishing or non-vanishing is reduced to that of the ordinary L-function because none of the factors occurring in front of $L(s, \chi_{\mathbf{Z}})$ vanishes at integers $1 - m$ with $m \geqq 1$. The assertion of the corollary is then clear from the standard properties of the classical L-function.

For the record, we also give the explicit value at negative integers.

Corollary 2. *Let m be an integer $\geqq 1$. Then*

$$B_{m,\chi,k,T} = N^{1-k} \frac{S(\chi, T)}{S_{\mathbf{Z}}(\chi_{\mathbf{Z}})} \prod_{\substack{p \mid N \\ p \nmid c}} \left(1 - \frac{\bar{\chi}(p)}{p^m}\right) B_{m,\chi_{\mathbf{Z}}}$$

except when $\chi_{\mathbf{Z}}$ is trivial, and $m = 1$.

Proof. We merely use the classical value of the Dirichlet L-series, cf. for instance Theorem 2.3 of Chapter XIV, [L 7].

The above formula is applied to the value of the L-function,

$$\boxed{L(1 - m, \chi, T) = -\frac{1}{m} B_{m,\chi,k,T}.}$$

§6. Universal Distributions

Let

$$Z_N = \frac{1}{N} \mathfrak{o}/\mathfrak{o} \approx \frac{1}{N} \mathbf{Z}^k/\mathbf{Z}^k$$

as before. Let

$$g_1 : Z_N \to A_1 \quad \text{and} \quad g_2 : Z_N \to A_2$$

be distributions of the same degree s, which we assume is an integer $\geqq 0$. Then a homomorphism from one to the other is a homomorphism $A_1 \to A_2$ which makes the following diagram commutative.

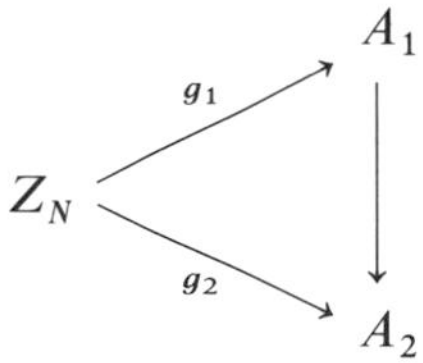

It is then clear that there is a universal distribution of degree s on Z_N, obtained as follows. Let $\mathbf{F}_N$ be the free abelian group generated by the elements of Z_N. We also call $\mathbf{F}_N$ the group of **divisors** on Z_N. We let $\mathbf{R}_N$ be the subgroup generated by all divisors of the form

$$(N/M)^s \sum_{(N/M)y=x} (y) - (x) \quad \text{with } x \in Z_M$$

for all divisors M of N. We call $\mathbf{R}_N$ the subgroup of **distribution relations**. Then the factor group

$$\mathbf{U}(N) = \mathbf{F}_N/\mathbf{R}_N$$

together with the natural map of Z_N into $\mathbf{U}(N)$ is a universal distribution of level N, and degree s. Any other universal distribution is isomorphic to this one by a unique isomorphism.

We are interested in the structure of $\mathbf{U}(N)$ as an abelian group, given first in [Ku 3], [Ku 4]. See also Yamamoto [Ya 1], [Ya 2]. It turns out to be free, e.g. when $s = 0$. We shall exhibit a basis. If

$$g : Z_N \to A$$

is a distribution, then by its **rank** we shall mean the rank of the abelian group generated by the image $g(Z_N)$. If A_N denotes this group, then the rank is the $\mathbf{Q}$-dimension of $A_N \otimes \mathbf{Q}$. We shall see that the rank of the universal distribution is $|C(N)|$.

Let us write the factorization of N with indices:

$$N = \prod_{i \geqq 1} p_i^{n_i}.$$

Then

$$\frac{1}{N}\,\mathfrak{o}/\mathfrak{o} = \bigoplus \frac{1}{p_i^{n_i}}\,\mathfrak{o}/\mathfrak{o}$$

and any element has a partial fraction decomposition

$$\frac{a}{N} = \sum \frac{a_i}{p_i^{n_i}} \bmod \mathfrak{o},$$

where a_i is well defined mod $p_i^{n_i}$, while a is well defined mod N. We let:

V_N = set of elements a/N as above, such that for every i, a_i is prime to p_i and $a_i \neq 1$, or $a_i = 0$.

It is then clear that V_N has cardinality $c(N)$, where

$$c(N) = |C(N)|$$

is the order of the Cartan group.

Theorem 6.1. *The images of the elements of V_N in the universal distribution of degree 0 form a free basis for* $\mathbf{U}(N)$.

Proof. We shall first show that if g is any distribution, and A_N is the abelian group generated by $g(Z_N)$, then $g(V_N)$ generates A_N. We do this by induction on the number of prime factors of N.

Let

$$\sum \frac{b_i}{p_i^{n_i}}$$

be an arbitrary element of Z_N. Write $b_1 = p_1^r a_1$ where a_1 is 0 or prime to p_1. If $a_1 = 0$ then we are through by induction, so we can assume that a_1 is prime to p_1 and $1 \leq r < n_1$. Then:

$$\begin{aligned} g\left(\frac{p_1^r a_1}{p_1^{n_1}} + \sum_{i \geq 2} \frac{b_i}{p_i^{n_i}}\right) &= g\left(p_1^r\left(\frac{a_1}{p_1^{n_1}} + \sum_{i \geq 2} \frac{c_i}{p_i^{n_i}}\right)\right) \\ &= \sum_{j \bmod p_1^r} g\left(\frac{a_1}{p_1^{n_1}} + \frac{j}{p_1^r} + \sum_{i \geq 2} \frac{c_i}{p_i^{n_i}}\right) \end{aligned}$$

by the distribution relation. Since $r < n_1$ it follows that

$$\frac{a_1}{p_1^{n_1}} + \frac{j}{p_1^r} = \frac{a_1'}{p_1^{n_1}},$$

where a_1' is prime to p_1.

Inductively, we may now repeat the same argument with respect to p_2, $p_3, \ldots$. It merely suffices to observe the following. In the first step of the argument, when we factored out p_1^r, thus changing b_i to c_i, if b_i is prime to p_1 then c_i is prime to p_1. Thus performing the same argument inductively on the other primes does not destroy the desired property for those primes which have already been taken care of. This concludes the first part of the proof.

Secondly, we show that we can recover those elements a/N for which a_i may be equal to 1 from the prescribed set V_N. Let

$$N' = N/p_1^{n_1}, \qquad y \in \frac{1}{N'}\mathfrak{o}/\mathfrak{o}.$$

From the distribution relation, we find:

$$\sum_{j \bmod p_1^{n_1}} g\left(\frac{j}{p_1^{n_1}} + y\right) = g(p_1^{n_1}y)$$

$$\sum_{k \bmod p_1^{n_1-1}} g\left(\frac{k}{p_1^{n_1-1}} + y\right) = g(p_1^{n_1-1}y).$$

Subtracting yields

$$\sum_{\substack{j \bmod p_1^{n_1} \\ (j,p)=1}} g\left(\frac{j}{p_1^{n_1}} + y\right) \equiv 0 \bmod A_{N'},$$

where $A_{N'}$ is the group generated by $g(Z_{N'})$. This yields

$$-g\left(\frac{1}{p_1^{n_1}} + y\right) \equiv \sum_{\substack{a_1 \neq 1 \\ (a_1,p_1)=1}} g\left(\frac{a_1}{p_1^{n_1}} + y\right) \bmod A_{N'}.$$

Observe that the same quantity y occurs on both sides of this relation. We may now repeat the procedure inductively on the partial fraction decomposition of y. If we write

$$y_1 = y = \sum_{i \geq 2} \frac{a_i}{p_1^{n_i}},$$

and say $a_2 = 1$, we get a similar congruence

$$-g\left(\frac{a_1}{p_1^{n_1}} + y_1\right) \equiv \sum_{\substack{a_2 \neq 1 \\ (a_2,p_2)=1}} g\left(\frac{a_1}{p_1^{n_1}} + \frac{a_2}{p_2^{n_2}} + y_2\right) \bmod A_{N''},$$

where $N'' = N/p_2^{n_2}$. In this way we reduce the proof to the case when N contains fewer prime factors, and then can apply induction with respect to the number of prime factors to conclude the proof.

If we can exhibit one distribution of degree 0 whose rank is $c(N)$, then the canonical homomorphism from the universal distribution has to be an isomorphism in light of the above fact that the elements of V_N generate the image of any distribution. In fact, we use the following stronger version.

Theorem 6.2. *Let*

$$g : \frac{1}{N}\mathfrak{o}/\mathfrak{o} \to A$$

be a distribution of degree 0. Let K be a field of characteristic 0. Assume that the distribution obtained by following g with the natural homomorphism

$$A \to A \otimes K$$

has K-rank $c(N)$, in the sense that the dimension of the vector space generated by the image of Z_N has dimension $c(N)$. Then g is the universal distribution.

Proof. The rank of the image is at most $c(N)$ by what we have proved. If the vector space generated by the image has that rank, then the above generators must remain free under g and the tensor product, so they must be linearly independent over **Z** in the abelian group generated by

$$g(Z_N) = g\left(\frac{1}{N}\mathfrak{o}/\mathfrak{o}\right).$$

Hence the canonical homomorphism from the universal distribution to g must be an isomorphism, as was to be shown.

We shall now construct a distribution of degree 0 on $\mathbf{K}/\mathfrak{o}$, having this maximal rank. We consider the Fourier series for each integer $m > 1$ given by

$$f_m(t) = \sum_{n=1}^{\infty} \frac{e^{2\pi i n t}}{n^m}.$$

It is immediately verified that f_m defines a distribution of degree $m - 1$. In fact, the distribution relation is satisfied for the variable in $\mathbf{R}/\mathbf{Z}$ rather than $\mathbf{Q}/\mathbf{Z}$. However, we shall now consider the associated Stickelberger distribution, and for this, we have to work on $\mathbf{Q}/\mathbf{Z}$.

We let

$$T: \prod \mathfrak{o}_p \to \prod \mathbf{Z}_p$$

be a surjective map as in §4. Then $f_k \circ T$ is a distribution of degree 0 on $\mathbf{K}/\mathfrak{o}$.

Theorem 6.3. *Let $k \geqq 2$. The Stickelberger distribution associated with $f_k \circ T$ has K-rank equal to $c(N)$ on Z_N, and is the universal distribution of degree 0 on $\mathbf{K}/\mathfrak{o}$.*

Proof. We apply Theorem 3.1 which gives the dimensions of the χ-eigenspaces for each character χ. It is immediately verified that the sum $S(\chi, f_k \circ T)$

has the form

$$S(\chi, f_k \circ T) = \lambda L(k, \chi, T)$$

where λ is a constant factor which is $\neq 0$, and $L(k, \chi, T)$ is the value of the L-function defined in §5, at the positive integer $k \geqq 2$. Theorem 5.3 then gives us the value of this sum for any complex variable, and in particular at the positive integers, where it is clear that the sum is $\neq 0$, since it is a product of a non-zero constant factor and the value of the ordinary L-series $L(k, \chi_{\mathbf{Z}})$. This proves the theorem.

The proof has been given rapidly because in the applications we shall deal with even distributions and $k = 2$. In that case, let us make things more explicit. We shall make no use of Theorem 6.3 in the applications.

Let $\mathbf{B}_k$ be the k-th Bernoulli polynomial (cf. for instance [L 8], Chapter 2, §2). When $k = 2$, then

$$\mathbf{B}_2(X) = X^2 - X + \tfrac{1}{6}.$$

This case will suffice for the applications. Then it is an elementary property of these polynomials, following immediately from their definition in terms of their generating function

$$\frac{te^{tX}}{e^t - 1} = \sum \mathbf{B}_k(X) \frac{t^k}{k!}$$

that the function

$$h_k : t \mapsto \frac{1}{k} \mathbf{B}_k(\langle t \rangle)$$

is a distribution of degree $k - 1$ on $\mathbf{Q}/\mathbf{Z}$. (For $k = 1$, define $h_1(0) = 0$.) Consequently, the lifted distribution

$$h_k \circ T$$

is a distribution of degree 0 on $\mathbf{K}/\mathfrak{o}$. It is even or odd according as k is even or odd. Again this is an immediate consequence of the formula

$$\mathbf{B}_k(1 - X) = (-1)^k \mathbf{B}_k(X),$$

which also follows directly from the generating series. Therefore the Stickelberger distribution associated to $h_k \circ T$, called the k-th **Bernoulli distribution** on $\mathbf{K}/\mathfrak{o}$, cannot be universal.

However, suppose we consider distributions with values in abelian groups where multiplication by 2 is invertible. Then every distribution has a unique expression as a sum of an even and an odd distribution. Furthermore, we may define the **universal even** (resp. **odd**) distribution into such groups. Adding the even (resp. odd) relations to the distribution relations immediately shows that this universal distribution exists.

The images of the elements in $V_N/\pm 1$ in Theorem 6.1 form free generators for the universal even (resp. odd) distribution into abelian groups where multiplication by 2 is invertible. In particular, these universal distributions are free. Theorem 6.2 can then also be formulated in this context. An even (resp. odd) distribution of maximal rank is then universal into abelian groups as above. Reference to Theorem 6.2 will then be made for the even (resp. odd) cases as well.

Theorem 6.4. *Let g be the universal even* (resp. *odd*) *distribution of degree 0 on $Z_N = (\mathbf{K}/\mathfrak{o})_N$ into abelian groups where multiplication by 2 is invertible. Let A_N be the group generated by $g(Z_N)$. Then A_N is free abelian, and has $c(N)/2$ generators, except in the case $N = 2$.*

Proof. This is obvious from Theorem 6.1, by identifying the free generators for the universal distribution according to the even or odd relation.

Theorem 6.5. *Let k be an integer $\geqq 1$. Let h_k be the k-*th *Bernoulli distribution on $\mathbf{Q}/\mathbf{Z}$. Then the Stickelberger distribution associated with $h_k \circ T$ is the universal even* (resp. *odd*) *distribution, according as k is even* (resp. *odd*), *with values in abelian groups where multiplication by 2 is invertible.*

Proof. Suppose k is even. By Theorem 6.2 (even), it suffices to verify that for each even character χ on the Cartan group we have

$$S(\chi, h_k \circ T) \neq 0.$$

By the definition of the Bernoulli-Cartan numbers, this sum is just what we called $B_{k,\chi,k,T}$, which is $\neq 0$ by Corollary 1 of Theorem 5.2. This concludes the proof.

CHAPTER 2

Modular Units

We introduce the explicit units in the modular function field which form the basis of everything that follows. We assume that the reader is acquainted with the elementary theory of modular forms and functions, and especially with the groups of automorphisms, as described in Shimura [Sh] or Lang [L 5], Chapters 1 through 6.

It turns out that the functions on the modular curves which have zeros and poles only at infinity (the cusps, as the saying goes) play an important role both in the function field theory and applications to number theory. Such functions can be called units, in analogy with units in number fields, and in fact they are units in the affine ring of the modular curves. This affine ring can be taken over the integers **Z**, or over **Q**. In the former case, these modular units when specialized to number fields yield units in the sense of Dirichlet. We shall consider such specializations in the last chapters of the book.

Here we are concerned with establishing the general theory in the function field. It turns out that this theory is very analogous to cyclotomic theory. Whereas in the classical cyclotomic theory the Galois group is the 1-dimensional Cartan $\mathbf{Z}(N)^*$, in the modular theory, it is the 2-dimensional Cartan which plays a role. This amounts to a "twist" which was conjectured in [KL 7] to be connected with the classical questions concerning class numbers of cyclotomic fields. See also [L8], p. 53. The precise connection was established by Wiles [Wi] and Mazur-Wiles [Ma-W].

In any case, we can obtain class numbers expressed here in terms of $B_{2,\chi}$ instead of $B_{1,\chi}$. These applications occur in Chapters 5 and 6.

Here we establish the link between the function theory and the abstract theory of Chapter 1, by constructing a system of units which form a distribution on $\mathbf{Q}^2/\mathbf{Z}^2$, and such that their divisors form the Stickelberger distribution associated with the second Bernoulli distribution. These units are given in terms of modular forms due to Klein, who introduced this particular normalization of the sigma function because it is easier than theta functions to handle with respect to transformations in $SL_2(Z)$ (cf. Klein [Kl 2], p. 204; Fricke [Fr], pp. 450, 451, 452 or Klein-Fricke [K-F], p. 22). They give an easier approach to the Siegel functions and the construction of units. The Siegel functions [Sie 2] and the associated fundamental theta function have been used to construct units by Ramachandra [Ra], Novikov [No 1], Robert [Ro 1] and Stark [St 2] among others.

Note that the distribution relations in the present case are the 2-dimensional analogues of the relations determined by Bass [Ba] for the cyclotomic units.

Other authors more recently have considered functions in the modular function field of what is usually denoted by $X_0(N)$, which have divisors concentrated at the cusps, e.g. Newman [Ne] and [Ogg]. These functions were obtained as quotients of the discriminant function $\Delta(m\tau)/\Delta(\tau)$. Working with the Klein forms goes farther, especially since one can profit from the Galois action of the whole modular function field. Aside from that, we shall see later that the Siegel functions generate all the units, except possibly for 2-torsion.

Units can also be constructed as quotients of Weierstrass functions. This was the point of view first mentioned in [L 5], Chapter 18, p. 257. Independently, such quotients on special elliptic curves were used in the work of Demjanenko and Kubert for applications to certain diophantine questions on elliptic curves (bounding torsion points over number fields), cf. their papers in the bibliography, especially [Ku 1], or Chapter 8. The starting point of our whole investigation came by putting together these two trends in [KL 1]. The fact that the differences of Weierstrass forms

$$\wp_a - \wp_b = -\frac{\sigma_{a+b}\sigma_{a-b}}{\sigma_a^2\sigma_b^2}$$

has an expression like that on the right is seen to amount formally to a quadratic expression. The systematic study of these quadratic relations was then carried out in [Ku 2], and is reproduced in the next chapter.

In this chapter, we merely give the construction of the appropriate units by this method, including some special units which will be important for the structure of the cuspidal divisor class group obtained in Chapter 5, §7.

§1. The Klein Forms and Siegel Functions

We recall some elementary definitions from the theory of modular forms. We let:

$\Gamma(1) = SL_2(\mathbf{Z})$

$\Gamma(N) =$ subgroup of elements

$$\alpha = \begin{pmatrix} a & b \\ c & d \end{pmatrix}$$

such that $\alpha \equiv 1 \bmod N$. By 1 we mean the unit 2×2 matrix, so this congruence condition is equivalent with

$$a \equiv d \equiv 1 \bmod N \quad \text{and} \quad c \equiv b \equiv 0 \bmod N.$$

Let k be an integer. By a **form of degree** k (**weight** $-k$) we mean a function

$$h\begin{pmatrix}\omega_1\\ \omega_2\end{pmatrix} = h(W) \quad \text{where } W = \begin{pmatrix}\omega_1\\ \omega_2\end{pmatrix},$$

of two complex variables, with $\mathrm{Im}(\omega_1/\omega_2) > 0$, satisfying the homogeneity property

MF 1. $$h\left(\lambda\begin{pmatrix}\omega_1\\ \omega_2\end{pmatrix}\right) = \lambda^k h\begin{pmatrix}\omega_1\\ \omega_2\end{pmatrix}, \qquad \lambda \in \mathbf{C}^*.$$

Let Γ be a subgroup of $\mathrm{SL}_2(\mathbf{Z})$, of finite index. We say that a form h as above is **modular on** Γ, or **with respect to** Γ, if it satisfies the additional properties:

MF 2. $$h\left(\alpha\begin{pmatrix}\omega_1\\ \omega_2\end{pmatrix}\right) = h\begin{pmatrix}\omega_1\\ \omega_2\end{pmatrix} \quad \textit{for all } \alpha \in \Gamma.$$

MF 3. *For τ in the upper half plane,* i.e. $\mathrm{Im}\,\tau > 0$, *the function $h(\tau, 1)$ is meromorphic at infinity, meaning that it has a convergent Laurent series expansion for some N:*

$$h(\tau, 1) = \sum_{n=n_0}^{\infty} a_n q^{n/N} \quad \textit{for some } n_0 \in \mathbf{Z},$$

where $q = e^{2\pi i\tau}$. A similar expansion should hold for the function $h \circ \alpha$ with any $\alpha \in \mathrm{SL}_2(\mathbf{Z})$.

Given a modular form h, the composite functions $h \circ \alpha$ with $\alpha \in \mathrm{SL}_2(\mathbf{Z})$ are called the **conjugates** of h. We often write $h(\tau)$ instead of $h(\tau, 1)$.

A form of weight 0 is called a **modular function**.

If a form is modular on $\Gamma(N)$, then we also say that the form has **level** N.

Given a lattice L, we have the usual Weierstrass functions $\sigma(z, L)$ and

$$\zeta(z, L) = \sigma'/\sigma(z, L).$$

From the partial fraction decomposition (Mittag-Leffler) obtained by logarithmic differentiation of the Weierstrass product for $\sigma(z, L)$, one sees that $\zeta(z, L)$ is homogeneous of degree -1, that is

$$\zeta(\lambda z, L) = \lambda^{-1}\zeta(z, L) \quad \text{for } \lambda \in \mathbf{C}^*.$$

There is a function $\eta(z, L)$ which is $\mathbf{R}$-linear in z, such that for $\omega \in L$,

$$\zeta(z + \omega, L) = \zeta(z, L) + \eta(\omega, L).$$

In the usual notation,

$$\eta_1 = \eta(\omega_1, L), \qquad \eta_2 = \eta(\omega_2, L)$$

and for *real* a_1, a_2,

$$\eta(a_1\omega_1 + a_2\omega_2, L) = a_1\eta_1 + a_2\eta_2.$$

The function $\eta(z, L)$ is homogeneous of degree -1, that is

$$\eta(\lambda z, \lambda L) = \lambda^{-1}\eta(z, L).$$

This is immediately proved from the corresponding property for $\zeta(z, L)$. We shall use the dot product notation

$$z = a_1\omega_1 + a_2\omega_2 = a \cdot W, \quad \text{where } W = \begin{pmatrix} \omega_1 \\ \omega_2 \end{pmatrix}.$$

We also write

$$W_\tau = \begin{pmatrix} \tau \\ 1 \end{pmatrix}.$$

We define the **Klein forms**

$$\mathfrak{k}(z, L) = e^{-\eta(z,L)z/2}\sigma(z, L).$$

We also write

$$\mathfrak{k}(z, L) = \mathfrak{k}_a(W).$$

Then the Klein forms satisfy the following properties. First, they are homogeneous of degree 1:

K 0. $\qquad \mathfrak{k}(\lambda z, \lambda L) = \lambda\mathfrak{k}(z,L) \quad$ *and* $\quad \mathfrak{k}_a(\lambda W) = \lambda\mathfrak{k}_a(W).$

This is obvious from the homogeneity property of the sigma function and the Weierstrass zeta function. The next property is also clear.

K 1. *If* $\alpha = \begin{pmatrix} a & b \\ c & d \end{pmatrix}$ *is in* $\mathrm{SL}_2(\mathbf{Z})$, *then*

$$\mathfrak{k}_a(\alpha W) = \mathfrak{k}_{a\alpha}(W).$$

We recall the standard transformation property of the sigma function with respect to the periods. Let $\omega = b_1\omega_1 + b_2\omega_2$ with integers b_1, b_2. Then

$$\sigma(z+\omega, L) = (-1)^{b_1b_2+b_1+b_2}e^{\eta(\omega,L)(z+\omega/2)}\sigma(z, L).$$

Then we find:

K 2. $$\mathfrak{k}_{a+b}(W) = \varepsilon(a,b)\mathfrak{k}_a(W)$$

where $\varepsilon(a,b)$ has absolute value 1, and is given explicitly by

$$\varepsilon(a,b) = (-1)^{b_1b_2+b_1+b_2}e^{-2\pi i(b_1a_2-b_2a_1)/2}.$$

This follows easily from the Legendre relation

$$\eta_2\omega_1 - \eta_1\omega_2 = 2\pi i.$$

We leave the computation to the reader. Cf. Theorem 1.3, Chapter 10.

So far, we needed no further assumption on a_1, a_2. Assume now that they are rational numbers, with denominators dividing an integer $N > 1$, say

$$a_1 = r/N \quad \text{and} \quad a_2 = s/N.$$

Let

$$\alpha = \begin{pmatrix} a & b \\ c & d \end{pmatrix} \equiv \begin{pmatrix} 1 & 0 \\ 0 & 1 \end{pmatrix} \bmod N$$

be in $\Gamma(N)$, and write

$$ar + cs = r + \left(\frac{a-1}{N}r + \frac{c}{N}s\right)N, \qquad br + ds = s + \left(\frac{b}{N}r + \frac{d-1}{N}s\right)N.$$

Then we find from **K 2**:

K 3. $$\mathfrak{k}_a\left(\alpha\begin{pmatrix}\omega_1\\ \omega_2\end{pmatrix}\right) = \mathfrak{k}_{a\alpha}\begin{pmatrix}\omega_1\\ \omega_2\end{pmatrix} = \varepsilon_a(\alpha)\mathfrak{k}_a\begin{pmatrix}\omega_1\\ \omega_2\end{pmatrix}$$

where $\varepsilon_a(\alpha)$ is a $(2N)$th root of unity, given precisely by

$$\varepsilon_a(\alpha) = \varepsilon(\alpha) = -(-1)^{\left(\frac{a-1}{N}r+\frac{c}{N}s+1\right)\left(\frac{b}{N}r+\frac{d-1}{N}s+1\right)}e^{2\pi i(br^2+(d-a)rs-cs^2)/2N^2}.$$

From this transformation law, we get:

Theorem 1.1. *Let $a \in (1/N)\mathbf{Z}^2$ but $a \notin \mathbf{Z}^2$. Then $\varepsilon_a(\alpha)^{2N} = 1$. Hence $\mathfrak{k}_a$ is a modular form on $\Gamma(2N^2)$, and $\mathfrak{k}_a^{2N}$ is on $\Gamma(N)$. If N is odd, then $\mathfrak{k}_a^N$ is on $\Gamma(N)$.*

The proof is immediate, by considering the cases when r, s are both even, or one of them is odd, or both are odd, and using $ad - bc = 1$, so that for instance, not both c, d are even.

From **K 3**, it is also an elementary matter to give the precise conditions under which a product

$$f = \prod \mathfrak{k}_a^{m(a)}$$

of Klein forms is modular of level N. We could state this result now, but it fits better in the next chapter, Theorem 4.1.

We now take $\omega_1 = \tau$ and $\omega_2 = 1$, so that

$$z = a_1\tau + a_2.$$

Using the q-expansion for the sigma function, given in most books on elliptic functions (for instance [L 5], Chapter 18, §2) we can easily derive the q-product for the Klein forms. We let the **Siegel functions** be defined by

$$g_a(\tau) = \mathfrak{k}_a(\tau)\,\Delta(\tau)^{1/12},$$

where $\Delta(\tau)^{1/12}$ is the square of the Dedekind eta function, namely the natural q-product for the 12th root of Δ, which is

$$\text{Dedekind } \eta(\tau)^2 = 2\pi i \cdot q^{1/12} \prod_{n=1}^{\infty} (1 - q^n)^2,$$

where $q^{1/12} = e^{2\pi i\tau/12}$. We shall use the notation

$$q = q_\tau = e^{2\pi i\tau} \quad \text{and} \quad q_z = e^{2\pi i z}.$$

Then from the q-product for the sigma function (see. e.g. [L 5], Chapter 18, §2), we obtain the q-product for the Siegel functions:

K 4. $$g_a(\tau) = -q_\tau^{(1/2)\mathbf{B}_2(a_1)} e^{2\pi i a_2(a_1 - 1)/2}(1 - q_z) \prod_{n=1}^{\infty} (1 - q_\tau^n q_z)(1 - q_\tau^n/q_z),$$

where $\mathbf{B}_2(X) = X^2 - X + \frac{1}{6}$ is the second Bernoulli polynomial.

Remarks. If we change a by an integral vector in $\mathbf{Z}^2$, then **K 2** shows that g_a changes by a root of unity. We can always make such a change so that a representative in the class mod $\mathbf{Z}^2$ has coordinates $a = (a_1, a_2)$ such that

$$0 \leqq a_1 < 1 \quad \text{and} \quad 0 \leqq a_2 < 1.$$

These are the standard representatives $\langle a_1 \rangle$ and $\langle a_2 \rangle$. In that case, the terms in the q-expansion of g_a can be analyzed as follows.

(1) The terms in the product on the right. Each expression

$$q_\tau^n q_z \quad \text{and} \quad q_\tau^n / q_z$$

is equal to a positive rational power of q_τ times some root of unity. Therefore the product on the right is invertible in

$$\mathbf{Z}[\boldsymbol{\mu}_N][[q_\tau^{1/N}]].$$

(2) The term $1 - q_z$ has the form

$$1 - q_z = 1 - q_\tau^{a_1} e^{2\pi i a_2}.$$

We still write

$$a_1 = \frac{r}{N} \quad \text{and} \quad a_2 = \frac{s}{N}$$

with $0 \leqq r, s < N$.

Case 1. $a_1 > 0$. Then again this term is a series in $q_\tau^{1/N}$, invertible over the ring $\mathbf{Z}[\boldsymbol{\mu}_N][[q_\tau^{1/N}]]$.

Case 2. $a_1 = 0$. Then this term has the form

$$1 - \zeta^s$$

where $\zeta = e^{2\pi i/N}$. Thus this term is a constant, and its divisibility in the cyclotomic field is known and elementary. It is a unit (in the sense of Dirichlet) if N is composite, and has the obvious order at p if $N = p^n$ is a prime power.

(3) The term

$$-e^{2\pi i a_2 (a_1 - 1)/2}$$

is a root of unity.

In each one of the above cases we have found that the power series in $q_\tau^{1/N}$ is invertible in $\mathbf{Q}(\boldsymbol{\mu}_N)[[q_\tau^{1/N}]]$, and therefore does not contribute to the order at q_τ. This leaves us with the final term.

(4) The term

$$q_\tau^{(1/2)\mathbf{B}_2(a_1)},$$

which is a rational power of q_τ, and gives us the order of the Siegel function g_a at q_τ, having assumed $0 \leqq a_1 < 1$.

From the above analysis, we conclude:

For any $a \in \mathbf{Q}^2$, $a \notin \mathbf{Z}^2$ we have

$$\boxed{\operatorname{ord}_{q_\tau} g_a = \tfrac{1}{2}\mathbf{B}_2(\langle a_1 \rangle).}$$

This formula will be used in a fundamental way in analyzing the divisor of g_a in §3.

It is possible to derive transformation laws for the Dedekind eta function just as for the Klein forms. In fact, one can express the eta function in terms of Klein forms. We shall do this later. For our present purposes, combining Theorem 1.1 with the fact that $\Delta(L)$ is a modular form on $\mathrm{SL}_2(\mathbf{Z})$, we get:

Theorem 1.2. *Assume that a has denominator dividing N. Then the Siegel functions g_a are modular functions, and g_a^{12N} is on $\Gamma(N)$. Furthermore, g_a has no zeros or poles on the upper half plane.*

Remark. Because the Dedekind function $\eta(\tau)$ is not a function of lattices, but depends on the choice of a basis, we could not define the Siegel functions in terms of a lattice. However, if we take the 12th powers, then we may define for any complex number z, the functions

$$g^{12}(z, L) = \mathfrak{k}^{12}(z, L)\,\Delta(L).$$

These give rise to modular forms when

$$z \in \frac{1}{N} L \quad \text{and} \quad z \notin L,$$

for some positive integer N.

We conclude this section by brief remarks on modular forms which describe a structure prevalent in all that we shall do.

Let F be a modular form on $\Gamma(N)$, defined as a function of two variables

$$F(W) = F\begin{pmatrix}\omega_1\\ \omega_2\end{pmatrix}$$

Let $\Gamma_1(N)$ be the subgroup of elements

$$\gamma = \begin{pmatrix}a & b\\ c & d\end{pmatrix} \in \mathrm{SL}_2(\mathbf{Z})$$

such that $c \equiv 0 \bmod N$, and $a \equiv d \equiv 1 \bmod N$. If $F(\gamma W) = F(W)$ for all $\gamma \in \Gamma_1(N)$, then we say that F is on $\Gamma_1(N)$. Such a form gives rise to functions $f(t, L)$, where L is a lattice, and t is a point of order exactly N with respect to L, as follows. Given a pair (t, L), we select any element ω_2 of L such that

$$t = \omega_2/N,$$

and then any element ω_1 of L such that ω_1/ω_2 lies in the upper half plane, and ω_1, ω_2 is a basis of L. We define

$$f(t, L) = F(W).$$

Conversely, given a function of pairs $f(t, L)$, we may define

$$F(W) = F\begin{pmatrix}\omega_1\\ \omega_2\end{pmatrix} = f\left(\frac{\omega_2}{N}, [\omega_1, \omega_2]\right).$$

The correspondence $F \leftrightarrow f$ gives alternative ways of defining modular forms on $\Gamma_1(N)$. We shall give one more way of prescribing equivalent data for a modular form.

Consider families $\{f_a\}$ of forms on $\Gamma(N)$ (always taken of some fixed weight) indexed by $a \in (1/N)\mathbf{Z}^2$ with a primitive mod $\mathbf{Z}^2$, satisfying the condition

FR 1. $\quad f_a(\gamma W) = f_{a\gamma}(W) \quad$ for all $\gamma \in \mathrm{SL}_2(\mathbf{Z})$.

There is a bijection between such families and forms on $\Gamma_1(N)$.

The bijection is obtained as follows.

Given $f(t, L)$, we define $f_a(W) = f(a \cdot W, [W])$, where $[W]$ is the lattice generated by ω_1, ω_2.

Conversely, given a family $\{f_a\}$, let $L = [W]$ for some W, write a point t of order N as $t = a \cdot W$, and define

$$f(t, L) = f_a(W).$$

It is a simple exercise to show that the above associations give rise to the stated bijection.

We shall call families $\{f_a\}$ as above **Fricke families** if they have one more property described as follows. Let

$$f_a(\tau) = \sum c_n q_\tau^{n/N}$$

be the $q_\tau^{1/N}$-expansion of f_a. We first assume that the coefficients c_n lie in the field $\mathbf{Q}(\boldsymbol{\mu}_N)$. Let d be an integer prime to N, and let σ_d be the automorphism of $\mathbf{Q}(\boldsymbol{\mu}_N)$ giving the usual action on the roots of unity, extended to the power series field by operating on the coefficients, and leaving $q_\tau^{1/N}$ fixed. Then the required property is that

FR 2. $$\sigma_d f_{a_1,a_2} = f_{a_1,da_2}.$$

In the applications, the form $f(z, [\tau, 1])$ is holomorphic in z, τ and depends only on z mod $[\tau, 1]$, so that it has an expansion

$$\sum a_{mn} q_z^m q_\tau^n, \quad \text{with } a_{mn} \in \mathbf{Q}(\boldsymbol{\mu}_N),$$

where $a_{mn} = a_{mn}(a_1, a_2)$ satisfies $a_{mn}(a_1, a_2)^{\sigma_d} = a_{mn}(a_1, da_2)$. Let $z = a_1\tau + a_2$. Then this last property is obviously satisfied.

The modular form f associated with a family of Fricke type will also be called of **Fricke type**, or a **Fricke form**.

Proposition 1.3. *The functions $\{g_a^{12N}\}$ form a Fricke family.*

Proof. Property **FR 1** follows from **K 3**, and **FR 2** follows from **K 4** and the above remarks.

The next proposition is inserted for completeness. We shall not use it in this book.

Proposition 1.4. *Let f be a Fricke form on $\Gamma_1(N)$, of weight 0 (so a function). Let L be a lattice, and let t be a complex number of period exactly N with respect to L. Let a bar denote complex conjugation. Then*

$$\overline{f(t, L)} = f(\bar{t}, \bar{L}).$$

Proof. Let $L = [\omega_1, \omega_2]$ and write

$$t = a_1\omega_1 + a_2\omega_2 \quad \text{with } a_1, a_2 \in \mathbf{Q}.$$

Let $\tau = \omega_1/\omega_2$. Then

$$f(t, L) = f(a_1\omega_1 + a_2\omega_2, [\omega_1, \omega_2]) = f_a(\tau) = f_{a_1,a_2}(\tau).$$

Let the q-expansion be

$$f_{a_1,a_2}(\tau) = \sum c_n e^{2\pi i n\tau}.$$

Then

$$\begin{aligned}\overline{f_{a_1,a_2}(\tau)} &= \sum \overline{c_n} e^{-2\pi i n\bar{\tau}} \\ &= (\sigma_{-1} f_{a_1,a_2})(-\bar{\tau}) \\ &= f_{a_1,-a_2}(-\bar{\tau}).\end{aligned}$$

Therefore

$$\begin{aligned}\overline{f_a(\tau)} &= f(-a_1\bar{\tau} - a_2, [-\bar{\tau}, 1]) \\ &= f(a_1\bar{\omega}_1 + a_2\bar{\omega}_2, [-\bar{\omega}_1, \bar{\omega}_2]) \\ &= f(\bar{t}, \bar{L}).\end{aligned}$$

This concludes the proof.

§2. Units in the Modular Function Field

Let $\mathfrak{H}$ denote the upper half plane. Let Γ be a subgroup of $\Gamma(1)$, of finite index. One can given the quotient space $\Gamma\backslash\mathfrak{H}$ the structure of a Riemann surface, which can be completed to a complete non-singular curve denoted by X_Γ. If $\Gamma = \Gamma(N)$, then this curve is denoted by $X(N)$, and the open part corresponding to $\Gamma\backslash\mathfrak{H}$ is denoted by Y_Γ or $Y(N)$ if $\Gamma = \Gamma(N)$. We do not cover these foundations in the present book. Cf. for instance Shimura [Sh], Chapter 1, or most other introductions to modular functions. For our present purposes, we do not need anything about the algebraization of this quotient space, except the following.

A form of weight 0 is called a **modular function**. The set of modular functions on Γ is a function field in one variable, i.e. has transcendence degree 1 and is finitely generated. Cf. Shimura as above, or [L 5]. It is also called the function field on X_Γ.

We say that a modular form is **defined over a field** K if the coefficients of its q-expansion lie in K, and similarly for every conjugate of the form. In our applications, $K = \mathbf{Q}(\zeta)$ for some root of unity ζ. Let:

F_N = function field on $X(N)$ consisting of those functions which are defined over $\mathbf{Q}(\boldsymbol{\mu}_N)$.

Then there is a natural isomorphism

$$\mathrm{Gal}(F_N/F_1) \approx \mathrm{GL}_2(N)/\pm 1, \quad \text{where } \mathrm{GL}_2(N) = \mathrm{GL}_2(\mathbf{Z}/N\mathbf{Z}).$$

Cf. Shimura [Sh] or Lang [L 5]. Indeed, the group $\mathrm{SL}_2(\mathbf{Z}/N\mathbf{Z})/\pm 1$ operates on a modular function by composition with the natural operation of $\mathrm{SL}_2(\mathbf{Z})$ on the upper half plane,

$$\alpha(\tau) = \frac{a\tau + b}{c\tau + d}.$$

In addition the group of matrices

$$\begin{pmatrix} 1 & 0 \\ 0 & d \end{pmatrix}$$

operates on this field as follows. We let σ_d be the automorphism of $\mathbf{Q}(\boldsymbol{\mu}_N)$ such that

$$\sigma_d \zeta = \zeta^d.$$

Then σ_d extends to the power series field $K((q^{1/N}))$, with $K = \mathbf{Q}(\boldsymbol{\mu}_N)$ by operating on the coefficients:

$$\sigma_d\left(\sum a_n q^{n/N}\right) = \sum (\sigma_d a_n) q^{n/N}.$$

It can be shown that this automorphism of the power series field induces an automorphism of F_N leaving F_1 fixed. Then $\mathrm{SL}_2(\mathbf{Z}/N\mathbf{Z})$ and the above group of automorphisms σ_d generate $\mathrm{GL}_2(\mathbf{Z}/N\mathbf{Z})$, and the kernel of the representation as automorphisms of F_N is ± 1. Cf. Shimura [Sh] and [L 5]. The correspondence between $\mathrm{GL}_2(N)$ and $\mathrm{Gal}(F_N/F_1)$ will be denoted

$$\alpha \mapsto \sigma_\alpha.$$

From the elementary theory of the modular function field, we assume also known that

$$F_1 = \mathbf{Q}(j).$$

We let R_N be the integral closure of $\mathbf{Z}[j]$ in the function field F_N of $X(N)$, and $\mathbf{Q}R_N$ the integral closure of $\mathbf{Q}[j]$. Elements of $(\mathbf{Q}R_N)^*$ will be called **units**, and elements of R_N^* will be called **units over Z**.

The only pole of j is at infinity. The points on the modular curve $X(N)$ lying above $j = \infty$ are called the **cusps.** One of these points corresponds to the embedding of F_N in the power series field $\mathbf{Q}(\boldsymbol{\mu}_N)((q^{1/N}))$, which to each modular function on the upper half plane associates its q-expansion. This point or cusp will be called the **standard cusp**, or **standard point at infinity**. We may view $q^{1/N}$ as a local parameter at the standard cusp.

From the standard algebraic characterization of integral closure as the intersection of all valuation subrings containing the given ring, we see that the units in F_N are precisely those modular functions in F_N which have zeros and poles only at the cusps. From the construction with Klein forms and the delta function, we then see that the Siegel functions are units (at the appropriate level), and furthermore, the functions

$$g_a^{12N}$$

with $a \in (1/N)\mathbf{Z}^2$, $a \notin \mathbf{Z}^2$, are units in $(\mathbf{C}R_N)^*$. However, the coefficients of the q-expansion lie in the field $\mathbf{Q}(\boldsymbol{\mu}_N)$. Consequently these functions actually lie in $(\mathbf{Q}R_N)^*$. Thus we have exhibited a group of units.

The unit group modulo constants is isomorphic to its group of divisors, and is embedded in the free abelian group generated by the cusps. The fact that a function has as many zeros as poles shows that it is embedded in the subgroup of degree 0. Hence the rank of the unit group modulo constants is at most equal to

$$\text{number of cusps on } X(N) - 1.$$

We shall prove later that the Siegel functions (modulo constants) have precisely this rank.

For any $\alpha \in \mathrm{GL}_2(N)$, we have

$$\sigma_\alpha g_a^{12N} = g_{a\alpha}^{12N},$$

from Proposition 1.3. Thus the conjugates of g_a^{12N} are again of the same form. In other words, the Siegel functions (raised to the appropriate power) are permuted by the elements of the Galois group $\mathrm{Gal}(F_N/F_1)$.

Let f be a unit in $(\mathbf{Q}R_N)^*$, and let

$$f = \sum a_n q^{n/N}$$

be its q-expansion, which may have a finite number of terms with $n < 0$. Let $a_m \neq 0$ be the lowest non-zero coefficient. We also write

$$c(f) = a_m.$$

The following criterion will be used constantly.

Lemma 2.1. *If the coefficients $a_n(f \circ \alpha)$ are algebraic integers, and the lowest coefficient $c(f \circ \alpha)$ is a unit in $\mathbf{Z}[\boldsymbol{\mu}_N]$ for all $\alpha \in \mathrm{SL}_2(N)$, then f is a unit over $\mathbf{Z}$.*

Proof. If one forms the polynomial

$$\prod_{\alpha} (T - f \circ \alpha)$$

where the product is taken for all $\alpha \in L_2(N)$, then one obtains a polynomial in T, whose coefficients are of level 1, and have q-expansions with coefficients in $\mathbf{Z}[\boldsymbol{\mu}]$. The q-expansion principle shows that they must be polynomials in $\mathbf{Z}[\boldsymbol{\mu}_N][j]$. This proves that f is integral over $\mathbf{Z}[\boldsymbol{\mu}_N][j]$. The same argument can be applied to f^{-1}, thus proving the lemma.

Theorem 2.2. *Let $a \in (1/N)\mathbf{Z}^2$, and suppose a is primitive* mod $\mathbf{Z}^2$ *(that is, has exact period N* mod $\mathbf{Z}^2$*).*

(i) *If N is composite, then g_a^{12N} is a unit over $\mathbf{Z}$.*
(ii) *If $N = p^n$ is a prime power, then g_a^{12N} is a unit in $R_N[1/p]$.*
(iii) *If $c \in \mathbf{Z}$, $c \neq 0$, c is prime to N, then $(g_{ca}/g_a)^{12N}$ is a unit over $\mathbf{Z}$.*

Proof. Write

$$a = (a_1, a_2) = \left(\frac{r}{N}, \frac{s}{N}\right)$$

with integers r, s which are not both $\equiv 0 \bmod N$. If

$$r \not\equiv 0 \bmod N$$

then the q-expansion of g_a begins with a root of unity times a pure power of q. Hence the only leading coefficients which are not units in the cyclotomic field occur when $r \equiv 0 \bmod N$. In that case, they are powers of $1 - \zeta^s$, and ζ^s is a primitive N-th root of 1. These have the following divisibility properties according as N is a prime power or not. When ζ is a primitive N-th root of unity, and N is composite then $1 - \zeta$ is a unit in $\mathbf{Z}[\boldsymbol{\mu}_N]$. On the other hand, if $N = p^n$, then $1 - \zeta$ is a unit in $\mathbf{Z}[\boldsymbol{\mu}_N, 1/p]$. The theorem then follows from Lemma 2.1. Note that when we take g_{ca}/g_a, we insure that the leading coefficient is a unit in every case, thus taking care of (iii).

§3. The Siegel Units as Universal Distribution

We begin with some general remarks on the Cartan group. Let p be a prime number, $G_p = \mathrm{GL}_k(\mathbf{Z}_p)$ with an arbitrary integer $k \geqq 2$ for the moment, and let C_p be the Cartan subgroup of Chapter 1, §1. We let G_p operate on vertical

k-tuples on the left. Let

$$e_1 = {}^t(1, 0, \ldots, 0)$$

be the first vertical unit vector. Let $G_{p,\infty}$ be the isotropy group of e_1. An element $r = {}^t(r_1, \ldots, r_k) \in \mathbf{Z}_p^k$ is called **primitive** if not all its coordinates are divisible by p. Then C_p operates simply transitively on the primitive elements, as already noted in Chapter 1, §1. Consequently, we have a decomposition

$$G_p = C_p G_{p,\infty}$$

such that every element of G_p has a unique expression as a product of an element in C_p and an element of the isotropy group $G_{p,\infty}$.

The case of interest to us is when $k = 2$, $e_1 = \begin{pmatrix} 1 \\ 0 \end{pmatrix}$. Then $G_{p,\infty}$ consists of the matrices

$$\begin{pmatrix} 1 & b \\ 0 & d \end{pmatrix}.$$

We may then form the product for all p, and also do the same thing at finite level N, i.e. after reduction mod N, for an arbitrary positive integer N. Thus $\mathrm{GL}_k(N)$ acts on $\mathbf{Z}(N)^k$ on the left. We again have a unique decomposition

$$\mathrm{GL}_k(N) = C(N)G_\infty(N)$$

where $G_\infty(N)$ is the isotropy group of e_1. When $k = 2$, it consists of the same matrices already written down, but with integers b, d mod N. The isotropy group will be abbreviated $G_\infty(N)$. Thus

$$G_\infty(N) = \prod_p G_{\infty,p}(p^{n(p)})$$

if $N = \prod p^{n(p)}$ is the prime factor decomposition of N.

We shall also have to deal with the Galois group of the modular function field F_N. In that case, this group is isomorphic to $\mathrm{GL}_2(N)/\pm 1$, and we have the unique decomposition

$$G(N)(\pm) = C(N)(\pm)G_\infty(N),$$

where the $(\pm)$ signifies factoring out ± 1. Of course, we also use the notation $G(N)/\pm 1$, or $C(N)/\pm 1$, or $G(N)^{\pm}$.

By the **standard prime at infinity** of the function field F_N, denoted by P_∞, we mean the place (discrete valuation) induced by the embedding

$$F_N \to \mathbf{Q}(\boldsymbol{\mu}_N)((q^{1/N}))$$

which to each modular function associates its q-expansion. We measure the order at P_∞ in terms of the absolute local parameter q, and abbreviate this order by $\operatorname{ord}_{P_\infty}$. This order may then be a rational number.

For each automorphism σ of F_N over F_1 we have the prime σP_∞, which is such that

$$\operatorname{ord}_{\sigma^{-1}P_\infty}(g) = \operatorname{ord}_{P_\infty}(\sigma g)$$

for every $g \in F_N$. From the q-expansions, we see that the isotropy group of P_∞ consists precisely of the matrices

$$\begin{pmatrix} 1 & b \\ 0 & d \end{pmatrix}.$$

The matrices

$$\begin{pmatrix} 1 & b \\ 0 & 1 \end{pmatrix}$$

constitute the "geometric" part of the isotropy group, over the complex numbers. The diagonal matrices

$$\begin{pmatrix} 1 & 0 \\ 0 & d \end{pmatrix}$$

constitute an "arithmetic" part, which operates on the coefficients of the q-expansion, and thus do not change the order of the function in terms of the parameter q.

Let $\mathscr{D}_N$ be the free abelian group generated by the primes at infinity (cusps), and let $\mathscr{F}_N$ be the subgroup of divisors of units. The factor group

$$\mathscr{C}_N = \mathscr{D}_N/\mathscr{F}_N$$

is called the group of **cuspidal divisor classes** on $X(N)$. We sometimes write $\mathscr{D}(N)$, $\mathscr{F}(N)$, and $\mathscr{C}(N)$ for these groups.

The q-expansion of the Siegel functions shows that

$$\boxed{\operatorname{ord}_{P_\infty}(g_a) = \tfrac{1}{2}\mathbf{B}_2(\langle a_1 \rangle),}$$

for $a = (a_1, a_2) \in (1/N)\mathbf{Z}^2$ and $a \notin \mathbf{Z}^2$. Of course, g_a does not lie in F_N, but it is modular (of higher level), some power g_a^{12N} does lie in F_N, and its q-expansion is still expressible in terms of $q^{1/N}$, so we commit the abuse of notation in writing the order in terms of g_a rather than its $12N$-power.

By **K 2** of §1, we see that $\mathrm{ord}_{P_\infty}(g_a)$ depends only on the class of a mod $\mathbf{Z}^2$, and also on the class of a mod ± 1. For $a \in (\mathbf{Q}/\mathbf{Z})_N$ the function g_a is well defined up to a multiple by a root of unity, and similarly, g_a and g_{-a} differ by a root of unity. Since we shall take divisors of these functions, these roots of unity will disappear.

We shall now apply the general situation of Chapter 1, §4. We define the map T by

$$T : (a_1, a_2) \mapsto a_1.$$

we may then write the order of Siegel functions at the conjugates of P_∞ in the form

$$\mathrm{ord}_{\sigma_\beta^{-1} P_\infty}(g_a) = \frac{1}{N} \mathbf{B}_2\left(\left\langle \frac{T(a\beta)}{N} \right\rangle\right)$$

Finally, the **divisor of** g_a may be written as

$$\boxed{\mathrm{div}(g_a) = \sum_\beta \tfrac{1}{2}\mathbf{B}_2\left(\left\langle \frac{T(a\beta)}{N} \right\rangle\right)(\sigma_\beta^{-1} P_\infty).}$$

The sum is taken over $\beta \in C(N)/\pm 1$.

Since $C(N)/\pm 1$ operates simply transitively on the primes at infinity of F_N, we have an isomorphism of the group ring

$$\mathbf{Z}[C(N)/\pm 1] \to \mathscr{D}$$

with the group of cuspidal divisors, and similarly an isomorphism

$$\mathbf{Q}[C(N)/\pm 1] \to \mathscr{D} \otimes \mathbf{Q}$$

with the free vector space generated by the primes at infinity over the rational numbers. We may then view $\mathrm{div}(g_a)$ as an element of this vector space, and we omit the P_∞ from the divisor notation.

The group ring of the factor group $C(N)/\pm 1$ injects naturally in the group ring of $C(N)$, by mapping

$$\sigma_{\pm\beta} \mapsto \sigma_\beta + \sigma_{-\beta}.$$

(This holds if $N > 2$. For $N = 2$, there is no need to do anything.) Thus we may view the divisor group as injected in the group ring of the full Cartan group, and also, as we have seen in Chapter 1, §3, in the injective limit for all levels N. We now realize that the association

$$a \mapsto \operatorname{div}(g_a)$$

exactly fits the pattern of the Stickelberger distribution, except that g_0 is not defined.

Given a mapping

$$h : \mathbf{Q}^k/\mathbf{Z}^k - \{0\} \to \text{abelian group}$$

which is defined except at 0, and satisfies the distribution relation except at 0, we may call this a **punctured distribution**.

If h is odd, we then define $h(0) = 0$.

If h is even, and for each N we have

$$\sum_{\substack{x \in Z_N \\ x \neq 0}} h(x) = 0,$$

then we define $h(0)$ to have any value.

It is then immediately verified that with this definition, the function on all of $\mathbf{Q}^k/\mathbf{Z}^k$ again satisfies the distribution relation. This may be called the prolongation of h to 0, or to all of $\mathbf{Q}^k/\mathbf{Z}^k$.

In particular, we can define

$$\operatorname{div}(g_0) = \sum_\beta \tfrac{1}{2}\mathbf{B}_2(0)\sigma_\beta^{-1}.$$

Then we obtain:

Theorem 3.1. *The function*

$$a \mapsto \operatorname{div}(g_a)$$

is isomorphic to the Stickelberger distribution associated with $\frac{1}{2}\mathbf{B}_2 \circ T$. It is the universal even distribution on $\mathbf{Q}^2/\mathbf{Z}^2$ into abelian groups on which 2 is invertible. The rank of the group generated by the Siegel functions modulo constants,

$$g_a \quad \textit{for} \quad a \in \frac{1}{N}\mathbf{Z}^2/\mathbf{Z}^2 \bmod \pm 1,$$

is equal to $|C(N)/\pm 1| - 1$, which is $\frac{1}{2}c(N) - 1$ except for $N = 2$.

Proof. This is clear from the formula for $\operatorname{div}(g_a)$. The number giving the rank of the units is one less than $\frac{1}{2}c(N)$ because of the single value at 0 which contributes 1 to the rank of the Stickelberger distribution. Furthermore, two functions have the same divisor if and only if their quotient is a constant. Thus the present situation is a special case of Theorem 6.5 of Chapter 1.

If we view the values g_a as lying in the multiplicative group of the function field modulo constants, then we may also say that the association

$$a \mapsto g_a$$

is the universal even punctured distribution on $\mathbf{Q}^2/\mathbf{Z}^2$.

Theorem 3.2. *Let* $N = p^n$ *be a prime power. Then the Siegel functions* g_a (mod constants), *for a primitive in* $Z_N^*/\pm 1$ *have rank*

$$|C(N)/\pm 1| - 1.$$

Proof. This is immediate from Theorem 3.1 and the structure theorem for a free basis of the universal distribution, Theorem 6.1 of Chapter 1.

§4. The Precise Distribution Relations

We let g be the Siegel functions. For each pair of rational numbers $a = (a_1, a_2)$ not both integral,

$$g_a(\tau) = \mathfrak{k}(a_1\tau + a_2, [\tau, 1])\,\Delta(\tau)^{1/12}.$$

The 12-th root is the "natural" one, and can be written as $\eta(\tau)^2$ where $\eta(\tau)$ is the Dedekind eta function. If we raise the right hand side to the 12-th power, then we may write the value in terms of lattices. Indeed, let t be a point of period exactly N with respect to a lattice L. Then we put

$$\boxed{g^{12}(t, L) = \mathfrak{k}^{12}(t, L)\,\Delta(L).}$$

The value of $\mathfrak{k}^{12}(t, L)$ changes by an N-th root of unity when t changes (additively) by an element of L. Raising further to the N-th power makes the expression on the right hand side dependent only on the value of t mod L. The following proposition from Robert [Ro 1] extends results of Klein and Ramachandra [Ra].

Theorem 4.1. *Let $L' \supset L$ be two lattices, and let N be the smallest positive integer such that $NL' \subset L$. Let*

$$t_1 = 0, \ldots, t_n$$

be a complete system of coset representatives of L'/L. Then

$$\text{(i)} \qquad \prod_{i=2}^{n} g^{12N}(t_i, L) = \left(\frac{\Delta(L')}{\Delta(L)}\right)^N.$$

If t is a complex number such that $Mt \in L$, $t \notin L'$, and

$$m = \text{l.c.m.}(N, M),$$

then

$$\text{(ii)} \qquad \prod_{i=1}^{n} g^{12m}(t + t_i, L) = g^{12m}(t, L').$$

Proof. The proof is based on translating the relation into a relation about modular functions. In order to prove that two modular functions of level m are equal it suffices to prove:

(1) They have the same orders of zeros and poles.
(2) At the standard prime at infinity, the leading coefficient of their q-expansions coincide.

Indeed, from (1) we conclude that their quotient is a constant, and this constant must be equal to 1 by (2). In our applications, we deal with modular functions having zeros or poles only at the points at infinity, so we need only check the orders of zeros and poles of the conjugates of the given modular functions over $\mathbf{C}(j)$ at the standard prime at infinity to verify (1). In fact, we may do so for a fixed power of the two modular functions, since for such a power, the orders get multiplied by the same integer.

The integer m in the theorem is chosen to be least such that each point $t + t_i$ has period m with respect to L. It is a function of the index $(L' : L)$ and of the period of t. If the assertion of the theorem is proved for a triple (L', L, t) and m, then it follows when m is replaced by any multiple, from the definition of g.

For simplicity of notation, we shall sometimes write

$$g_m(t + t_i, L) \quad \text{instead of } g^{12m}(t + t_i, L).$$

As to (i), we can reduce the proof to the case when $(L' : L)$ is prime. Indeed, let

$$L' \supset L_1 \supset L.$$

Let u_i $(i = 1, \ldots, p)$ be representatives of L' mod L_1 and let v_j $(j = 1, \ldots)$ be representatives of L_1 mod L, taking

$$u_1 = v_1 = 0.$$

Then, using induction, we get

$$\begin{aligned}\prod_{i=2}^{n} g^{12N}(t_i, L) &= \prod_{i \neq 1} \prod_{j} g^{12N}(t_i + u_j, L) \prod_{j \neq 1} g^{12N}(u_j, L) \\ &= \prod_{i \neq 1} g^{12N}(t_i, L_1)\left(\frac{\Delta(L_1)}{\Delta(L)}\right)^N \\ &= \left(\frac{\Delta(L')}{\Delta(L)}\right)^N.\end{aligned}$$

This reduces the formula to the case when $(L' : L) = p$, prime, which we assume.

In this case, we can take $L' = [\tau, 1]$ and L to be any one of the lattices

$$[p\tau, 1], \qquad [\tau + b, p] \quad \text{with } 0 \leqq b \leqq p - 1.$$

Coset representatives of L'/L are given by:

$$\begin{aligned} e\tau, &\quad \text{with } 0 \leqq e \leqq p - 1, \quad \text{if } L = [p\tau, 1] \\ e, &\quad \text{with } 0 \leqq e \leqq p - 1, \quad \text{if } L = [\tau + b, p]. \end{aligned}$$

Let us deal with the first case. We have to show that

$$\prod_{e=1}^{p-1} g^{12p}(e\tau, [p\tau, 1]) = \left(\frac{\Delta(\tau)}{\Delta(p\tau)}\right)^p.$$

or equivalently, replacing $p\tau$ by τ,

$$\prod_{e=1}^{p-1} g^{12p}\left(\frac{e\tau}{p}, [\tau, 1]\right) = \left(\frac{\Delta(\tau/p)}{\Delta(\tau)}\right)^p.$$

But the left hand side is equal to

$$\prod_{e=1}^{p-1} q^{12p(1/2)\mathbf{B}_2(e/p)}(1 - q_\tau^{e/p})^{12p} \prod_{n=1}^{\infty} (1 - q_\tau^n q_\tau^{e/p})^{12p}(1 - q_\tau^n/q_\tau^{e/p})^{12p}.$$

By the distribution relation we find

$$\sum_{e=0}^{p-1} p\mathbf{B}_2\left(\frac{e}{p}\right) = \mathbf{B}_2(0) = \tfrac{1}{6},$$

so

$$\sum_{e=1}^{p-1} 12p\tfrac{1}{2}\mathbf{B}_2\left(\frac{e}{p}\right) = 1 - p.$$

Using the standard product

$$\Delta(\tau) = q\prod(1 - q^n)^{24} \quad \text{and} \quad \Delta(\tau/p) = q^{1/p}\prod(1 - q^{n/p})^{24},$$

the desired relation falls out.

The second case is even easier and is left to the reader.

Example. Let us take $L = NL'$. Then we get the relation

$$\prod_{i=2}^{n} g^{12N}(t_i, L) = N^{12N},$$

so the right hand side is constant.

We now prove (ii). Let $L' \supset L_1 \supset L$ for some sublattice L_1 such that $(L_1 : L') = p$ is prime. Let

$$u_1, \ldots, u_{n'}$$

be coset representatives of L_1 in L' (with $u_1 = 0$), and let

$$t_1, \ldots, t_p$$

be coset representatives of L in L_1. Then

$$t_k + u_j \qquad (1 \leqq k \leqq p, 1 \leqq j \leqq n')$$

are coset representatives of L in L'. By induction on the number of prime factors of $(L' : L)$, we see that if we have proved our relation for prime index, then

$$\prod_{i=1}^{n} g_m(t + t_i, L) = \prod_{k=1}^{p} \prod_{j=1}^{n'} g_m(t + t_k + u_j; L)$$
$$= \prod_{k=1}^{p} g_m(t + t_i; L_1)$$
$$= g_m(t, L').$$

This reduces to proving the relation for the integer m when

$$(L' : L) = p,$$

which we assume.

In this case, we can take $L' = [\tau, 1]$, and L is any one of the lattices

$$[p\tau, 1], \qquad [\tau + b, p], \qquad 0 \leqq b \leqq p - 1.$$

Coset representatives of L'/L are given by:

$$\begin{aligned} e\tau, &\quad \text{with } 0 \leqq e \leqq p - 1, \quad \text{if } L = [p\tau, 1] \\ e, &\quad \text{with } 0 \leqq e \leqq p - 1, \quad \text{if } L = [\tau + b, p]. \end{aligned}$$

The point t can then be written

$$t = \frac{rp\tau + s}{M} \qquad \text{if } L = [p\tau, 1],$$

$$t = \frac{r(\tau + b) + sp}{M} \quad \text{if } L = [\tau + b, p].$$

Lemma 4.2. *For any r, s not both* $\equiv 0 \pmod{M}$, *the following pairs of modular functions have the same order at infinity,* i.e. *in their q-expansions:*

(1) $\displaystyle\prod_{e=0}^{p-1} g_m\left(\frac{r(\tau+b)+sp}{M}+e; [\tau+b,p]\right)$ *and* $\displaystyle g_m\left(\frac{r(\tau+b)+sp}{M}; [\tau,1]\right)$

(2) $\displaystyle\prod_{e=0}^{p-1} g_m\left(\frac{rp\tau + s}{M} + e\tau; [p\tau, 1]\right)$ *and* $\displaystyle g_m\left(\frac{rp\tau + s}{M}; [\tau, 1]\right).$

Proof. In the first case, we put

$$\tau' = \frac{\tau + b}{p}.$$

Then $q_{\tau'} = q_{\tau/p}\zeta_p^b$, and

$$\operatorname{ord}_{q_\tau} = \frac{1}{p} \cdot \operatorname{ord}_{q_{\tau'}}.$$

From the basic q-product for the Siegel functions, we see that each term on the left hand side has order at q equal to $1/p$ times the order at q of the right hand side. This proves case (1).

The second case depends on a different phenomenon. This time, we let

$$\tau' = p\tau,$$

so that

$$\operatorname{ord}_{q_{\tau'}} = \frac{1}{p} \cdot \operatorname{ord}_{q_\tau}.$$

The order at q of the left hand side is equal to the sum

$$p \sum_{e=0}^{p-1} 6m\mathbf{B}_2\left(\left\langle \frac{pr/M + e}{p} \right\rangle\right),$$

which, by the basic relation of the Bernoulli polynomial, is equal to

$$6m\mathbf{B}_2\left(\left\langle \frac{rp}{M} \right\rangle\right).$$

This is precisely equal to the order of the right hand side, and our lemma is proved.

The conjugates of a modular function $f(\tau)$ over $\mathbf{C}(j)$ are given by the action of $\mathrm{SL}_2(\mathbf{Z})$, and are of the form $f(\gamma\tau)$ for $\gamma \in \mathrm{SL}_2(\mathbf{Z})$. Under the action of γ, the point t changes, but one sees immediately that the lemma suffices to prove that the two functions occurring say in part (2) of the lemma have the same orders at all points at infinity.

Next we have to check the leading coefficients.

We proceed in an analogous manner. In this case, we have only to look at one conjugate of the modular functions under consideration, say corresponding to the sublattice $L' = [\tau, p]$. From the q-product **K 4** for the Siegel functions, we see that the leading coefficient of the q-expansion of the left hand side is given by the following values:

(a) $r \equiv 0 \pmod{M}$:

$$(1 - e^{2\pi i ps/N})^{12m} \prod_{k=0}^{p-1} e^{2\pi i 6m\frac{r}{M}\frac{s}{M} + \frac{k}{p}} = (1 - e^{2\pi i ps/N})^{12m} e^{2\pi i 6mrsp/M^2}.$$

(b) $r \not\equiv 0 \pmod{M}$: $e^{2\pi i 6mrsp/M^2}$.

The same q-product for

$$g_m\left(\frac{r\tau + ps}{M}; [\tau, 1]\right)$$

shows that the leading coefficient of the function on the right hand side has the same value, and our theorem is proved.

Remark. For most purposes, Theorem 4.1 suffices. In certain applications, if one wishes to get more precise exponents when $L' = L/2$, then one needs a refinement of this theorem, as stated next and also due to Robert. No use will be made of this other than in Lemma 5.2 of Chapter 11.

Theorem 4.3. *Let $L' = L/2$, so $N = 2$, and let t be a complex number such that $4t \in L$ and $2t \notin L$, so $M = 4$ in Theorem 4.1. Then*

$$\prod_{i=1}^{4} g^{12 \cdot 2}(t + t_i, L) = g^{12 \cdot 2}(t, L').$$

Proof. The proof is the same as in Theorem 4.1, except that in the last step, one must check that the leading coefficients of the q-expansions are equal in the present case, which is done by inspection.

Note that in the situation of Theorem 4.3, $m = 4$ but the product relationship is stated for the lower power 2.

§5. The Units over Z

Let S_N be the group generated by the functions g_a, where $Na \in \mathbf{Z}^2$ but $a \notin \mathbf{Z}^2$. In other words, it is the group of functions

$$g = \prod_a g_a^{m(a)}$$

with a satisfying the above conditions. Let

W = group generated by all elements $\zeta, 1 - \zeta$, where $\zeta \in \boldsymbol{\mu}$.

If some element g above is constant, then this constant is the constant term of the q-expansion, and consequently lies in W. Thus the group S_N modulo constants is precisely equal to the group

$$S_N W/W.$$

Let $S_N(\mathbf{Z})$ be the subgroup of S_N consisting of those functions which are units over **Z**. We are interested in the factor groups

$$S_N(\mathbf{Z})W/W \quad \text{and} \quad S_N(\mathbf{Z})\boldsymbol{\mu}/\boldsymbol{\mu}.$$

These are essentially the groups of units over **Z**, modulo constants. To study these groups, we index the functions g_a by elements

$$a \in \left(\frac{1}{N}\mathbf{Z}^2/\mathbf{Z}^2\right)\Big/ \pm 1 = Z_N/\pm 1, \qquad a \neq 0.$$

For this we use formulas **K 0** and **K 2** of §1 which tell us that up to the stated indeterminacy of a, the functions g_a differ only by a root of unity.

The analysis following the q-expansion of the Siegel functions shows that we can write g_a in the form

$$g_a = c(a)g_a^*,$$

where g_a^* is a q-series all of whose coefficients are algebraic integers, and having leading coefficient 1, while the constant $c(a)$ has the following properties:

If N is composite, then $c(a)$ is a unit (as an algebraic number).

If $N = p^n$ is a prime power, then

$$\operatorname{ord}_p c(a) = \begin{cases} 0 & \text{if } \langle a_1 \rangle \neq 0 \\ \operatorname{ord}_p(1-\zeta) & \text{if } \langle a_1 \rangle = 0, \end{cases}$$

and ζ is a p-power root of unity. Furthermore, $c(a) \in W$. If $a_2 = s/N$ with s prime to p, then ζ is a p^n-th root of unity.

We define

$$c(g) = \prod c(a)^{m(a)}.$$

We may call $c(g)$ the **leading coefficient** of g in the q-expansion. It is well defined up to a root of unity, given the indeterminacy of $a \in Z_N/\pm 1$.

The function g is a unit over **Z**, i.e. a unit in the integral closure of $\mathbf{Z}[j]$ in the modular function field F_N if and only if it is a unit at every divisor extending a minimal prime in $\mathbf{Z}[j]$. These primes are of two types: geometric and arithmetic, involving reduction modulo some prime number. Of these, obviously only those involving reduction mod p are relevant, and we have the following criterion:

Lemma 5.1. *The function g is a unit over* **Z** *if and only if $c(g \circ \alpha)$ is a unit for every conjugate $g \circ \alpha$ of g over* $\mathbf{Q}(j)$.

Write

$$g = g_{\text{comp}} \prod_p g_{(p)},$$

where g_{comp} is the product taken over all composite a, and $g_{(p)}$, for each prime p, is the product taken over those a whose denominator is a power of p. By the distribution relation, we may assume that such a are primitive of the same order $p^{n(p)}$. Furthermore, by Theorem 2.2, we know that g_{comp} is a unit over $\mathbf{Z}$.

Lemma 5.2. *The function g is a unit over $\mathbf{Z}$ if and only if $g_{(p)}$ is a unit over $\mathbf{Z}$ for each prime p.*

Proof. Immediate from Theorem 2.2 or Lemma 5.1.

We note that the group $(\mathbf{Z}/N\mathbf{Z})^* = \mathbf{Z}(N)^*$ operates by multiplication on Z_N^*. We use this for $N = p^n$, equal to a prime power.

Theorem 5.3. *Let $N = p^n$* be a prime power. Let

$$g = g_{(p)} = \prod g_a^{m(a)}$$

where the product is taken for $a \in Z_N^*/\pm 1$. Then g is a unit over $\mathbf{Z}$ if and only if for each orbit of $\mathbf{Z}(p^n)^*$ we have

$$\sum_{a \in \text{orbit}} m(a) = 0.$$

Proof. By Lemma 5.2, we know that g is a unit over $\mathbf{Z}$ if and only if

$$\sum_a \operatorname{ord}_p c(a)^{m(a)} = 0.$$

The analysis of the constant $c(a)$ shows that its order at p is $\neq 0$ for exactly one orbit of $\mathbf{Z}(p^n)^*/\pm 1$, and that it is constant for that orbit. Applying this to each conjugate of g yields the theorem, because the orbits are permuted transitively by the elements of the Cartan group.

§6. The Weierstrass Units

The units which we construct in this section are useful for special applications, but the section may be omitted by the reader eager to reach as fast as possible the theorems describing the structure of all units and the cuspidal divisor class group. On the other hand, the section may also serve as motivation for the next chapter.

Let $\wp$ be the Weierstrass elliptic function.

If $a = (a_1, a_2) \in (1/N)\mathbf{Z}^2$ and is primitive mod $\mathbf{Z}^2$, then we let

$$\wp_a(\tau) = \wp(a_1\tau + a_2, [\tau, 1]).$$

The classical factorization of the $\wp$-function in terms of the sigma function can also be expressed in terms of the Klein forms, because the quadratic term in the exponential factor drops out. Thus we have:

$$\wp_a - \wp_b = -\frac{\mathfrak{k}_{a+b}\mathfrak{k}_{a-b}}{\mathfrak{k}_a^2\mathfrak{k}_b^2}.$$

Since we know the q-product for the Klein form, we obtain immediately the q-product for such differences of Weierstrass forms. There is no need to write it down.

Theorem 6.1. *Let*

$$a,\, b,\, c,\, d \in \frac{1}{N}\mathbf{Z}^2$$

be primitive mod $\mathbf{Z}^2$, *and assume that* $a \pm b$ *and* $c \pm d \not\equiv 0$ mod $\mathbf{Z}^2$. *Then the function*

$$\frac{\wp_a - \wp_b}{\wp_c - \wp_d}$$

is a unit in $\mathbf{Q}R_N$.

Proof. The conjugates of this function are again of the same form, as is obvious from the elementary theory of the automorphisms of the modular function field, cf. [Sh] or [L 5]. The differences of Weierstrass functions have no zeros or poles on the upper half plane, so the theorem is clear, since the coefficients of the q-expansion lie in the cyclotomic field.

In certain cases, quotients as in Theorem 6.1 are units in R_N, that is units over $\mathbf{Z}$. To determine these cases, we tabulate the behavior of the leading term of the q-expansion.

Let us identify $\mathbf{Z}^2$ with $\mathbf{Z}\tau + \mathbf{Z}$, where τ is a variable in the upper half plane. If

$$a = a_1\tau + a_2$$

we also write $\wp(a) = \wp(a_1\tau + a_2)$. We recall the **Klein form** whose q-expansion is given by formula **K 4** of Chapter 2, §1:

$$\mathfrak{k}(a) =$$

$$-2\pi i q_\tau^{(1/2)a_1(a_1-1)} e^{2\pi i a_2(a_1-1)/2}(1-q_a)\prod_{n=1}^{\infty}(1-q_\tau^n q_a)(1-q_\tau^n/q_a)(1-q_\tau^n)^{-2}.$$

Let

$$\wp(a,b) = \wp(a) - \wp(b).$$

As we have already mentioned, we have the formula

$$\wp(a,b) = -\frac{\mathfrak{k}(a+b)\mathfrak{k}(a-b)}{\mathfrak{k}(a)^2\mathfrak{k}(b)^2},$$

Suppose that $a, b \in (1/N)(\mathbf{Z}\tau + \mathbf{Z})$. We say that the pair (a, b) is N-**admissible** if a, b, $a+b$ and $a-b$ are primitive of period N mod $\mathbf{Z}\tau + \mathbf{Z}$. If N is fixed during a discussion, we say more briefly that the pair is admissible.

We write

$$\langle \pm a_1 \rangle = \min(\langle a_1 \rangle, \langle -a_1 \rangle)$$

where $\langle a_1 \rangle$ is the representative mod $\mathbf{Z}$ lying in the interval $[0, 1)$.

Lemma 6.2. *Let (a, b) be an N-admissible pair. Then*

$$\operatorname{ord}_q \wp(a,b) = \min(\langle \pm a_1 \rangle, \langle \pm b_1 \rangle).$$

Proof. Since $\wp$ is an even function, one can replace a and b by $-a$ and $-b$ if necessary. Furthermore, if $m \in \mathbf{Z}\tau + \mathbf{Z}$, then we know from **K 2** of §1 that

$$\mathfrak{k}(a+m)/\mathfrak{k}(a) \quad \text{is a root of unity.}$$

We may then assume that

$$0 \leqq a_1 \leqq \tfrac{1}{2} \quad \text{and} \quad 0 \leqq b_1 \leqq \tfrac{1}{2},$$

and, say, that $0 \leqq b_1 \leqq a_1$. The assertion is then immediate by looking at the leading terms which are pure powers of g, and give the order in the q-product for the Klein forms.

If $f_1(q)$ and $f_2(q)$ are two power series in q (possibly with fractional exponents), then we write

$$f_1(q) \approx f_2(q)$$

to mean that $f_1(q) = \lambda f_2(q)u(q)$, where λ is a root of unity, and

$$u(q) = 1 + \sum_{n=1}^{\infty} c_n g^{n/N} \quad \text{with } c_n \in \mathbf{Z}[\boldsymbol{\mu}_N].$$

Lemma 6.3. *Let a, b be an N-admissible pair. Let*

$$e = \min(\langle \pm a_1 \rangle, \langle \pm b_1 \rangle).$$

Then the following table gives the $\approx$-equivalence for $(2\pi i)^2 \wp(a, b)$. We let ζ denote a primitive N-th root of unity.

	$(2\pi i)^2 \wp(a, b) \approx$
Case 1. $\langle a_1 \rangle = 0$ or $\langle b_1 \rangle = 0$	$(1 - \zeta)^2$
In all other cases, $\langle a_1 \rangle \neq 0$, $\langle b_1 \rangle \neq 0$	
Case 2. $\langle a_1 + b_1 \rangle \neq 0$ and $\langle a_1 - b_1 \rangle \neq 0$	$q^e, \quad \frac{1}{N} \leqq e < \frac{1}{2}$
Case 3. Precisely one of $\langle a_1 + b_1 \rangle$ and $\langle a_1 - b_1 \rangle = 0$	$(1 - \zeta)q^e$
Case 4. Both $\langle a_1 + b_1 \rangle$ and $\langle a_1 - b_1 \rangle = 0$. (Equivalent with $\langle a_1 \rangle = \langle b_1 \rangle = \frac{1}{2}$.)	$(1 - \zeta)^2 q^{1/2}$

Proof. The table is immediate from the analysis of the q-expansions for the Siegel functions made in §1.

Theorem 6.4. *Let a be a primitive point of order N. Let r, r', s, s' be integers prime to N such that $r + r'$ and $s + s'$ are also prime to N. Then*

$$\frac{\wp(ra) - \wp(r'a)}{\wp(sa) - \wp(s'a)}$$

is a unit over $\mathbf{Z}$, i.e. *a unit in* R_N. *Its q-expansion has coefficients in* $\mathbf{Z}[\boldsymbol{\mu}_N]$, *and begins with a unit in* $\mathbf{Z}[\boldsymbol{\mu}_N]$ *times a fractional power of q. Its conjugates have the same form.*

Proof. This is immediate from the table.

One may raise the question as to the precise nature of the group of units (modulo constants) generated by the quotients in Theorem 6.1. This question is completely answered in the next chapter.

Here, we go on with further results on quotients of Weierstrass functions. For what follows, it is convenient to use the Weber function

$$f(z, L) = \frac{g_2(L)g_3(L)}{\Delta(L)}\wp(z, L),$$

which gives a way of getting a function of weight 0 out of the Weierstrass function.

For the rest of this section, we assume that N **is prime** $\geqq 5$.

It is also convenient to express this function more algebraically in terms of an elliptic curve A and a point P of order N on A. Suppose A is the "generic" elliptic curve with invariant $j(\tau)$. We consider pairs $(P, \mathfrak{C})$ consisting of a point P of exact order N, and a cyclic group $\mathfrak{C}$ of order N such that P is also of order N mod $\mathfrak{C}$. Given two such pairs $(P, \mathfrak{C})$ and $(P', \mathfrak{C}')$, there exists $\alpha \in \mathrm{GL}(A_N)$ such that

$$\alpha\mathfrak{C} = \mathfrak{C}' \quad \text{and} \quad \alpha P = P'.$$

If

$$A_{\mathbf{C}} \approx \mathbf{C}/[\tau, 1],$$

and P is represented by $(r\tau + s)/N$ in $\mathbf{C}$, then we have an algebraic function f_A on A such that

$$f_A(P) = f\left(\frac{r\tau + s}{N}; [\tau, 1]\right).$$

We shall write the left hand side also as $f(P, A)$. Strictly speaking, we should also indicate the (algebraic) differential of first kind into this notation, which fixes the lattice in an isomorphism class (multiplication by non-zero scalars) but we omit it for simplicity of notation.

Let $P_{\mathfrak{C}}$ be the residue class of P in $A/\mathfrak{C}$. We may form the function

$$f(P_{\mathfrak{C}}, A/\mathfrak{C}) = f_{A/\mathfrak{C}}(P_{\mathfrak{C}}),$$

which we can abbreviate by $f_{A/\mathfrak{C}}(P)$. If $[\tau, 1]$ is the lattice of A, and $\mathfrak{C}$ corresponds to the lattice

$$L \supset [\tau, 1] \quad \text{such that } L/[\tau, 1] \quad \text{is cyclic of order } N,$$

$$f_{A/\mathfrak{C}}(P) = f\left(\frac{r\tau + s}{N}; L\right).$$

Such lattices L are easily classified, and a generating point P also. They are of two types:

Case 1. $L = [\tau, 1/N]$; P represented by $r\tau/N$ with $(r, N) = 1$.

Case 2. $L = [(\tau + v)/N, 1]$; P represented by s/N with $(s, N) = 1$, and $0 \leqq v \leqq N - 1$.

Let r be an integer prime to N. Then the conjugates of

$$f_{A/\mathfrak{C}}(rP)$$

over $\mathbf{Q}(j)$ are of the same form, namely

$$f_{A/\mathfrak{C}'}(rP'),$$

if the pair $(P', \mathfrak{C}')$ is conjugate to $(P, \mathfrak{C})$. In other words, for every automorphism σ of F_N over $\mathbf{Q}(j)$, we have

$$(f_{A/\mathfrak{C}}(rP))^\sigma = f_{A/\sigma\mathfrak{C}}(rP^\sigma),$$

because we can take A defined over $\mathbf{Q}(j)$, fixed by σ.

We used the function f for convenience. In practice, when we take quotients and the factor $g_2 g_3/\Delta$ disappears, it is then customary to deal directly with the $\wp$-function and the modular form of weight 2, so we write

$$\wp(P, A) = \wp\left(\frac{r\tau + s}{N}, [\tau, 1]\right),$$

if $A_{\mathbf{C}} = \mathbf{C}/[\tau, 1]$.

A pair of points $P, Q \in A_N$ is said to be **admissible** for $\mathfrak{C}$ if it satisfies the following conditions.

The cyclic groups (P) and (Q) generated by P, Q are equal. The points P, Q, $P \pm Q$ are not in $\mathfrak{C}$.

Since N is prime $\geqq 5$, admissible pairs obviously exist. We define:

$$\wp[P, Q; \mathfrak{C}] = \wp(P, A/\mathfrak{C}) - \wp(Q, A/\mathfrak{C}).$$

Write $Q = rP$ for some integer r (necessarily prime to N). The association

$$(P, \mathfrak{C}; A) \mapsto \wp[P, rP; \mathfrak{C}] = \wp(P, A/\mathfrak{C}) - \wp(rP, A/\mathfrak{C})$$

defines a modular form on $X(N)$.

Theorem 6.5. *Let N be prime $\geqq 5$. For (P_1, Q_1) admissible for $\mathfrak{C}_1$, and (P_2, Q_2) admissible for $\mathfrak{C}_2$, the expression*

$$u = \frac{\wp[P_1, Q_1; \mathfrak{C}_1]}{\wp[P_2, Q_2; \mathfrak{C}_2]}$$

defines a modular function on $X(N)$, having the following properties.

(i) *It is a unit in $\mathbf{Q}R_N$. If $\mathfrak{C}_1 = \mathfrak{C}_2$, then it is a unit in R_N,* i.e. *it is a unit over* $\mathbf{Z}$.
(ii) *Its conjugates over $\mathbf{Q}(j)$ are of the same type as u.*
(iii) *Its divisor on $X(N)$ is the N-*th *multiple of a divisor.*

Proof. The first statement follows as usual from the fact that the function, expressed in terms of τ, has no zero or pole on the upper half plane. If $\mathfrak{C}_1 = \mathfrak{C}_2$, then there are integers a, b, c, d prime to N, such that

$$a \pm b \quad \text{and} \quad c \pm d \quad \text{are prime to } N;$$

and there is an integer r prime to N such that one of the conjugates of u has the form

$$u(\tau) = \frac{f(ar\tau, [N\tau, 1]) - f(br\tau, [N\tau, 1])}{f(cr\tau, [N\tau, 1]) - f(dr\tau, [N\tau, 1])}.$$

Put $\tau' = N\tau$ so $\tau = \tau'/N$. Arguing as in Theorem 6.4, we see that u is integral over $\mathbf{Z}[j(\tau')] = \mathbf{Z}[j(N\tau)]$. Since $j(N\tau)$ is integral over $\mathbf{Z}[j(\tau)]$, it follows that u is integral over $\mathbf{Z}[j]$, whence u is a unit in R_N because u and u^{-1} have the same shape.

Now let us return to the case when $\mathfrak{C}_1$, $\mathfrak{C}_2$ are not necessarily equal. When we take the quotient, we may do so either with the Weierstrass function or the Weber function. Note that the classical q-expansions for g_2, g_3,

Δ as functions of τ or of

$$\frac{\tau + \nu}{N}$$

as in Case 2 all start with a non-zero constant term, and consequently that the factor $g_2 g_3/\Delta$ does not contribute to the zeros or poles at infinity of the unit u, or any of its conjugates.

Consider the function

$$f[P, Q; \mathfrak{C}] = f(P, A/\mathfrak{C}) - f(Q, A/\mathfrak{C}).$$

We look at the q-expansion of this function and its conjugates, according to Case 1 or Case 2 above.

Case 1. No fractional power of q occurs in that expansion, or in other words, only integral powers of q occur.

Case 2. Fractional powers of q may occur, but the power series begins with a non-zero constant term.

These assertions are immediately verified, either by using the q-expansions for the Klein forms, or using directly the q-expansion for the Weierstrass function, cf. [L 5], Chapter 4. In Case 1, this implies that the order of the function at this particular conjugate is divisible by N, since we measure the order in terms of the parameter $q^{1/N}$ on the modular curve $X(N)$. In Case 2, the order is equal to 0, so also divisible by N. This proves assertions (ii) and (iii), and concludes the proof.

CHAPTER 3

Quadratic Relations

In Chapter 2 we gave the simple transformation law of the Klein forms. It is quite easy from this to deduce when a power product of Klein forms is modular of a given level. We also want to determine when power products of other functions are modular of level N, for instance forms which are differences of Weierstrass $\wp$-functions. A standard elementary formula from the theory of elliptic functions tells us that

$$\wp(u) - \wp(v) = -\frac{\sigma(u+v)\sigma(u-v)}{\sigma(u)^2\sigma(v)^2}.$$

Formally, the right hand side is a "parallelogram". This leads into a purely algebraic formalism about quadratic relations and quadratic relations mod N. This formalism reduces analytic questions concerning modular forms to algebraic questions concerning universal quadratic relations. We begin with the algebraic theory of quadratic forms over the integers, giving two descriptions of universal relations. We then apply these to forms of various types. We shall also see that all the various groups of units which we can generate by means of Weierstrass forms, Klein forms, etc. essentially give the same group.

This entire chapter closely follows Kubert's paper [Ku 2].

§1. Formal Quadratic Relations

We shall deal with quadratic forms on $\mathbf{Z} \times \mathbf{Z}$. For notation, we let

$$r = (r_1, r_2) \in \mathbf{Z} \times \mathbf{Z}$$

denote pairs of integers, not both equal to 0. Let:

F = free abelian group generated by the elements of $(\mathbf{Z} \times \mathbf{Z})/\pm 1$.

For this chapter, we call F the group of **divisors** on $(\mathbf{Z} \times \mathbf{Z})/\pm 1$, or divisors for short. If $r \in \mathbf{Z} \times \mathbf{Z}$, we denote by (r) the divisor having $\pm r$ with multiplicity 1, and all other elements of $(\mathbf{Z} \times \mathbf{Z})/\pm 1$ with multiplicity 0. We shall eventually apply all this to lattices, so viewing τ as a variable, we sometimes write

$$r = r_1\tau + r_2,$$

in which case we also write $(r) = (r_1\tau + r_2)$. In other words, we identify $\mathbf{Z} \times \mathbf{Z}$ with $\mathbf{Z}\tau + \mathbf{Z}$, as a 2-dimensional free module over the integers.

Let

$$\alpha = \sum_{(r)} m(r)(r)$$

be a divisor. We say that α satisfies the **quadratic relation** if it satisfies

QUAD. $\qquad \sum m(r)r_1^2 = \sum m(r)r_2^2 = \sum m(r)r_1 r_2 = 0.$

The subgroup of F generated by all divisors satisfying the quadratic relations will be called the subgroup of **quadratic relations**, and will be denoted by K.

For each pair of elements $r, s \in \mathbf{Z} \times \mathbf{Z}$ not both zero, we let the **parallelogram** associated with r, s be the element of F given by

$$p(r, s) = (r + s) + (r - s) - 2(r) - 2(s).$$

Theorem 1.1. *The group of quadratic relations is generated by the parallelograms and the relations*

$$c^2(r) - (cr) \quad \textit{for } c \in \mathbf{Z} \textit{ and all } r.$$

Proof. This theorem will be a consequence of a stronger one as follows.

The **degree** of a divisor is the integer

$$\deg \alpha = \sum m(r),$$

equal to the sum of its coefficients. We let F^+ be the subgroup of divisors of even degree. We let

$$K^+ = K \cap F^+$$

be the subgroup of K consisting of divisors of even degree.

Theorem 1.2. *The subgroup K^+ is generated by the parallelogram elements.*

It is clear that Theorem 1.2 implies Theorem 1.1, because a relation $4(r) - (2r) = 0$ has degree 3, parallelograms have degree -2, so by subtracting we get an element of degree 1. Given any element of K of odd degree, we can subtract a combination as above to transform it into an element of even degree, and then apply Theorem 1.2 to conclude the proof of Theorem 1.1.

Let us now prove Theorem 1.2. Let K_0^+ be the subgroup of K^+ generated by the parallelograms. We have to show that

$$K_0^+ = K^+.$$

We define the **height**

$$h(r_1, r_2) = |r_1| + |r_2|,$$

and

$$h(\sum m(r)(r)) = \max h(r) \quad \text{for } m(r) \neq 0.$$

Given an element $\beta \in K^+$, we first show that we can find $\alpha \in K_0^+$ such that

$$h(\beta - \alpha) \leqq 2.$$

Write the element β in the form $\beta = r\tau + s$, where r, s are integers, and suppose $h(\beta) \geqq 3$. We find integers (r', s') such that

$$\begin{aligned} |r - r'| + |s - s'| &< |r| + |s| \\ |r - 2r'| + |s - 2s'| &< |r| + |s| \\ |r'| + |s'| &< |r| + |s| \\ (r', s') &\neq (r - r', s - s'). \end{aligned}$$

This is trivially done. We then let

$$a = r'\tau + s' \quad \text{and} \quad b = (r - r')\tau + (s - s').$$

Then $a + b = r\tau + s$, and $\beta - p(a, b)$ has lower height than β. This reduces the proof to the case when $h(\beta) \leqq 2$, and therefore to a finite number of explicit cases, which can be handled explicitly one at a time, as follows.

We note that

$$(\tau + 1) + (\tau - 1) \equiv 2(\tau) + 2(1) \qquad \text{mod } K_0^+,$$

as one sees by using the parallelogram $p(a, b)$ where $a = (\tau)$ and $b = (1)$. Similarly,

$$(2\tau) + (2) \equiv 2(\tau + 1) + 2(\tau - 1) \qquad \text{mod } K_0^+,$$

as one sees by using the parallelogram $p(a, b)$ where

$$a = (\tau - 1) \quad \text{and} \quad b = (\tau + 1).$$

Considering all possible cases of height $\leqq 2$, we are therefore reduced to proving that if a linear combination of

$$(\tau), \qquad (1), \qquad (\tau + 1), \qquad (2\tau)$$

lies in K^+, then it lies in K_0^+. We need a lemma.

Lemma 1.3. *The element* $8(\tau) - 2(2\tau)$ *is in* K_0^+.

Proof. We can write this element in the form

$$\alpha - \beta - 2\gamma,$$

where:

$$\begin{aligned} \alpha &= (5\tau) + (\tau) - 2(2\tau) - 2(3\tau) - [(5\tau) + (3\tau) - 2(\tau) - 2(4\tau)] \\ \beta &= (3\tau) + (\tau) - 2(\tau) - 2(2\tau) \\ \gamma &= (4\tau) + (2\tau) - 2(\tau) - 2(3\tau), \end{aligned}$$

and α, β, γ obviously lie in K_0^+. This proves the lemma.

Now suppose that we have an element $\xi \in K^+$, written in the form

$$\xi = m_1(\tau) + m_2(1) + m_3(2\tau) + m_4(\tau + 1).$$

By definition,

$$m_1 + 4m_3 + m_4 = 0, \qquad m_2 + m_4 = 0, \quad \text{and} \quad m_4 = 0.$$

It follows that $m_2 = 0$, so that m_1 is even. Furthermore

$$m_1 + 4m_3 = 0.$$

But $m_1 + m_3$ is even by assumption, so m_3 is even, $m_3 = 2n_3$, and $m_1 = -8n_3$. We can now use the lemma to conclude the proof.

§2. The Even Primitive Elements

We want a theorem also giving various generators for the group generated by primitive elements. Thus we are led to making further definitions.

Let $\alpha \in K$. We shall say that α is **primitive** (of level N) if any element $r\tau + s$ occurring with non-zero multiplicity in α is primitive mod N, that is has period N in $\mathbf{Z}^2/N\mathbf{Z}^2$. We let:

$K^+_{\mathrm{pr}}(N) =$ subgroup of K^+ consisting of primitive elements.

$K_{\mathrm{sp}}(N) =$ subgroup generated by primitive parallelograms, i.e. parallelograms

$$(r + s) + (r - s) - 2(r) - 2(s)$$

such that $r + s$, $r - s$, r, s are primitive. For simplicity of notation, we omit the N, and write simply K^+_{pr} and K_{sp}. We call the elements of K_{sp} **special**, and call K_{sp} the **special subgroup**, with respect to N. Obviously K_{sp} is contained in K^+_{pr}.

Theorem 2.1. *We have* $K_{\mathrm{sp}}(N) = K^+_{\mathrm{pr}}(N)$.

It is obvious that this theorem implies the previous one, by considering primitive elements with respect to all divisors of N.

The desired inclusion $K^+_{\mathrm{pr}} \subset K_{\mathrm{sp}}$ will be proved by a recursive procedure, depending on a descent with respect to two norms. We work with both coordinates of an element of $\mathbf{Z} \times \mathbf{Z}$, and to omit indices, it is now convenient to let r, s denote integers. We also write our "elementary" divisors in the form $r\tau + s$. We define

$$h(r\tau + s) = |r| + |s| \quad \text{and} \quad h_2(r\tau + s) = |s|.$$

Lemma 2.2. *Let β be primitive in F, such that $h_2(\beta) \geqq 2$. Then there exists a special element α such that*

$$h_2(\beta - \alpha) = 0 \quad \textit{or} \quad 1.$$

Proof. We may assume that β consists of a single element, say

$$\beta = (r\tau + s), \quad \text{with } |s| \geqq 2.$$

We may assume that $h_2(\beta) = s > 0$. (Remember that an element is identified with its negative.) We shall use a special element α of the form

$$(r\tau + s) + ((r - 2x)\tau + (s - 2)) - 2(x\tau + 1) - 2((r - x)\tau + (s - 1)),$$

where x is to be determined, in order to insure primitivity. Once we have found x, it is then clear that we have achieved our purposes since $h_2(\beta - \alpha)$ has decreased by at least one unit, and we can repeat the procedure to obtain the statement of the lemma.

Write $N = \prod p^{n(p)}$. To achieve the desired primitivity of the elements occurring in α, it suffices to do so mod $p^{n(p)}$ for each p dividing N.

If $p = 2$ and r is even (the worst case), we let $x_p = r - 1$. Since s must be odd, each one of our elements has the period $2^{n(2)}$ mod $2^{n(2)}$.

In all other cases, the equation

$$(r - x_p)(r - 2x_p) \equiv 0 \pmod{p}$$

has at most two solutions, so we can find x_p which is not a solution.

Finally, we pick $x \equiv x_p (\mathrm{mod}\, p^{n(p)})$ for all p, thereby finishing the proof of the lemma.

We have therefore reduced the proof of the theorem to the cases when $s = 0$ or $s = \pm 1$.

Lemma 2.3. *For any $r \in \mathbf{Z}$ we have*

$$(r\tau \pm 1) \equiv 0 \bmod K_{\mathrm{sp}}, (\tau), (\tau + 1), (1).$$

The congruence on the right is of course meant modulo the group generated by K_{sp} and the indicated elements.

Proof. Since elements of $(\mathbf{Z} \times \mathbf{Z})/\pm 1$ are identified mod ± 1, we may assume without loss of generality that $r > 0$. We shall prove that if $r \geqq 2$, then there exists a special α such that

$$h(\beta - \alpha) < h(\beta).$$

Indeed, we take

$$\alpha = (r\tau \pm 1) + ((r - 2)\tau \pm 1) - 2(\tau) - 2((r - 1)\tau \pm 1).$$

We may therefore proceed again recursively to prove our lemma, using the element

$$(\tau + 1) + (\tau - 1) - 2(\tau) - 2(1)$$

to show that $(\tau - 1)$ lies in $K_{\mathrm{sp}}, (\tau), (\tau + 1), (1)$.

Lemma 2.4. *Let $(r\tau)$ be primitive.*

(i) *If N is even or if r is odd, then*

$$(r\tau) \equiv 0 \bmod K_{sp}(N), (\tau), (\tau + 1), (1).$$

(ii) *If N is odd and r is even, then*

$$(r\tau) \equiv 0 \bmod K_{sp}(N), (\tau), (\tau + 1), (1), (2).$$

Proof. Let $\beta = (r\tau)$. If r is odd, $|r| > 3$, we write $r = 2x + 1$. We take the special element

$$\alpha = (r\tau) + (\tau + 2) - 2(-x\tau - 1) - 2((x + 1)\tau - 1).$$

Then $h(\beta - \alpha) < h(\beta)$. Recursively, we can therefore show that

$$\beta \equiv 0 \bmod K_{sp}, (\tau), (\tau + 1), (1), (\tau + 2).$$

However, the fact that

$$(\tau + 2) + (\tau) - 2(\tau + 1) - 2(1)$$

is special shows that the preceding congruence reduces to the one specified in the lemma, as was to be shown, in case (i).

Suppose next that r is even. Then N is odd because $r\tau$ is assumed primitive. Write $r = 2x$, and let

$$\alpha = (r\tau) + (2) - 2(x\tau + 1) - 2(x\tau - 1).$$

The same type of argument as before concludes the proof.

Putting the three lemmas together, we have shown:

Theorem 2.5. *If β is primitive of level N in F, then*

(i) $\beta \equiv 0 \bmod K_{sp}(N), (\tau), (\tau + 1), (1)$ *if N is even*
(ii) $\beta \equiv 0 \bmod K_{sp}(N), (\tau), (\tau + 1), (1), (2)$ *if N is odd.*

Observe that this theorem applies to any primitive element of F. In order to get a proof of Theorem 2.1, we must of course start using particular properties of K_{pr}^{+}, which we now do.

Suppose given an element

$$m_1(1) + m_2(\tau) + m_3(\tau + 1),$$

of which we need only assume that it satisfies the quadratic relations **QUAD**. Then

$$m_1 + m_3 = 0, \qquad m_2 + m_3 = 0, \qquad m_3 = 0.$$

Therefore all $m_i = 0$ and we are done in the first case of Theorem 2.5, applied to our special situation.

In the second case, assume that we have an element

$$m_1(1) + m_2(2) + m_3(\tau) + m_4(\tau + 1)$$

which satisfies the quadratic relations, and whose degree $\sum m_i$ is even. The third quadratic relation shows that $m_4 = 0$. Furthermore,

$$m_1 + 4m_2 = 0 \quad \text{and} \quad m_3 = 0.$$

Therefore m_1 is even, and it follows that m_2 is even because the degree is even. It will therefore suffice to prove the next lemma.

Lemma 2.6. *The element* $2(2) - 8(1)$ *lies in the group generated by the primitive parallelograms.*

Proof. We can write

$$8(1) - 2(2) = p(\tau - 2, 1) + 2p(\tau - 1, 1) + p(\tau, 1) - p(\tau - 1, 2).$$

Let $\alpha = \sum m(r)(r)$ be an element of K. We call α **non-trivial** mod N if for every (r) such that $m(r) \neq 0$, we have

$$r = (r_1, r_2) \neq 0 \qquad (\text{mod } N).$$

We let:

$K_{\text{nt}}(N) =$ subgroup of K generated by the elements non-trivial mod N.
$K_{\text{nt}}^{+}(N) =$ subgroup of K_{nt} consisting of elements of even degree.

Theorem 2.7. *The group* K_{nt}^{+} *is generated by the non-trivial parallelograms* mod N, i.e. *the parallelograms* $p(r, s)$ *with*

$$r + s,\ r - s,\ r,\ s \not\equiv 0 \qquad (\text{mod } N).$$

Proof. This is immediate from the theorem, because for non-trivial r mod N, $r_1\tau + r_2$ is primitive for some m dividing N, the two cases apply, and the argument of Theorem 2.5 applies.

§3. Weierstrass Forms

If not both r_1, r_2 are $\equiv 0 \bmod N$, we define

$$\sigma_r(\tau, 1) = \sigma_r(\tau) = \sigma\left(\frac{r_1\tau + r_2}{N}; [\tau, 1]\right),$$

where σ is the Weierstrass sigma function, whose addition formula was recalled in Chapter 2, §1. We call σ_r a **Weierstrass form.**

Theorem 3.1. *Let $\{m(r)\}$ be a family of integers, associated with pairs of integers $r = (r_1, r_2) \in \mathbf{Z}^2$, not both $\equiv 0 \bmod N$. Assume also that almost all $m(r) = 0$. The form*

$$f = \prod_r \sigma_r^{m(r)}$$

is modular with respect to $\Gamma(N)$ if and only if the family $\{m(r)\}$ satisfies the quadratic relations **QUAD**.

Proof. Let

$$\gamma = \begin{pmatrix} a & b \\ c & d \end{pmatrix}$$

be an element of $\Gamma(N)$, so that we can write

$$\begin{aligned} a - 1 &= xN & b &= yN \\ c &= zN & d - 1 &= wN \end{aligned}$$

with integers x, y, z, w. We use the addition formula of σ to see that for any given choice of exponents $m(r)$, the function f transforms under γ by picking up an exponential factor and a power of -1. We use the Legendre relation

$$\eta_2\tau = \eta_1 + 2\pi i$$

once on the exponential factor. We see immediately that this factor is equal to 1 for all $\gamma \in \Gamma(N)$ if and only if the three relations among the coordinates r_i, r_j are satisfied. Observe also that the first relation mod 2 is equivalent to

$$\sum m(r_i)r_i \equiv 0 \pmod 2,$$

because $r_i^2 \equiv r_i \pmod 2$. The explicit computation of the power of -1 shows that it also disappears under the stated conditions.

We want to translate Theorem 2.1 and the corollary into statements about forms. We introduce some notation. To each form

$$f = \prod \mathfrak{k}_r^{m(r)} \quad \text{or} \quad \prod \sigma_r^{m(r)}$$

we associate its **divisor** (depending on a given product representation as above)

$$(f) = \sum m(r)(r).$$

Conversely, if

$$\alpha = \sum m(r)(r)$$

is a divisor, we let

$$\sigma^{(\alpha)} = \prod \sigma_r^{m(r)} \quad \text{and} \quad \mathfrak{k}^{(\alpha)} = \prod \mathfrak{k}_r^{m(r)}.$$

We let:

$\mathfrak{S}(N)$ = subgroup of the group generated by the Weierstrass forms consisting of all the elements which have level N.

= group of forms expressed as products in Theorem 3.1 satisfying the quadratic relations **QUAD**.

$\mathfrak{S}^+(N)$ = subgroup of $\mathfrak{S}(N)$ consisting of elements of even weight.

$\mathfrak{S}_{\text{pr}}^+(N)$ = group generated by the elements $\sigma^{(\alpha)}$, where $\alpha \in K_{\text{pr}}^+(N)$.

Remark. It is frequently useful to know that -1 lies in $\mathfrak{S}^+(N)$. This is easily seen by observing that the divisor

$$(1) - (-1)$$

satisfies the quadratic relations, and

$$\frac{\sigma_{0,1}}{\sigma_{0,-1}} = -1.$$

Theorem 3.2. *The group $\mathfrak{S}^+(N)$ is generated by the forms*

$$\wp_r - \wp_s = -\frac{\sigma_{r+s}\sigma_{r-s}}{\sigma_r^2\sigma_s^2}.$$

The group $\mathfrak{S}_{\text{pr}}^+(N)$ is generated by these expressions where r, s, $r+s$, $r-s$ are primitive.

Note that -1 lies in the group generated by the forms $\wp_r - \wp_s$, since we can also use the difference $\wp_s - \wp_r$ and take the quotient to get -1.

§4. The Klein Forms

Let $\mathfrak{K}(N)$ be the group of forms which are generated by Klein forms, and are modular with respect to $\Gamma(N)$, or as we also say, on $\Gamma(N)$. We want to know when a product

$$f = \prod \mathfrak{k}_r^{m(r)}$$

is in $\mathfrak{K}(N)$.

Theorem 4.1. *A product f as above is modular of level N if and only if the family $\{m(r)\}$ satisfies the following conditions.*

***N* odd.**

QUAD(N)odd. $\sum m(r)r_1^2 \equiv \sum m(r)r_2^2 \equiv \sum m(r)r_1r_2 \equiv 0 \pmod N$

***N* even.**

QUAD(N)even.

$$\sum m(r)r_1^2 \equiv \sum m(r)r_2^2 \equiv 0 \pmod{2N}$$
$$\sum m(r)r_1r_2 \equiv 0 \pmod N.$$

Proof. It is easily checked from **K 3** of Chapter 2, §1 that the relations **QUAD(N)** in both cases imply that f is on $\Gamma(N)$. Cf. Theorem 1.1 of Chapter 2.

Conversely, we shall leave the easier case N odd to the reader. Suppose that N is even. For any

$$\alpha = \begin{pmatrix} a & b \\ c & d \end{pmatrix} \in \mathrm{SL}_2(\mathbf{Z})$$

we have

$$\frac{f\left(\alpha\begin{pmatrix}\omega_1\\ \omega_2\end{pmatrix}\right)}{f\begin{pmatrix}\omega_1\\ \omega_2\end{pmatrix}} = e^{\pi i \sum E(r_1, r_2; a, b, c, d)}$$

where $E(r_1, r_2; a, b, c, d)$ is a simple expression easily calculated from **K 3**, namely:

$$E(r_1, r_2; a, b, c, d) =$$

$$m(r_1, r_2)\left[\frac{ab}{N^2} r_1^2 + \frac{c(d-1)-c}{N^2} r_2^2 + \left(\frac{b}{N} + \frac{a-1}{N}\right) r_1 \right.$$

$$\left. + \left(\frac{d-1}{N} + \frac{c}{N}\right) r_2 + \left(\frac{cb}{N^2} + \frac{(a-1)(d-1)}{N^2}\right) r_1 r_2 + \frac{d-a}{N} r_1 r_2 \right].$$

If f is on $\Gamma(N)$, then $\sum E(r_1, r_2; a, b, c, d)$ is an even integer, for all $\alpha \in \Gamma(N)$. We shall use special values of α. First use

$$\alpha = \begin{pmatrix} 1 & N \\ 0 & 1 \end{pmatrix} \quad \text{and} \quad \alpha = \begin{pmatrix} 1 & 0 \\ N & 1 \end{pmatrix}.$$

Then we find

$$\sum m(r_1, r_2)(r_1 N + r_1^2) \equiv 0 \pmod{2N}$$
$$\sum m(r_1, r_2)(r_2 N + r_2^2) \equiv 0 \pmod{2N}.$$

Since N is even, we get

$$\sum m(r_1, r_2) r_1^2 \equiv 0 \pmod{2},$$

and since $r_1^2 \equiv r_1 \pmod{2}$, we also get

$$\sum m(r_1, r_2) r_1 \equiv 0 \pmod{2}.$$

Hence

$$\sum m(r_1, r_2) r_1^2 \equiv 0 \pmod{2N}.$$

The same goes for r_2^2 replacing r_1^2, so that two of the three desired relations are proved.

For the cross term involving $r_1 r_2$, we observe that in the sum

$$E(r_1, r_2; a, b, c, d)$$

the sum of all the terms, except possibly for the $r_1 r_2$ terms, are even integers, by using what we have just proved, so for instance

$$\frac{ab}{N^2} \sum m(r_1, r_2) r_1^2 \quad \text{is an even integer,}$$

and similarly for the other terms. To deal with the r_1r_2 terms, we use a third element α, namely

$$\begin{pmatrix} a & b \\ c & d \end{pmatrix} = \begin{pmatrix} 1-N & N \\ -N & 1+N \end{pmatrix}$$

Then

$$\sum m(r_1, r_2)\left(\frac{cb}{N^2} + \frac{(a-1)(d-1)}{N^2}\right) r_1 r_2 \quad \text{is an even integer.}$$

Hence

$$\sum m(r_1, r_2)\frac{(d-a)}{N^2} r_1 r_2 = \text{even integer},$$

so that

$$\sum m(r_1, r_2)\frac{2}{N} r_1 r_2 = \text{even integer},$$

and finally

$$\sum m(r_1, r_2) r_1 r_2 \equiv 0 \pmod{N},$$

as was to be shown.

Theorem 4.2. $\mathfrak{S}(N) \subset \mathfrak{K}(N)$.

Proof. Suppose

$$f = \prod \sigma_r^{m(r)} \in \mathfrak{S}(N).$$

Then

$$f = \prod \mathfrak{k}_r^{m(r)},$$

because the quadratic relations **QUAD** show that the extra exponential factor introduced to get the Klein forms from the Weierstrass forms will disappear under those conditions. This proves the desired inclusion.

More substantially, we have the converse for elements of even degree.

Theorem 4.3. $\mathfrak{K}^+(N) \subset \mathfrak{S}^+(N)$, *and therefore* $\mathfrak{K}^+(N) = \mathfrak{S}^+(N)$.

Proof. To begin, we observe that

$$\wp_r - \wp_s = -\frac{\mathfrak{k}_{r+s}\mathfrak{k}_{r-s}}{\mathfrak{k}_r^2\mathfrak{k}_s^2}.$$

(The extra factors distinguishing the Klein forms from the Weierstrass forms cancel.) Thus the "parallelograms" built out of Klein forms or Weierstrass forms coincide.

Lemma 4.4. *If α satisfies the quadratic relations* **QUAD**, *then*

$$\sigma^{(\alpha)} = \mathfrak{k}^{(\alpha)}.$$

Proof. Obvious, because the Klein forms and Weierstrass forms differ by a term which is quadratic in the exponent.

Let $f \in \mathfrak{R}^+(N)$. Then Theorem 4 shows that

$$(f) \equiv 0 \bmod \sum_{D|N} K_{\mathrm{sp}}(D), (\tau), (\tau + 1), (1) \qquad \text{if} \quad N = 2$$

$$(f) \equiv 0 \bmod \sum_{D|N} K_{\mathrm{sp}}(D), (\tau), (\tau + 1), (1), (2) \quad \text{if} \quad N \neq 2$$

Therefore, to prove the theorem, it suffices to prove the next lemma.

Lemma 4.5. *Let $f \in \mathfrak{R}^+(N)$. If $N \neq 2$, and*

$$(f) \equiv 0 \bmod (\tau), (\tau + 1), (1), (2),$$

then f lies in $\mathfrak{S}^+(N)$. If $N = 2$, then $(f) \equiv 0 \bmod (\tau), (\tau + 1), (1)$.

Proof. Write

$$(f) = m_1(1) + m_2(2) + m_3(\tau) + m_4(\tau + 1),$$

such that $\sum m_i$ is even, and satisfy **QUAD(N)**.

Since f has even weight, it suffices to prove $f \in \mathfrak{S}(N)$. We distinguish cases.

***N* odd.** The quadratic relations **QUAD(N)** give us

$$m_1 + 4m_2 \equiv 0 \pmod N$$
$$m_3 \equiv 0 \pmod N$$
$$m_4 \equiv 0 \pmod N.$$

We can therefore rewrite

$$(f) = (m_1 + 4m_2)(1) + m_2((2) - 4(1)) + m_3(\tau) + m_4(\tau).$$

Let $n = m_1 + 4m_2$. Then $n \equiv 0 \pmod{N}$. By Lemma 4.4, the second term above is in $\mathfrak{S}(N)$. We now show that $N(1)$ is in $\mathfrak{S}(N)$. If we can show the existence of integers q_1, q_2 such that

$$\alpha = q_1(1) + (N - q_1)(1 + N) + q_2(2) - q_2(2 + N)$$

and q_1, q_2 satisfy the quadratic relations, then

$$\mathfrak{k}^{N(1)} = \mathfrak{k}^{\alpha} \in \mathfrak{S}(N) \quad \text{by Lemma 4.4}$$

and we have what we want. The quadratic relations amount to

$$q_1 + (N - q_1)(N + 1)^2 + 4q_2 - q_2(N + 2)^2 = 0,$$

i.e.

$$q_1(N + 2) + q_2(N + 4) = (N + 1)^2,$$

which can be solved with $q_1, q_2 \in \mathbf{Z}$ since N is odd.

By symmetry, it also follows that $N(\tau), N(\tau + 1)$ are in $\mathfrak{S}(N)$, thereby concluding the odd case.

***N* even, $N \neq 2$.** We write again

$$(f) = m_1(1) + m_2(2) + m_3(\tau) + m_4(\tau + 1).$$

Then

$$\begin{aligned} n_1 &= m_1 + 4m_2 + m_4 \equiv 0 \pmod{2N} \\ n_3 &\equiv \qquad\qquad\; m_3 + m_4 \equiv 0 \pmod{2N} \\ n_4 &= \qquad\qquad\qquad\;\; m_4 \equiv 0 \pmod{N}. \end{aligned}$$

We also put $n_2 = m_2$. We rewrite (f) in the form

$$\begin{aligned} (f) &= (m_1 + 4m_2 + m_4)(1) + m_2((2) - 4(1)) + (m_3 + m_4)(\tau) \\ &\quad + m_4((\tau + 1) - (1) - (\tau)) \\ &= n_1(1) + n_2((2) - 4(1)) + n_3(\tau) + n_4((\tau + 1) - (1) - (\tau)). \end{aligned}$$

By Lemma 4.4, the second term lies in $\mathfrak{S}(N)$, actually in $\mathfrak{S}^+(N)$ because m_4 is even, so m_1 is even, m_3 is even, and since

$$m_1 + \cdots + m_4 \quad \text{is even,}$$

we get $m_2 = n_2$ is even and the second term has even weight.

It now suffices to show that $2N(1)$, $2N(\tau)$, $N((\tau + 1) - (1) - (\tau))$ are in $\mathfrak{S}^+(N)$, or in other words

$$\mathfrak{k}_{0,1}^{2N}, \qquad \mathfrak{k}_{1,0}^{2N}, \qquad \left(\frac{\mathfrak{k}_{1,1}}{\mathfrak{k}_{0,1}\mathfrak{k}_{1,0}}\right)^N$$

are in $\mathfrak{S}^+(N)$. We first do this for $\mathfrak{k}_{0,1}^{2N}$. We let

$$\alpha = q_1(1) + (2N - q_1)(1 + N) + q_2(2) - q_2(2 + N),$$

where we select q_1, q_2 to satisfy the quadratic relations, that is

$$q_1 + (2N - q_1)(N + 1)^2 + 4q_2 - q_2(N + 2)^2 = 0,$$

or equivalently

$$q_1(N + 2) + q_2(N + 4) = 2(N + 1)^2.$$

We distinguish the cases when $N \equiv 0 \pmod 4$ and $N \equiv 2 \pmod 4$, but we can again find a solution in the same trivial way. Then using the basic property **K 3** of the Klein forms, we see at once that

$$\sigma^{(\alpha)} = \mathfrak{k}^{(\alpha)} = (-1)^{q_1 + q_2}\mathfrak{k}_{0,1}^{2N}.$$

Furthermore, the divisor

$$(1) - (-1)$$

satisfies the quadratic relations, and

$$\frac{\sigma_{0,1}}{\sigma_{0,-1}} = -1.$$

If necessary we multiply $\sigma^{(\alpha)}$ by this last factor to get the desired expression of $\mathfrak{k}_{0,1}^{2N}$ in $\mathfrak{S}^+(N)$. This proves in the case $N \neq 2$ that $\mathfrak{k}_{0,1}^{2N}$ lies in $\mathfrak{S}^+(N)$.

The other form $\mathfrak{k}_{1,0}^{2N}$ can be handled in the same way, or by observing that their expressions can be obtained by applying a suitable element of $GL_2(\mathbf{Z}/N\mathbf{Z})$ to the expression for $\mathfrak{k}_{0,1}^{2N}$, i.e. making a conjugation by an element of the Galois group.

We now handle

$$\left(\frac{\mathfrak{k}_{1,1}}{\mathfrak{k}_{0,1}\mathfrak{k}_{1,0}}\right)^N.$$

It suffices to show that there exist integers q_1, q_2 such that

$$\begin{aligned}\alpha = (N+1)(\tau+1) - (\tau+1+N) - q_1(1) - (N-q_1)(1+N)\\ - N(\tau) + q_2(2) - q_2(2+N)\end{aligned}$$

satisfies the quadratic relations, because this expression mod N is just

$$N((\tau+1) - (1) - (\tau)),$$

and then by **K 2**, we get

$$\sigma^{(\alpha)} = \pm\mathfrak{k}^{(\alpha)}$$

is the desired form up to a sign, which we have seen can be taken care of.

To satisfy the desired conditions on q_1, q_2 we note that

$$\sum m(r_1, r_2)r_1^2 = 0$$
$$\sum m(r_1, r_2)r_1 r_2 = 0$$

are satisfied for any choice of q_1, q_2.

Expanded out, the third quadratic condition amounts to solving

$$q_1(N+2) - (N+4)q_2 = (1+N)(N+2).$$

One has to consider the cases $N \equiv 2 \pmod 4$ and $N \equiv 0 \pmod 4$, which is easy and left to the reader.

***N* = 2.** In this case, one checks that ± 1 and

$$\wp_1 - \wp_\tau, \qquad \wp_1 - \wp_{\tau+1}, \qquad \wp_\tau - \wp_{\tau+1}$$

generate $\mathfrak{R}^+(N)$. This concludes the proof in all cases.

Remark. One observes that the above proof in the case when N is a prime power actually allows us to obtain a free basis for $\mathfrak{R}^+(N)$.

§5. The Siegel Group

We recall that η is defined to be $\Delta^{1/24}$ where

$$\Delta = (2\pi i)^{12} q_\tau \prod (1 - q_\tau^n)^{24}.$$

If $a = (a_1, a_2) = (r/N, s/N)$, we let $\mathfrak{k}_a = \mathfrak{k}_{r,s}$ and we define

$$g_a = \mathfrak{k}_a \eta^2.$$

This is a modular function, i.e. it has weight 0. Let U be the group generated by the functions g_a, $a \in \mathbf{Q}^2$, $a \notin \mathbf{Z}^2$. For any positive integer N, we let:

$U_N =$ the subgroup of U generated by the functions g_a such that $Na \in \mathbf{Z}^2$.

We set

$$U(N) = U \cap F_N,$$

where F_N is the modular function field of level N. In this section we wish to determine explicitly when an element of U_N belongs to $U(N)$. We use formula **K 3** of Chapter 2, §1, together with **QUAD(N)** of §4. What we have to check is when $\prod g_a^{m(a)}$ is modular with respect to $\Gamma(N)$. We first establish the modularity of the powers of η^2. From **K 3** and **QUAD(N)** we easily see the following.

Lemma 5.1. *η^2 is modular with respect to $\Gamma(12)$, and in fact*

$$\eta^2 = \lambda[\mathfrak{k}_{(1/2,0)}\mathfrak{k}_{(0,1/2)}\mathfrak{k}_{(1/2,1/2)}][\mathfrak{k}_{(1/3,0)}\mathfrak{k}_{(0,1/3)}\mathfrak{k}_{(1/3,1/3)}\mathfrak{k}_{(1/3,-1/3)}]^{-1}$$

for some constant λ.

$\eta^4 = (\eta^2)^2$ is modular with respect to $\Gamma(6)$.

η^6 is modular with respect to $\Gamma(3)$, and in fact

$\eta^6 = \lambda_1[\mathfrak{k}_{(1/2,0)}\mathfrak{k}_{(0,1/2)}\mathfrak{k}_{(1/2,1/2)}]^{-1}$ for some constant λ_1.

$\eta^{12} = (\eta^2)^6$ is modular with respect to $\Gamma(2)$.

$\eta^{24} = \Delta$ is modular with respect to $\Gamma = \mathrm{SL}_2(\mathbf{Z})$.

We associate to η^2 a 1-dimensional character of Γ. If $\alpha \in \Gamma$ we have

$$\eta^2\left(\alpha\begin{pmatrix}\omega_1\\ \omega_2\end{pmatrix}\right) = \psi(\alpha)\eta^2\begin{pmatrix}\omega_1\\ \omega_2\end{pmatrix}$$

where ψ is a character whose image is contained in $\boldsymbol{\mu}_{12}$ (the group of 12-th roots of unity). Moreover,

$$\psi\left(\begin{pmatrix} 1 & 1 \\ 0 & 1 \end{pmatrix}\right) = e^{2\pi i/12} \quad \text{and} \quad \psi\left(\begin{pmatrix} 0 & 1 \\ -1 & 0 \end{pmatrix}\right) = e^{2\pi i/4} = i.$$

Since $\begin{pmatrix} 1 & 1 \\ 0 & 1 \end{pmatrix}$ and $\begin{pmatrix} 0 & 1 \\ -1 & 0 \end{pmatrix}$ generate Γ, we see that ψ is determined completely by these values. We have the exact sequence

$$0 \to \operatorname{Ker} \psi \to \Gamma \to \boldsymbol{\mu}_{12} \to 0.$$

Since $\boldsymbol{\mu}_{12}$ is abelian, $\Gamma' = [\Gamma, \Gamma] \subset \operatorname{Ker} \psi$. But we know that

$$(\Gamma : \Gamma') = 12,$$

so $\Gamma' = \operatorname{Ker} \psi$. From Lemma 1 we have that $\Gamma(12) \subset \Gamma'$.

Suppose now that we are given $g \in U_N$ and

$$g = \prod g_a^{m(a)}, \qquad a \in \frac{1}{N}\mathbf{Z}^2, \qquad a \notin \mathbf{Z}^2.$$

If $\alpha \in \Gamma(N)$, then

$$g(\alpha\tau) = \prod_a \varepsilon_a(\alpha)^{m(a)} \psi(\alpha)^{\Sigma m(a)} g(\tau),$$

where ε_a is the root of unity defined in **K 3**. Let

$$\varepsilon_{g,a} = \prod_a \varepsilon_a(\alpha)^{m(a)} \psi(\alpha)^{\Sigma m(a)}.$$

We note that $g \in U(N)$ is equivalent to the statement that for each $\alpha \in \Gamma(N)$ we have $\varepsilon_{g,\alpha} = 1$.

Theorem 5.2. *Let N be prime to 6. Let*

$$g = \prod g_a^{m(a)},$$

as above. Then g is modular of level N if and only if the family $\{m(a)\}$ satisfies the relations **QUAD(N)**, *and 12 divides $\sum m(a)$.*

Proof. Let $d = \sum m(a)$. Since N is prime to 6, it follows that N is odd. To kill the N-part of $\varepsilon_{g,\alpha}$ we must kill the N-part of

$$\prod \varepsilon_a(\alpha)^{m(a)}$$

since $\psi(\alpha)$ has no N-part. But from the argument given in **QUAD(N) odd**, we see that

$$\sum m(a)(Na_1)^2 \equiv \sum m(a)(Na_2)^2 \equiv \sum m(a)(Na_1)(Na_2) \equiv 0 \pmod{N}$$

for this to occur. Then according to **QUAD(N) odd**, it follows that $\prod \mathfrak{k}_a^{m(a)}$ is modular with respect to $\Gamma(N)$. So η^{2d} must be modular with respect to $\Gamma(N)$. Since $(N, 6) = 1$, we can find $\gamma \in \Gamma(N)$ such that

$$\gamma \equiv \begin{pmatrix} 1 & 1 \\ 0 & 1 \end{pmatrix} \pmod{12}.$$

But then

$$\psi(\gamma) = \psi\left(\begin{pmatrix} 1 & 1 \\ 0 & 1 \end{pmatrix}\right) = e^{2\pi i/12},$$

so 12 divides d, and η^{2d} is modular with respect to Γ.

The case $(N, 6) = 1$ will be referred to as **Case 1**. We shall now treat the other cases, which are tedious.

Theorem 5.3. *In the other cases, g is modular of level N if and only if the family $\{m(a)\}$ satisfies the conditions for each of the following cases.*

Case 2. $(N, 12) = 3$: *$\{m(a)\}$ satisfies* **QUAD(N)** *and* $4 \mid \sum m(a)$.

Case 3. $(N, 12) = 4$: *$\{m(a)\}$ satisfies* **QUAD(N)** *and* $3 \mid \sum m(a)$.

Case 4. $(N, 12) = 2$: *$\{m(a)\}$ satisfies* **QUAD(N)** *and* $6 \mid \sum m(a)$.

or

Case 5. $(N, 12) = 2$: *$\{m(a)\}$ satisfies*:

$$\left(\sum m(a), 6\right) = 3,$$

$$N \Big| \sum m(a)(Na_i)^2 \quad \textit{for } i = 1, 2 \quad \textit{and} \quad \frac{1}{N} \sum m(a)(Na_i)^2 \quad \textit{is odd}.$$

$$N \Big| \sum 2m(a)(Na_1)(Na_2) \quad \textit{and} \quad \frac{1}{N} \sum 2m(a)(Na_1)(Na_2) \quad \textit{is odd}$$

Case 6. $(N, 12) = 12$: $\{m(a)\}$ *satisfies* **QUAD(N)**.

Case 7. $(N, 12) = 6$: $\{m(a)\}$ *satisfies* **QUAD(N)** *and* $2 | \sum m(a)$.

or

Case 8. $(N, 12) = 6$: $\{m(a)\}$ *satisfies*

$$(\textstyle\sum m(a), 2) = 1,$$

$$N \Big| \sum m(a)(Na_i)^2 \quad \textit{and} \quad \frac{1}{N} \sum m(a)(Na_i)^2 \quad \textit{is odd}$$

$$N \Big| \sum 2m(a)(Na_1)(Na_2) \quad \textit{and} \quad \frac{1}{N} \sum 2m(a)(Na_1)(Na_2) \quad \textit{is odd.}$$

Proof. We give the proof according to the cases. We let

$$d = \sum m(a).$$

Case 2. $(N, 6) = 3$. We note first that if N is odd, then ε_a has no 2-component by immediate inspection. So ψ^d restricted to $\Gamma(N)$ must have no 2-component. We may find $\gamma \in \Gamma(N)$ such that

$$\gamma \equiv \begin{pmatrix} 1 & 3 \\ 0 & 1 \end{pmatrix} \pmod{12}.$$

Then $\psi^d(\gamma) = e^{2\pi i d/4}$. So $4 | d$ and η^{2d} is modular with respect to $\Gamma(3)$. Therefore η^{2d} is modular with respect to $\Gamma(N)$, so

$$\prod \mathfrak{k}_a^{m(a)}$$

is also modular with respect to $\Gamma(N)$.

Case 3. $(N, 6) = 2$. Then $\prod \varepsilon_a^{m(a)}$ has no 3-component, so ψ^d restricted to $\Gamma(N)$ must have no 3-component. Suppose $4 | N$. We may then find $\gamma \in \Gamma(N)$ such that

$$\gamma \equiv \begin{pmatrix} 1 & 4 \\ 0 & 1 \end{pmatrix} \pmod{12}.$$

Then $\psi^d(\gamma) = e^{2\pi i/3}$. Hence $3 | d$ and η^{2d} is modular with respect to $\Gamma(4)$, hence with respect to $\Gamma(N)$. So

$$\prod \mathfrak{k}_a^{m(a)}$$

is also modular with respect to $\Gamma(N)$. If $4 \nmid N$ we may find $\gamma \in \Gamma(N)$ such that

$$\gamma \equiv \begin{pmatrix} 1 & 2 \\ 0 & 1 \end{pmatrix} \pmod{12}.$$

Thus $3 | d$ again. If $6 | d$ then η^{2d} is modular with respect to $\Gamma(2)$, hence with respect to $\Gamma(N)$. Otherwise, up to a constant factor,

$$\eta^6 = [\mathfrak{k}_{(1/2,0)}\mathfrak{k}_{(0,1/2)}\mathfrak{k}_{(1/2,1/2)}]^{-1};$$

and $\prod g_a^{m(a)}$ is modular with respect to $\Gamma(N)$ is equivalent to:

$$\sum m(a)(Na_i)^2 - \frac{d}{3}\frac{N^2}{2} \equiv 0 \pmod{2N} \quad \text{for } i = 1, 2$$

$$\sum m(a)(Na_1)(Na_2) - \frac{d}{3}\frac{N^2}{4} \equiv 0 \pmod{N}.$$

So

$$N \Big| \sum m(a)(Na_i)^2 \quad \text{and} \quad \sum m(a)(Na_i)^2/N + 1 \equiv 0 \pmod 2,$$

whence $\sum m(a)(Na_i)^2/N$ is odd, and

$$\frac{N}{2} \Big| \sum m(a)(Na_1)(Na_2) \quad \text{and} \quad \sum 2m(a)(Na_1)(Na_2)/N + 1 \equiv 0 \pmod 2,$$

whence $\sum 2m(a)(Na_1)(Na_2)/N$ is odd.

Case 4. $(N, 6) = 6$. If $12 | N$ then η^2 is $\Gamma(N)$-modular, and

$$\prod \mathfrak{k}_a^{m(a)}$$

is also $\Gamma(N)$-modular. If $2 | d$ then η^{2d} is $\Gamma(N)$-modular, and so is $\prod \mathfrak{k}_a^{m(a)}$. Suppose then $12 \nmid N$ and $2 \nmid d$. We check the condition for modularity from **QUAD(N)**, and find that

$$\sum m(a)(Na_i)^2 + dN^2/6 \equiv 0 \pmod{2N}$$
$$\sum m(a)(Na_1)(Na_2) + dN^2/4 \equiv 0 \pmod{N}.$$

So

$$N \Big| \sum m(a)(Na_i)^2 \quad \text{and} \quad \sum m(a)(Na_i)^2/N + 1 \equiv 0 \pmod 2,$$

whence $\sum m(a)(Na_i)^2/N$ is odd, and

$$\frac{N}{2}\Big|\sum m(a)(Na_1)(Na_2) \quad \text{and} \quad \sum 2m(a)(Na_1)(Na_2)/N + 1 \equiv 0 \pmod 2,$$

whence $\sum 2m(a)(Na_1)(Na_2)/N$ is odd. The theorem follows.

We note that in all cases above except in Case 1, the function η^{2d} may be expressed by Klein forms with level dividing N. In Case 1, the function η^{2d} is a power of Δ which may not necessarily be expressed by Klein forms with level dividing N.

CHAPTER 4

The Siegel Units Are Generators

In this chapter we essentially prove that the Siegel units generate all units, and we also get a precise description of all the units of a given level.

We observe right away that the Siegel units g_a with $a \in (1/N)\mathbf{Z}^2$ do not have level N. The first thing we prove is that essentially the group generated by such units has no cotorsion in the group of all units of all levels (the only exception occurs when N is composite, even, and for 2-torsion). If we then combine this result with the characterization by means of quadratic relations, we get the desired description of the units at level N.

The method of proof consists in reducing multiplicative relations among the q-expansions of units to additive relations of an appropriately chosen Fourier coefficient (more or less, the leading coefficient after the constant term). In analysing a modular function such that some power can be expressed as a product of Siegel functions, we are led to taking roots of q-products. By a theorem of Shimura, we know that a suitably reduced form of the q-product has integral coefficients, and we see that this can happen only if the modular function itself is already expressible as a power product of Siegel functions.

The additive relations of divisibility occur among elements of the cyclotomic fields, and hence we devote a section to describing an appropriate basis for the integers of these fields, which allow us to recognize easily when the coefficients are integral.

§1. Statement of Results

For this chapter, we are only interested in modular functions up to constant factors. Hence for $a \in \mathbf{Q}^2/\mathbf{Z}^2$ we mean by g_a the class of functions modulo constants. By abuse of language, we still speak of g_a as a function. We say that such a function is **modular** if it lies in the modular function field F_N for some level N (strictly speaking, if some representative function and therefore any representative function lies in a modular function field). We shall investigate when a root of a power product of Siegel functions is modular.

To begin with, we note that the Siegel functions themselves are modular. It is clear from the explicit transformation law for a Klein form $\mathfrak{k}_a$ that it is modular. (See formula **K 3**.) It is well known for η^2, and can be proved in the same style. Indeed, up to a constant factor we have (e.g. from q-expansions) as in Chapter 3, §5:

$$\eta^{-6} = \mathfrak{k}_{(1/2,0)}\mathfrak{k}_{(1/2,1/2)}\mathfrak{k}_{(0,1/2)}$$
$$\eta^{-8} = \mathfrak{k}_{(1/3,0)}\mathfrak{k}_{(0,1/3)}\mathfrak{k}_{(1/3,1/3)}\mathfrak{k}_{(1/3,-1/3)},$$

and therefore $\eta^2 = \eta^8/\eta^6$ is modular whence g_a is modular.

For convenience of expression, it will be useful to adopt the notation g_a above only when $2a \neq 0$. If $2a = 0$, let us put

$$h_a = \mathfrak{k}_a\eta^2.$$

We see from the distribution relation that

$$h_{(1/2,0)} = \lambda h^2_{(1/4,0)}h^2_{(1/4,1/2)},$$

where λ is a constant. Thus it is clear that $h^{1/2}_{(1/2,0)}$ is modular. Likewise, if $2a = 0$, it is clear that $h_a^{1/2}$ is modular. In this light, if $2a = 0$, we **define**

$$g_a = h_a^{1/2} = (\mathfrak{k}_a\eta^2)^{1/2},$$

so that also in this case g_a is modular.

The next two theorems give a more precise result than just describing the group of modular units. We let U be the group generated by the Siegel functions g_a for all $a \in \mathbf{Q}^2/\mathbf{Z}^2$, $a \neq 0$, and the constants $\neq 0$. Let F be the modular function field, equal to the union of all the modular fields F_N of all levels N. Let $\mathbf{Q}R$ be the integral closure of $\mathbf{Q}[j]$ in F. In view of the fullness of the group of Siegel units (Chapter 2, Theorem 3.2 and the discussion preceding Lemma 2.1 of that chapter), it follows that the factor group of modular units $(\mathbf{Q}R)^*$ modulo U is a torsion group. It turns out that they are equal except for 2-torsion. Furthermore, let U_N be the subgroup of U generated by the Siegel functions g_a such that $Na = 0$, and the non-zero constants.

Theorem 1.1. *Let $N = p^r$ be a prime power. Then U_N is equal to its own division group in the group of modular units. In other words, if g is a modular function such that some positive power of g lies in U_N, then g also lies in U_N.*

Theorem 1.2. *Let N be an arbitrary integer > 1. Let l be an odd integer. If g is a modular function such that g^l lies in U_N, then g lies in U_N.*

In the proof of Theorem 1.1 it will clearly be sufficient to assume that $g^l \in U_N$ for some prime l, and in the proof of Theorem 1.2 it will suffice to assume that l is an odd prime.

The rest of the chapter gives the proofs of the above theorems. We begin by some lemmas giving an appropriate basis for the cyclotomic integers. The main part of the proof consists in reducing the study of multiplicative relations among the units to additive relations among such integers, by projecting on the first coefficient of the q-expansion. Using a theorem of Shimura (see Lemma 3.1) to the effect that the Fourier coefficients of modular forms are algebraic integers, we can formulate versions of Theorems 1.1, and 1.2 involving only formal power series arguments, and no other property of modular forms. This is done in Theorems 4.3 and 5.2.

We conclude this first section by combining the above results with the main theorem of Chapter 3, to get a characterization of the units at a given level.

Theorem 1.3. *Let F_N be the modular function field of level N. Assume that $N = p^n$ is a prime power with $p \neq 2, 3$. Then the units in F_N (modulo constants) consist of the power products*

$$\prod g_a^{m(a)}$$

of the Siegel functions with $a \in (1/N)\mathbf{Z}^2$, $a \notin \mathbf{Z}^2$, such that the family $\{m(a)\}$ satisfies the quadratic relations **QUAD(N)**, *and* $\sum m(a) \equiv 0 \bmod 12$.

Proof. By Theorem 3.2 of Chapter 2 we know that powers of the Siegel functions generate a group of maximal rank in the unit group of F_N. Consequently any unit has some power which lies in the group generated by the Siegel functions. By Theorem 5.1 of Chapter 3 and Theorem 1.1 of Chapter 4, such units lie in U_N. One merely has to impose the additional level condition to prove the theorem.

The combination of Theorems 1.1, 1.2 and Theorem 5.2, 5.3 of the preceding chapter yield a similar characterization of the units of given level in F_N, but the statements become more involved because of the possibility of 2-torsion. In any case, for prime power including powers of 2 and 3, we have obtained a complete characterization of the units. In the composite case, we have a characterization only up to 2-torison. For an analysis of this phenomenon, see Kubert [Ku 6].

§2. Cyclotomic Integers

Let $\mathbf{Q}_N = \mathbf{Q}(e^{2\pi i/N})$, and let $\mathfrak{o}_N$ be the algebraic integers in $\mathbf{Q}_N$. It is standard that

$$\mathfrak{o}_N = \mathbf{Z}[e^{2\pi i/N}].$$

We want an appropriate $\mathbf{Z}$-basis for $\mathfrak{o}_N$.

Suppose first that $N = p^r$ is a prime power. Let $\boldsymbol{\mu}_N$ as usual be the group of N-th roots of unity. Let $S_j (j = 1, \ldots, N/p)$ be the cosets of $\boldsymbol{\mu}_p$ in $\boldsymbol{\mu}_N$. For each coset, choose arbitrarily an element ζ_j in S_j.

Lemma 2.1. *The set*

$$\bigcup (S_j - \{\zeta_j\})$$

forms a $\mathbf{Z}$*-basis for* $\mathfrak{o}_{p^r}$.

Proof. We note that the number of cosets is p^{r-1}, so that

$$\bigcup (S_j - \{\zeta_j\})$$

has $p^r - p^{r-1}$ elements, which is the right number. Since the elements of $\boldsymbol{\mu}_N$ generate $\mathfrak{o}_N$ over $\mathbf{Z}$, we need only show that ζ_j can be recovered. To do this, multiply the identity

$$\sum_{\zeta \in \boldsymbol{\mu}_p} \zeta = 0$$

by ζ_j. We get

$$\zeta_j + \sum_{\zeta \neq 1} \zeta_j \zeta = 0.$$

But the sum is the same as that taken over all elements of $S_j - \{\zeta_j\}$, thus proving the lemma.

We next extend the lemma to arbitrary N. We let

$$N = \prod p^{n(p)}$$

be the prime factorization of N. For each element $\zeta \in \boldsymbol{\mu}_N$ we let

$$\zeta = \prod \zeta_p$$

be the factorization into prime power roots of unity. For each prime p we form the set

$$S(p) = \bigcup (S_j^{(p)} - \zeta_j^{(p)})$$

as in the preceding lemma.

Lemma 2.2. *Let S be the set of N-th roots of unity ζ such that $\zeta_p \in S(p)$ for each $p \mid N$. Then S is a* **Z**-*basis for $\mathfrak{o}_N$.*

Proof. Again, the cardinality of S is the right one. Let ω be an arbitrary N-th root of unity. Write

$$\omega = \prod_{\omega_p \in S(p)} \omega_p \prod_{\omega_p \notin S(p)} \omega_p.$$

As shown in Lemma 2.1, we may write $\omega_p \notin S(p)$ as an integral linear combination of elements in $S(p)$. We make this substitution and expand out the product by distributivity to prove the lemma.

Next we wish to determine when sums of the form

$$\alpha = \sum a_\zeta \zeta, \qquad a_\zeta \in \mathbf{Z},$$

taken over all N-th roots of unity, are divisible by a prime number l. We begin with the prime power case.

Lemma 2.3. *Let $N = p^r$. A prime l divides $\sum a_\zeta \zeta$ if and only if, for each pair of elements ω, ξ in the same coset of $\boldsymbol{\mu}_p$, l divides $a_\omega - a_\xi$.*

Proof. Let S_j as before be the cosets of $\boldsymbol{\mu}_p$. Since

$$-\zeta_J = \sum_{S_j - \{\zeta_j\}} \zeta$$

We obtain

$$\sum a_\zeta \zeta = \sum_j \sum_{\zeta \in S_j - \{\zeta_j\}} (a_\zeta - a_{\zeta_j})\zeta.$$

Since the union of the sets $S_j - \{\zeta_j\}$ forms a basis for $\mathfrak{o}_N$ over **Z**, the lemma is obvious.

Next we deal with arbitrary N. We first describe some notation. Let ω, ξ be N-th roots of unity such that for each $p|N$, ω_p, and ξ_p lie in the same coset of $\boldsymbol{\mu}_p$ and $\omega_p \neq \xi_p$. Let

$$\Phi(\omega, \xi) = \text{the set of } N\text{-th roots of unity } \varphi \text{ such that } \varphi_p = \omega_p \text{ or } \varphi_p = \xi_p.$$

For $\varphi \in \Phi(\omega, \xi)$ we define

sign $\varphi = (-1)^k$, where k is the cardinality of the set of primes p such that $\varphi_p = \omega_p$.

Lemma 2.4. *Let l be a prime number. The following two conditions are equivalent:*

(i) *l divides $\alpha = \sum a_\zeta \zeta$.*

(ii) *For every pair of N-th roots of unity ω, ξ such that $\omega_p/\xi_p \in \boldsymbol{\mu}_p$, and $\omega_p \neq \xi_p$ for all $p|N$, the prime l divides*

$$\sum_{\varphi \in \Phi(\omega,\xi)} (\text{sign } \varphi) a_\varphi.$$

Proof. Assume (i). We express $\sum a_\zeta \zeta$ in terms of the special basis of Lemma 2.2, making the choice of ξ_p as the element which has been excluded from its coset relative to $\boldsymbol{\mu}_p$. By definition of the basis S in Lemma 2.2, we conclude that $\omega \in S$, i.e. ω is a basis element. It follows at once that

$$\sum_{\varphi \in \Phi(\omega,\xi)} (\text{sign } \varphi) a_\varphi$$

is the coefficient of ω, in the expression of $\alpha = \sum a_\zeta \zeta$ in terms of the basis S. It is then clear that (i) implies (ii).

Conversely, suppose that for every choice of ω, ξ, condition (ii) is satisfied. Write $N = \prod p^{r(p)}$. We choose one element

$$\xi_{p,j(p)} \in S_{j_p}(p)$$

from each coset of $\boldsymbol{\mu}_p$ in $\boldsymbol{\mu}_{p^{r(p)}}$, and we form all possible elements

$$\xi = \prod_p \xi_{p,j(p)}.$$

Given $\omega \in S$, there will be precisely one element ξ above satisfying the hypothesis of condition (ii), and

$$\sum_{\varphi \in \Phi(\omega,\xi)} (\text{sign } \varphi) a_\varphi$$

is the coefficient of ω in the expression of $\sum a_\zeta \zeta$ in terms of the basis S. Thus each coefficient with respect to this basis is divisible by l, whence α is divisible by l, thereby proving the lemma.

§3. Remarks on q-Expansions

As mentioned already, we shall reduce multiplicative properties of modular units to additive properties of some of their Fourier coefficients. We recall a theorem of Shimura [Sh], Theorem 3.52, p. 85.

The space of cusp forms for $\Gamma(N)$ over $\mathbf{C}$ is generated by forms whose Fourier coefficients are rational integers.

As a consequence, we obtain

Lemma 3.1. *Let f be a modular function with divisor concentrated at the cusps. If the q-expansion of f at some cusp has algebraic coefficients in a number field K, then these coefficients have bounded denominators, and are integral at all but a finite number of primes of K.*

Proof. A conjugate g of f will have the given q-expansion at infinity. Then $\Delta^n g$ is a cusp form for suitably large n. Let $\Delta^n g = h$. The Fourier coefficients of h are a finite linear combination over K of a $\mathbf{Q}$-basis for modular forms of weight $12n$. By Shimura's theorem, it is clear that h has bounded denominators, and $g = h/\Delta^n$. Since $q^{-1}\Delta$ is invertible as a power series over $\mathbf{Z}$, the lemma follows.

Let K be a field with a valuation, assumed non-archimedean. We can extend the valuation to the power series whose coefficients have bounded valuation, by the maximal value of the coefficients (Gauss lemma). In particular, if f is such a power series, n is a positive integer, and f^n has integral coefficients for the valuation, so does f. This will be applied to coefficients of modular forms.

We let $q_{\tau/N} = q^{1/N} = e^{2\pi i\tau/N}$. We shall consider the power series field

$$\mathbf{Q}_N((q^{1/N})), \qquad \mathbf{Q}_N = \mathbf{Q}(\boldsymbol{\mu}_N).$$

If

$$f = \sum \alpha_n q_N^n, \qquad \alpha_n = \alpha_n(f),$$

and $\alpha_r q^{r/N}$ is the lowest term ($\alpha_r \neq 0$), then we **define**

$$f^* = \frac{f}{\alpha_r q^r}.$$

We call f^* the **reduced power series**, or **reduced form**, of f. If f is a modular function, then of course f^* is usually not a modular function. It is a power series in $q^{1/N}$, whose lowest term is 1. We give examples of this by recalling the q-products for the Siegel functions.

Writing $q_a = q_\tau^{a_1} e^{2\pi i a_2}$, we have

$$g_a = q_\tau^{(1/2)\mathbf{B}_2\langle a_1\rangle}(1 - q_a) \prod_{n=1}^{\infty} (1 - q_\tau^n q_a)(1 - q_\tau^n/q_a), \quad \text{up to a constant factor,}$$

where $\langle x\rangle$ is the smallest real number $\geqq 0$ in the residue class of x mod $\mathbf{Z}$, and

$$B_2(X) = X^2 - X + \tfrac{1}{6}.$$

Then for (a_1, a_2) not of period 2 mod $\mathbf{Z}^2$, we have:

$$\begin{aligned} g_a^* &= (1 - q_a) \prod (1 - q_\tau^n q_a)(1 - q_\tau^n/q_a) && \text{if } \langle a_1\rangle \neq 0, \\ g_a^* &= \prod (1 - q_\tau^n q_a)(1 - q_\tau^n/q_a) && \text{if } \langle a_1\rangle = 0. \end{aligned}$$

On the other hand, if $2a = 0$, then with the definition of this chapter,

$$g_a^* = \prod_{n=1}^{\infty} (1 - q_\tau^n/q_a).$$

In writing the above products, it is understood that a_1, a_2 are chosen equal to their representatives $\langle a_1\rangle$, $\langle a_2\rangle$. Furthermore, since $g_a = g_{-a}$ (up to a constant factor), we shall always assume that a_1 has the property

$$0 \leqq a_1 \leqq N/2.$$

Thus we consider only representative elements a of $(\mathbf{Q}^2/\mathbf{Z}^2)/\pm 1$.

Let N be the least common multiple of the denominators of a_1, a_2. We call N the **primitive denominator** of a, and say that a has **primitive period** N. If we have merely $Na = 0$, then we call N a **denominator**, and say that N is a **period** of a.

Suppose that a has primitive period N. We shall be interested in the coefficient of $q^{1/N}$ in the q-expansion of g_a^*. Let α_1 be this coefficient, so that

$$g_a^* = 1 + \alpha_1 q^{1/N} + \cdots.$$

Then $\alpha_1 \neq 0$ only in case $a_1 = 1/N$ [remember we just excluded the case when $a_1 = (N-1)/N$ for $N > 2$]. Furthermore,

$$\textit{if} \quad a = \left(\frac{1}{N}, a_2\right) \quad \textit{then} \quad \alpha_1 = -e^{2\pi i a_2}.$$

This is obvious from the q-products, and is valid also for $N = 2$ according to the special definition of §1.

From the q-products for the Siegel functions, we see that the reduced power series g_a^* have coefficients in the ring of cyclotomic integers $\mathfrak{o}_N$. It follows at once that any element g of the group generated by the Siegel functions and the non-zero constants also has a reduced form g^* of the same type, i.e. with coefficients in $\mathfrak{o}_N$.

Furthermore, if g is a power series such that some power g^l lies in that group, then g^* also has this property, in view of the Gauss lemma for power series recalled above.

We shall be interested in power products of the functions g_a. Let

$$m : \mathbf{Q}^2/\mathbf{Z}^2 \to \mathbf{Z}$$

be a function such that $m(a) = 0$ for almost all a (all but a finite number). We call m **even** if $m(a) = m(-a)$, and normalized if $m(0) = 0$. *Throughout the sequel, we assume that such functions m are even and normalized, unless otherwise specified.*

We say that a denominator d **occurs** in m if there is some a such that $m(a) \neq 0$ and $da = 0$. We say that N is a **denominator for** m if $Na = 0$ for all a such that $m(a) \neq 0$. We say that a **occurs** in m if $m(a) \neq 0$.

We define

$$g(m) = \prod g_a^{m(a)},$$

where the product is taken over $a \in (\mathbf{Q}^2/\mathbf{Z}^2)/\pm 1$.

Then $g(m)$ is modular. Similarly we define the power series

$$g^*(m) = \prod g_a^{*m(a)},$$

whose leading term is 1. Since we can take roots of such power series formally by means of the binomial expansion, we may then also define the power series

$$g^*(m/n) = g^*(m)^{1/n} = \prod g_a^{*m(a)/n},$$

for any positive integer n.

Let

$$\mathbf{A}_{\mathbf{Z}} = \prod \mathbf{Z}_p$$

be the integral adeles. For any $\sigma \in \mathrm{GL}_2(\mathbf{A}_{\mathbf{Z}})$ we define σm by

$$(\sigma m)(a) = m(a\sigma^{-1}),$$

and

$$g^*(\sigma m) = \prod g_{a\sigma}^{*m(a)}.$$

If N is a denominator for m, then the effect of σ in this last formula is determined by the image of σ in $\mathrm{GL}_2(\mathbf{Z}/N\mathbf{Z})$.

With this notation, Lemma 3.1, and the above remarks (Gauss lemma), we have:

Lemma 3.2. *Let g be modular. Let l be prime. If*

$$g^l = \prod g_a^{m(a)},$$

then the power series $g^*(\sigma m/l)$ has l-integral coefficients for every element $\sigma \in \mathrm{GL}_2(\mathbf{A}_{\mathbf{Z}})$.

Finally we recall the

Distribution relation. *Given $b \in (\mathbf{Q}^2/\mathbf{Z}^2)/\pm 1$, and a positive integer D, there is a constant λ such that*

$$\prod_{Da=b} g_a = \lambda g_b.$$

This is merely the relation of Chapter 2, except that with our present definition adjusted for $2a = 0$, it also holds on

$$(\mathbf{Q}^2/\mathbf{Z}^2)/\pm 1.$$

§4. The Prime Power Case

The first lemma will be used both in this case and the general composite case, providing the beginning of induction arguments.

Lemma 4.1. *Let l be a prime number. Let g be modular, and suppose that*

$$g^l = \prod g_a^{m(a)}.$$

Assume that every a occurring in m has prime period. Then there exists a representation

$$g = \lambda \prod g_a^{m'(a)},$$

where λ is constant, and every denominator occurring in m' also occurs in m.

In the light of the remarks of §3, it will suffice to prove the following version.

Lemma 4.2. *Let l be a prime number. Assume that the power series*

$$g^*(\sigma m/l) = \prod g_{a\sigma}^{*m(a)/l}$$

has l-integral coefficients for every $\sigma \in \mathrm{GL}_2(\mathbf{A_Z})$. Assume also that every a occurring in m has prime period. Then there exists a representation

$$\prod g_a^{m(a)} = \lambda \prod g_a^{m'(a)}$$

where λ is a constant, and:

(i) *If a occurs in m' then l divides $m'(a)$.*
(ii) *Every denominator occurring in m' also occurs in m.*

Proof. For each prime p define

$$g_{(p)}^*(m/l) = \prod_{pa=0} g_a^{*m(a)/l}.$$

Then $g_{(p)}^*(m/l)$ is a power series in $q^{1/p}$ with leading term 1. Furthermore, we have

$$g^*(m/l) = \prod_p g_{(p)}^*(m/l).$$

It follows that the coefficients of $q^{1/p}$ in $g^*(m/l)$ and $g_{(p)}^*(m/l)$ are equal. The same also applies to $g^*(\sigma m/l)$ and $g_{(p)}^*(\sigma m/l)$.

Let us call a, b **independent** if the cyclic groups (a) and (b) generated by a and b have only 0 as their intersection.

We now prove:

If $pa = pb = 0$ then l divides $m(a) - m(b)$.

If a, b are not independent we choose c such that $pc = 0$, and such that c, a are independent. Then

$$m(a) - m(b) = m(a) - m(c) + m(c) - m(b),$$

and if l does not divide $m(a) - m(b)$, then l does not divide $m(a) - m(c)$ or $m(c) - m(b)$. Thus without loss of generality, we may assume that a, b are independent.

We then find an automorphism σ such that

$$a\sigma = \left(\frac{1}{p}, 0\right) \quad \text{and} \quad b\sigma = \left(\frac{1}{p}, \frac{1}{p}\right).$$

The coefficient of $q^{1/p}$ in $g^*_{(p)}(\sigma m/l)$ is equal to the coefficient of $q^{1/p}$ in the product

$$\prod_c \left[(1 - q_c) \prod_{n=1}^{\infty} (1 - q^n q_c)(1 - q^n/q_c)\right]^{m(c)/l},$$

where the product is taken over c satisfying

$$c = a\sigma + v(b\sigma - a\sigma), \qquad v = 0, \ldots, p - 1.$$

Therefore this coefficient is equal to

$$-\sum_c \frac{1}{l} m(c)\zeta_c, \quad \text{where } \zeta_c = e^{2\pi i v/p}.$$

By hypothesis, this coefficient is l-integral, and by Lemma 2.3, we conclude that l divides $m(a) - m(b)$, as desired.

Let us fix a primitive element d such that $pd = 0$. Write

$$\prod_{pa=0} g_a^{m(a)} = \prod_{pa=0} g_a^{m(a)-m(d)} \prod_{pa=0} g_a^{m(d)}.$$

The distribution relation tells us that

$$\prod_{pa=0} g_a = \lambda \text{ is constant.}$$

Having proved that l divides $m(a) - m(d)$ also proves the lemma.

Note that in the case $p = 2$, the product above is just

$$\prod_c (1 - q_c) \prod_{n=2}^{\infty} (1 - q^n q_c),$$

by our convention concerning the definition of g_c in the case of level 2. Again

$$-\sum \frac{1}{l} m(c)\zeta_c$$

is the appropriate coefficient, and the same proof works.

We shall now prove Theorem 1.1. and formulate a formal version concerning bounded denominators, as for the previous lemmas.

Theorem 4.3. *Let l be a prime number. Assume that the power series*

$$g^*(\sigma m/l) = \prod g_{a\sigma}^{*m(a)/l}$$

has l-integral coefficients for every $\sigma \in \mathrm{GL}_2(\mathbf{A}_{\mathbf{Z}})$, and that if a occurs in m, then a has prime power denominator. Then there exists an even function m' such that

$$\prod g_a^{m(a)} = \lambda \prod g_a^{m'(a)},$$

where λ is constant, and:

(i) *If a occurs in m' then l divides $m'(a)$.*
(ii) *Denominators for m' can be taken among prime power denominators for m.*

Proof. By induction on the largest prime power denominator. The prime case is taken care of by Lemma 4.2. Let p^r be the largest denominator, and assume the theorem for lower cases.

We prove first:

Let a, b have period p^r. If $p(a - b) = 0$, then l divides $m(a) - m(b)$.

For suppose not. If $(a - b) = (a)$, i.e. $(a) = (b)$, we choose c such that $p(a - c) = 0$, $(c) \neq (a)$ and c has period p^r. Then

$$m(a) - m(b) = m(a) - m(c) + m(c) - m(b),$$

and l does not divide one of the two terms on the right. Without loss of generality, we may therefore assume that $(a) \neq (b)$.

Choose t such that $p^{r-1}t = b - a$. Then a, t generate

$$\frac{1}{p^r}\mathbf{Z}^2/\mathbf{Z}^2.$$

By taking an appropriate conjugation σ, we may assume that

$$a = \left(\frac{1}{p^r}, 0\right) \quad \text{and} \quad t = \left(0, \frac{1}{p^r}\right).$$

We shall now see that in the expansion of $g^*(m/l)$, the coefficient of q^{1/p^r} is not l-integral, which will yield a contradiction.

Again we can restrict our attention to the primitive p^r-vectors a, and in fact to the set

$$a + vt, \quad \text{with } 0 \leqq v \leqq p^r - 1.$$

The coefficient of q^{1/p^r} is equal to

$$-\sum_v \frac{1}{l} m\left(\frac{1}{p^r}, \frac{v}{p^r}\right) e^{2\pi i v/p^r}.$$

We have $a = (1/p^r, 0)$ and $b = (1/p^r, 1/p)$. Lemma 2.3 shows that $l|(m(a) - m(b))$, as desired.

Given a coset B of $((1/p)\mathbf{Z}^2/\mathbf{Z}^2)/\pm 1$ in $((1/p^r)\mathbf{Z}^2/\mathbf{Z}^2)/\pm 1$, the distribution relation shows that

$$\prod_{b \in B} g_b = \lambda g_v, \quad \text{where } v = pb,\ b \in B.$$

Let $d \in B$. Then

$$\begin{aligned}\prod_{b \in B} g_b^{m(b)} &= \prod_{b \in B} g_b^{m(b) - m(d)} \prod_{b \in B} g_b^{m(d)} \\ &= \lambda^{m(d)} \prod_{b \in B} g_b^{m(b) - m(d)} g_v^{m(d)}.\end{aligned}$$

However,

$$\prod_B \prod_{b \in B} g_b^{(m(b) - m(d))/l}$$

has l-integral Fourier coefficients because we have seen that $l|(m(b) - m(d))$. Since v has denominator p^{r-1}, we are done by induction.

§5. The Composite Case

Theorem 5.1. *Let l be a prime number $\neq 2$. Let g be modular, and suppose that*

$$g^l = \prod g_a^{m(a)}.$$

Let N be a period for every a occurring in m. Then there exists a representation

$$g = \lambda \prod g_a^{m'(a)},$$

where λ is constant, and N is also a period for every a occurring in m'.

As in the prime power case, we can reformulate Theorem 5.1 only in terms of power series, because of Shimura's theorem (Lemma 3.1).

Theorem 5.2. *Let l be a prime $\neq 2$. Assume that the power series*

$$g^*(\sigma m/l) = \prod g_{a\sigma}^{*m(a)/l}$$

has l-integral coefficients for every $\sigma \in \mathrm{GL}_2(\mathbf{A}_{\mathbf{Z}})$. Let N be a period for every a occurring in m. Then there exists a representation

$$\prod g_a^{m(a)} = \lambda \prod g_a^{m'(a)},$$

where λ is constant and:

(i) *If a occurs in m' then l divides $m'(a)$.*
(ii) *If a occurs in m' then N is a period for a.*

Proof. Again by induction on the largest primitive denominator occurring in m. We let M be maximal such that there exists a with $m(a) \neq 0$ and a has primitive period M, with $M \mid N$.

It will be convenient to abbreviate

$$\frac{1}{n}\mathbf{Z}^2/\mathbf{Z}^2 = Z_n \quad \text{and} \quad Z_n^* = \text{subset of } Z_n \text{ of elements with primitive denominator equal to } n.$$

We shall carry out a development applying to the 2-dimensional situation of Z_n, similar to that carried out for roots of unity in §2. We let M_0 be the product of all the primes dividing M, taken to the first power. If a, a' are in the same coset of Z_{M_0} and $a_p \neq a'_p$ for all $p \mid M$, then we define

$$\Phi(a, a') = \text{set of } x \in Z_M^* \quad \text{such that } x_p = a_p \text{ or } a'_p \quad \text{for all } p \mid M.$$

Observe that the condition a, a' to be in the same coset of Z_{M_0} means that

$$p(a_p - a'_p) = 0 \quad \text{for all } p \mid M.$$

Lemma 5.3. *Let a, a' be in the same coset of Z_{M_0}, and assume $a_p \neq a'_p$ for all $p|M$. Let m be the function of Theorem 5.2. Then*

$$\sum_{x \in \Phi(a,a')} (\text{sign } x) m(x) \equiv 0 \pmod{l}.$$

Proof. Suppose first that $(a_p - a'_p) \cap (a_p) = 0$ for all $p|M$. We can find d primitive such that $(d) \cap (a) = 0$, and $a - a' \in (d)$. We put $b = a + d$. Then b has primitive denominator M, and

$$a' = a + k(b - a)$$

with some integer k. For some σ we have

$$a\sigma = \left(\frac{1}{M}, 0\right) \quad \text{and} \quad b\sigma = \left(\frac{1}{M}, \frac{1}{M}\right).$$

Then the coefficient α of $q^{1/M}$ in

$$g^*(\sigma m/l) = \prod_{x \in Z_N} g_{x\sigma}^{*m(x)/l} = \cdots + \alpha q^{1/M} + \cdots$$

is

$$\alpha = \sum_c \frac{1}{l} m(c) \zeta_c,$$

where $c = a + v(b - a)$, $0 \leqq v < M$, since M was chosen to be the largest primitive denominator. By Lemma 2.4 we conclude that

$$\sum_{x \in \Phi(a,a')} (\text{sign } x) m(x) \equiv 0 \pmod{l},$$

and the lemma is proved in this case.

We now wish to remove the condition that $(a_p - a'_p)$ and (a_p) are independent. This is done by induction. Given a, a' we induct on the cardinality of the number of primes such that $(a_p) = (a'_p)$. We assume the result for cardinality s, and consider the case $s + 1$. Let p_0 be a prime such that $(a_{p_0}) = (a'_{p_0})$. Choose d_{p_0} of the same primitive denominator as a_{p_0}, lying in the same coset of Z_{p_0}, i.e.

$$d_{p_0} \equiv a_{p_0} \bmod Z_{p_0},$$

and such that

$$(d_{p_0}) \neq (a_{p_0}) = (a'_{p_0}).$$

Let

$$d = \sum_{p \neq p_0} a_p + d_{p_0} \quad \text{and} \quad d' = \sum_{p \neq p_0} a'_p + d_{p_0}.$$

Then by induction, all congruences being mod l, we have

$$\begin{aligned} 0 &\equiv \sum_{x \in \Phi(a,d')} (\text{sign } x)m(x) \\ &\equiv \sum_{x_{p_0} = a_{p_0}} + \sum_{x_{p_0} = d_{p_0}} (\text{sign } x)m(x) \end{aligned}$$

and similarly for $x \in \Phi(d, a')$ instead of $x \in \Phi(a, d')$. Then

$$\sum_{\substack{x_{p_0} = a_{p_0} \\ x \in \Phi(a,d')}} \equiv - \sum_{\substack{x_{p_0} = d_{p_0} \\ x \in \Phi(a,d')}}$$

and

$$\sum_{\substack{y_{p_0} = a'_{p_0} \\ y \in \Phi(d,a')}} \equiv - \sum_{\substack{y_{p_0} = d_{p_0} \\ y \in \Phi(d,a')}}$$

The expressions on the right of these last two congruences are the negative of each other, as one sees at once from the definition of the sign. Hence

$$\sum_{\substack{x_{p_0} = a_{p_0} \\ x \in \Phi(a,d')}} \equiv - \sum_{\substack{y_{p_0} = a'_{p_0} \\ y \in \Phi(d,a')}}$$

The definitions now show that

$$\sum_{\substack{x_{p_0} = a_{p_0} \\ x \in \Phi(a,d')}} + \sum_{\substack{y_{p_0} = a'_{p_0} \\ y \in \Phi(d,a')}} = \sum_{x \in \Phi(a,a')} (\text{sign } x)m(x),$$

and the lemma is proved.

For the next lemma, we define some notation. Let $p|M$. We write

$$Z_M^*/Z_p$$

for the set of congruence classes of elements of Z_M^* modulo Z_p.

Lemma 5.4. *Let $m: Z_M^* \to A$ be a function into an abelian group, not necessarily even. Then the following two conditions are equivalent.*

(1) *For all pairs* $a, a' \in Z_M^*$ *satisfying* $a \equiv a' \bmod Z_{M_0}$ *and* $a_p \neq a'_p$ *for all* $p|M$, *we have*

$$\sum_{\substack{x \in Z_M^* \\ x \in \Phi(a,a')}} (\text{sign } x)m(x) = 0.$$

(2) *For each* $p|M$, *there exists a function*

$$\psi_p : Z_M^*/Z_p \to A$$

such that if $a \in Z_M^*$ *then*

$$m(a) = \sum_{p|M} \psi_p(r_p(a)),$$

where $r_p(a)$ *is the residue class of* $a \bmod Z_p$.

Proof. We first prove that the prime decomposition of the function m as in condition (2) implies that m satisfies the linear relations of condition (1). We write

$$\sum_{x \in \Phi(a,a')} (\text{sign } x)m(x) = \sum_{x \in \Phi(a,a')} \text{sign } x \sum_{p|M} \psi_p(r_p(x))$$
$$= \sum_{p|M} \sum_{x \in \Phi(a,a')} (\text{sign } x)\psi_p(r_p(x)).$$

Fix a prime p_0. To each $x \in \Phi(a, a')$ we associate x' such that

$$x_p = x'_p \quad \text{if } p \neq p_0 \quad \text{and} \quad x_{p_0} \neq x'_{p_0}.$$

Then

$$x \equiv x' \bmod Z_{p_0} \quad \text{and} \quad \text{sign } x' = -\text{sign } x.$$

Thus elements in the sum occur in pairs, giving contributions which cancel each other, and we have shown that (2) implies (1).

We shall now prove that (1) implies (2). We write M as a prime power product,

$$M = \prod p^{\nu(p)}.$$

We denote by $\{s_{j_p}\}$ the elements of $Z_{p^{\nu(p)}}^*/Z_p$. Thus j_p is an index for the residue classes of $Z_{p^{\nu(p)}}^*$ modulo Z_p. We fix an arbitrary choice of an element

$a_{j_p} \in S_{j_p}$ for each j_p. We define, relative to this choice,

> X_M = set of elements $x \in Z_M^*$ such that $x_p = a_{j_p}$ for at least one prime $p|M$, and $j_p = j_p(x)$ is the index for the residue class of x itself.

Lemma 5.5. *Let $m: Z_M^* \to A$ be a function into an abelian group, not necessarily even, and satisfying condition (1). If m_1 is another such function such that*

$$m(x) = m_1(x) \quad \textit{for all } x \in X_M,$$

then $m = m_1$. In other words, m is determined by its values on X_M.

Proof. Suppose $y \in X_M$ so $y_p \neq a_{j_p(y)}$ for all $p|M$. Let

$$a(y) = \sum_{p|M} a_{j_p(y)}.$$

Then

$$\sum_{z \in \Phi(a(y), y)} (\text{sign } z) m(z) = 0.$$

But

$$\Phi(a(y), y) = [\Phi(a(y), y) \cap X] \cup \{y\}.$$

This shows that $m(y)$ is determined by values of m on X_M, thereby proving the lemma.

In order to solve for the functions ψ_p of Lemma 5.2, we shall have to consider a filtration of X_M as follows. Let t be the number of prime factors of M. We let

> $X_k(M)$ = set of $x \in X_M$ such that the number of primes $p|M$ such that $x_p = a_{j_p(x)}$ is exactly equal to k.

For each prime $p_0|M$, we define a function

$$f_{p_0}: X_M/Z_{p_0} \to \{0, \ldots, t-1\}$$

by

$$f_{p_0}(y) = \text{number of primes } p|M,\ p \neq p_0 \text{ such that } y_p = a_{j_p(y)}.$$

We define a function

$$f: \bigcup_{p|M} X_M/Z_p \to \{0, \ldots, t-1\}$$

in the natural way, since the union is disjoint. For each prime $p|M$ we let

$[X_M/Z_p]_k =$ set of elements $S \in X_M/Z_p$ such that

$$f(y) = k \quad \text{for all } y \in S.$$

Then we have a disjoint union

$$X_M/Z_p = \bigcup_{k=0}^{t-1} [X_M/Z_p]_k.$$

We define ψ_p by descending induction, starting with

$$[X_M/Z_p]_{t-1}.$$

We have to satisfy

$$m(x) = \sum_{p|M} \psi_p(r_p(x)), \quad \text{all } x \in X_t(M).$$

We observe that each $r_p(x)$ occurs in only one of these equations. Consequently it is possible to solve for the ψ_p in this case. Inductively, we define ψ_p on

$$\bigcup_{k=i-1}^{t-1} [X_M/Z_p]_k$$

for all $p|M$ such that for all $x \in X_t(M) \cup \cdots \cup X_i(M)$ we have

$$m(x) = \sum_{p|M} \psi_p(r_p(x)).$$

If $x \in X_{i-1}(M)$ then

$$r_p(x) \in [X_M/Z_p]_k \quad \text{with } k = i - 1 \quad \text{or} \quad i - 2.$$

If $k = i - 1$, the definition of ψ_p is done by induction. If $k = i - 2$, we have the same phenomenon as in the first step, namely each $r_p(x)$ occurs in only one of the equations

$$m(y) = \sum_{p|M} \psi_p(r_p(y)), \quad \text{with } y \in X_t(M) \cup \cdots \cup X_{i-1}(M)$$

and therefore we can solve for the functions ψ_p on $[X_M/Z_p]_{i-1}$.

Finally, we define for any $x \in Z_M^*$,

$$m^*(x) = \sum_{p|M} \psi_p(r_p(x)).$$

Then m^* is a solution to the system of equations (1) on Z_M^* by the first part of the proof, i.e. (2), implies (1), and agrees with m on X_M. Therefore $m^* = m$, and the proof of the lemma is complete.

For the application, we of course want to deal with an even function m, in which case we also want the functions ψ_p to be even. This is taken care of by the next lemma.

Lemma 5.6. *Let l be a prime $\neq 2$, and let*

$$m : Z_M^* \to \mathbf{Z}/l\mathbf{Z}$$

be a function which is even (i.e. $m(a) = m(-a)$), *satisfying condition* (1) *of Lemma* 5.4. *Then the functions ψ_p of* (2) *can be selected to be even for each $p|M$.*

Proof. With the functions ψ_p of (2), we define

$$\psi_p^*(y) = \tfrac{1}{2}[\psi_p(y) + \psi_p(-y)].$$

This is an even function which again satisfies (2), as desired.

We come to the main part of the proof of Theorem 5.4. We write the product of Siegel functions in the form

$$\prod g_a^{m(a)} = \prod_{a \in Z_M^*} g_a^{m(a)} \cdot \text{other factors}.$$

We now view $m(a)$ as lying in the abelian group $\mathbf{Z}/l\mathbf{Z}$, and solve for the functions ψ_p mod l, so that we have for $a \in Z_M^*$,

$$m(a) \equiv \sum_{p|M} \psi_p(r_p(a)) \bmod l.$$

We lift the functions ψ_p to $\mathbf{Z}$ in any way. Then

$$\prod_{a \in Z_M^*} g_a^{m(a)} = \prod_{a \in Z_M^*} \prod_{p|M} g_a^{\psi_p(r_p(a))} f^l,$$

where f is modular. If $pa = pb$ then $r_p(a) = r_p(b)$, and therefore

$$\prod_{a \in Z_M^*} g_a^{m(a)} = \prod_{p|M} \prod_{a \in Z_M^*} g_a^{\psi_p(r_p(a))} f^l$$

breaks up into a product over equivalence classes of elements mod Z_p. In each such equivalence class, the exponent is constant. Thus for each equivalence class $S \in Z_M^*/Z_p$ we have

$$\prod_{a \in S} g_a^{\psi_p(r_p(a))} = \left(\prod_{a \in S} g_a\right)^{\psi_p(S)}$$

Set $t = pa$. Suppose first that $(M/p, p) \neq 1$. Then by the distribution relations,

$$\prod_{a \in S} g_a = \lambda g_t$$

for some constant λ. If $(M/p, p) \neq 1$, then there exists $t' \in Z_{M/p}^*$ such that $pt' = t$. Then by the distribution relations we have

$$\prod_{a \in S} g_a = \lambda g_{t'}/g_t.$$

Hence in either case,

$$\prod g_a^{m(a)}$$

has been expressed as an l-th power times a product of Siegel functions of lower level, and the proof of the theorem is complete by induction.

Theorem 5.7. *Suppose that a product*

$$\prod g_a^{m(a)} = \lambda$$

is constant, so modular. Let M be the largest integer which occurs in m as a primitive denominator. If a, $a' \in Z_M^*$ satisfy $a_p \equiv a'_p$ mod Z_p and $a_p \neq a'_p$ for all $p|M$, then

$$\sum_{x \in \Phi(a,a')} (\text{sign } x) m(x) = 0.$$

Proof. If this sum is $\neq 0$ for some a, a', then there exists a prime l such that the sum is $\not\equiv 0$ mod l. This contradicts Lemma 5.3.

§6. Dependence of Δ

We conclude with remarks about Δ. When does Δ belong to the group generated by the Klein forms? The question in the most general case is closely linked to the possibility of taking modular square roots from the group generated by the Siegel functions. We give a complete answer to the question in the prime power case.

Let $\mathfrak{K}_N$ be the group generated by the Klein forms $\mathfrak{k}_a$, where a has period N, modulo constants.

Theorem 6.1. *Suppose $N = p^r$ is a prime power. Then Δ belongs to $\mathfrak{K}_N$ if and only if $(p^2 - 1)/2$ divides 12.*

Proof. If $(p^2 - 1)/2$ divides 12, then $p \leqq 5$ and we can write Δ explicitly as a product of Klein forms as follows. We observe that for any prime p we have

$$\prod_{a \in Z_p^*/\pm 1} \mathfrak{k}_a = \begin{cases} \lambda\Delta^{-(p^2-1)/24} & \text{if } p \neq 2 \\ \lambda\Delta^{-3/12} & \text{if } p = 2, \end{cases}$$

where λ is constant. Indeed, the weights of both sides are equal. The expression on the left is invariant under $SL_2(\mathbf{Z})$, and if $(p^2 - 1)/2$ divides 12, then an appropriate integral power yields Δ.

Conversely we use induction on r. Suppose

$$\prod \mathfrak{k}_a^{m(a)} = \lambda\Delta,$$

with some constant λ, then the product $\prod g_a^{m(a)}$ is constant. Note that $\sum m(a) = 12$. It will suffice therefore to prove:

Lemma 6.2. *Suppose that*

$$\prod g_a^{m(a)} = \textit{constant},$$

and that every a occurring in m has period N where $N = p^r$ is a prime power. Then $(p^2 - 1)/2$ divides $\sum m(a)$.

Proof. By induction on r. Take first $r = 1$. By Theorem 5.7, for each pair $a \neq a' \in Z_p$ we have

$$m(a) = m(a') = m.$$

Thus

$$\sum m(a) = \sum_{a \in Z(p)^*} m = m(p^2 - 1)/2,$$

and the lemma is proved in this case. Assume the lemma for some r, we shall prove it for $N = p^{r+1}$. By Theorem 5.7, if

$$a \equiv a' \pmod{Z_p}$$

then $m(a) = m(a')$. Let $\{S\}$ be the elements of $Z^*_{p^{r+1}}/Z_p$. Then

$$\prod_{a \in Z^*_{p^{r+1}}} g_a^{m(a)} = \prod_S \prod_{a \in S} g_a^{m(a)} = \prod_S g_{pS}^{m(S)}.$$

However

$$\sum_{a \in S} m(a) - m(S) = (p^2 - 1)m(S) \equiv 0 \bmod p^2 - 1.$$

We are therefore done by induction.

Remark. In general, for the composite case, it can be shown that if N has prime factors p_i, $(i = 1, \ldots, t)$ and

$$\text{g.c.d. } (p_i^2 - 1)/2 \text{ divides } 12,$$

then Δ belongs to $\mathfrak{K}_N$. Conversely, if Δ belongs to $\mathfrak{K}_N$, the non-2-part of the above g.c.d. must divide 3. A complete statement in this case will involve determining exactly when the square root of a modular unit is a modular unit. Cf. Kubert [Ku 6].

§7. Projective Limits

This section will not be used in the sequel and can be omitted. In passing to the projective limit of the unit groups, we lose some structure, but it gives some insight to see roughly what the whole situation looks like in a setting analogous to that studied by Iwasawa in the cyclotomic case, e.g. [Iw 3] and [Iw 4], where Iwasawa deals with Γ-extensions, and $\Gamma \approx \mathbf{Z}_p$.

We can carry out a similar theory for the tower of function fields of the modular curves $X(p^n)$, where p is prime, taken to be > 3 for simplicity. The role of Γ is here played by a Cartan group (units in the unramified extension of degree 2 over $\mathbf{Z}_p$). Taking into account the results of this chapter, all the non-formal work has already been done, and all that remains to do is to put it together with the usual Kummer theory to obtain analogues of some of Iwasawa's results, but with much simpler formulations. This is the purpose of the present short section, which extracts this Kummer theory.

We may form the usual projective family of group rings, their limit R (Iwasawa algebra) which here turns out to be essentially the power series in two variables $\mathbf{Z}_p[[X_1, X_2]]$, and the Galois group G of the extension of the modular function field by p^n-th roots of units. Then G turns out to be a 1-dimensional free module over R.

Let K_n be the function field of the modular curve $X(p^n)$ with constant field $\mathbf{Q}(\boldsymbol{\mu}_{p^n})$. Let $K_\infty = \bigcup K_n$. The group $\mathrm{GL}_2(\mathbf{Z}_p)$ is represented as a group of automorphisms of K_∞, cf. Shimura [Sh] or [L 5], Chapter VI, §3. It contains the Cartan group C_p, isomorphic to the group of units in the unramified extension of degree 2 of $\mathbf{Z}_p$, in some fixed representation after a choice of basis over $\mathbf{Z}_p$.

Let U_n be the group generated by the 12-th powers of Siegel functions g_a^{12} with $a \in \mathbf{Q}^2$ and $a \notin \mathbf{Z}^2$ such that $p^n a \in \mathbf{Z}^2$. Alternatively, we need only take primitive indices a of exact order p^n mod $\mathbf{Z}^2$.

From Formulas **K 1** through **K 3** of Chapter 2, §1, one sees that the roots of unity $\boldsymbol{\mu}_{p^n}$ are contained in U_n. The distribution relations, Theorem 4.1 of Chapter 2, show that

$$p^{12} \in U_n,$$

and in fact the subgroup of U_n consisting of the constants is precisely

$$\{\boldsymbol{\mu}_{p^{2n}}, p^{12}\}.$$

Let:

$$\Gamma = C_p/\pm 1 = C_p(\pm 1), \qquad \Gamma_n = C_p(p^n)/\pm 1 = \Gamma \bmod p^n.$$

$$R_n = \mathbf{Z}(p^n)[\Gamma_n], \qquad R = \lim R_n.$$

The limit is the projective limit, and defines the **Iwasawa algebra** in the present case. Since C_p contains the units $\equiv 1 \pmod{p}$, which are isomorphic under the exponential map to $\mathbf{Z}_p^2$, it follows that R contains a subring of finite index, isomorphic to

$$\mathbf{Z}_p[[X_1, X_2]],$$

where γ_1, γ_2 are independent generators of the units $\equiv 1 \pmod{p}$, and as usual, $X_i = \gamma_i - 1$.

Let

$$U = \bigcup U_n.$$

In Theorem 1.1, we saw that U is the group of units in K_∞, except for the fact that we have taken 12-th powers here, which will be irrelevant since the theorem we want will be concerned with Kummer theory in extensions of p-power order. Let

$$\Omega = K_\infty(U^{1/p^\infty}).$$

Then Ω is a Kummer extension of K_∞ and is Galois over $\mathbf{Q}(j)$, and over each field K_n. Let

$$G = \mathrm{Gal}(\Omega/K_\infty).$$

Then G is a Γ-module, and also a R-module. The projective structure can be obtained by the cofinal system of finite Kummer extensions

$$\Omega_n = K_\infty(U_n^{1/p^n}), \quad \text{with Galois group } G_n = \mathrm{Gal}(\Omega_n/K_\infty).$$

The field diagram is as follows.

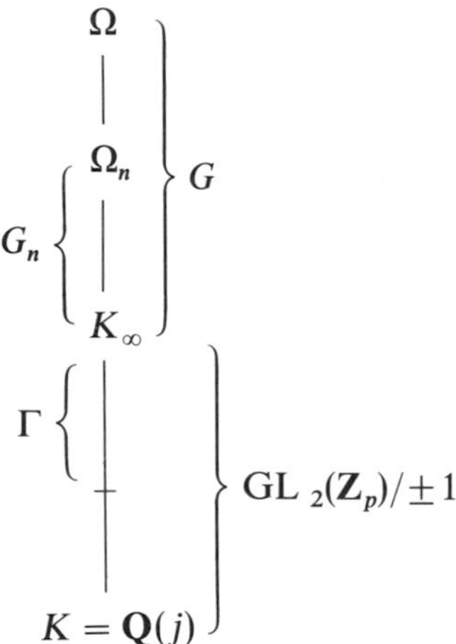

The standard Kummer theory of elementary algebra yields a duality for the finite abelian extension,

$$G_n \times U_n^{1/p^n} K_\infty^* / K_\infty^* \to \boldsymbol{\mu}_{p^n} \tag{1}$$

We have an isomorphism

$$U_n^{1/p^n} K_\infty^* / K_\infty^* \approx U_n^{1/p^n} / U_n^{1/p^n} \cap K_\infty^*.$$

But

$$U_n^{1/p^n} \cap K_\infty^* = U_n \boldsymbol{\mu}_{(n)}, \tag{2}$$

where $\boldsymbol{\mu}_{(n)}$ is the group of p^{2np^n}-th roots of unity. Indeed, the Galois group of $\mathbf{Q}(p^{1/p^n}, \boldsymbol{\mu}_{p^n})$ over $\mathbf{Q}$ is a semidirect product of the abelian Galois group of p^n-th roots of unity and a cyclic group of order p^n. The field of roots of unity is maximal subabelian in this Galois extension, and p^{1/p^n} generates an extension of degree p^n over $\mathbf{Q}(\boldsymbol{\mu}^{(p)}) = \mathbf{Q}_\infty$, the tower of roots of unity. Hence it generates an extension of the same degree over K_∞, because $\mathbf{Q}_\infty$ is the constant field of K_∞. By Theorem 1.1, as already mentioned, we conclude that (2) is true.

Therefore the Kummer theory gives a duality

$$G_n \times U_n^{1/p^n}/U_n \boldsymbol{\mu}_{(n)} \to \boldsymbol{\mu}_{p^n}. \tag{3}$$

Let $\sigma \in G_n$ and $u \in U_n^{1/p^n}$. The Kummer pairing

$$(\sigma, u) \mapsto \langle \sigma, u \rangle$$

is given by the formula

$$\langle \sigma, u \rangle = \sigma u/u = u^{\sigma - 1}.$$

The group Γ of course operates by conjugation on G, and operates on the roots of unity via the determinant. By first principles of functoriality, the symbol satisfies the formula

$$\langle \sigma^\gamma, u^\gamma \rangle = \langle \sigma, u \rangle^\gamma = \langle \sigma, u \rangle^{\det \gamma}, \tag{4}$$

for all $\gamma \in \Gamma$. By σ^γ we mean $\tilde{\gamma}\sigma\tilde{\gamma}^{-1}$, where $\tilde{\gamma}$ is a lifting of γ to an automorphism of Ω. Similarly $u^\gamma = \tilde{\gamma}u$. The value of the symbol is independent of the choice of lifting $\tilde{\gamma}$, because the symbol $\langle \sigma, u \rangle$ is equal to $\langle \sigma, u' \rangle$ if $u^{p^n} = u'^{p^n}$, and because Ω/K_∞ is abelian.

In the Kummer pairing, we want to take projective limits on the Galois groups, and injective limits on the p^n-th roots of units. For this purpose, we use the isomorphism

$$\frac{1}{p^n} \mathbf{Z}_p/\mathbf{Z}_p \approx \mathbf{Z}_p/p^n\mathbf{Z}_p = \mathbf{Z}_p(p^n)$$

obtained by multiplication with p^n. Then we use the isomorphism of the group ring of Γ_n with formal linear combinations of elements in Γ_n with coefficients in

$$\frac{1}{p^n}\mathbf{Z}_p/\mathbf{Z}_p,$$

namely we write

$$\mathbf{Z}(p^n)[\Gamma_n] \approx \frac{1}{p^n}\mathbf{Z}_p/\mathbf{Z}_p[\Gamma_n].$$

With this notation, ***we have an isomorphism of R-modules***

(5) $$\frac{1}{p^n}\mathbf{Z}_p/\mathbf{Z}_p[\Gamma_n] \to U_n^{1/p^n}/U_n\boldsymbol{\mu}_{p^{2n}}$$

by means of the mapping

$$\gamma \mapsto g_\gamma^{12/p^n}.$$

where g_γ is the Siegel function, and the map is well defined since we are allowed to take the p^n-th root mod roots of unity. Since Γ_n operates simply transitively, since the only relation among the Siegel functions is given by the distribution relation that

$$\prod_{\gamma \in C(p^n)} g_\gamma = p \text{ mod roots of unity},$$

and since the p^n-th roots of p generate an extension disjoint from Ω_n over K_∞, it follows that there cannot be any relation in the homomorphism from the group ring in (5) to the group on the right, and hence that the map is an isomorphism.

The group G_n can also be considered as R-module. By duality, ***we have an isomorphism***

(6) $$\mathbf{Z}(p^n)[\Gamma_n] \to G_n,$$

and the Kummer duality can be represented by the pairing

(7) $$\mathbf{Z}(p^n)[\Gamma_n] \times \frac{1}{p^n}\mathbf{Z}_p/\mathbf{Z}_p[\Gamma_n] \to \boldsymbol{\mu}_{p^n}$$

given by

$$\left(\sum x(\alpha)\alpha, \sum y(\beta)\beta\right) \mapsto \mathbf{e}\left(\sum x(\alpha)y(\alpha) \det \alpha\right),$$

where $\mathbf{e}(t) = e^{2\pi i t}$. We have an isomorphism of pairings

(8)
$$\begin{array}{ccccc} G_n & \times & U^{1/p^n}/U_n\boldsymbol{\mu}_{p^{2n}} & \longrightarrow & \boldsymbol{\mu}_{p^n} \\ \updownarrow \approx & & \updownarrow \approx & & \updownarrow = \\ \mathbf{Z}(p^n)[\Gamma_n] & \times & \dfrac{1}{p^n}\mathbf{Z}_p/\mathbf{Z}_p[\Gamma_n] & \longrightarrow & \boldsymbol{\mu}_{p^n}. \end{array}$$

This takes care of the story at finite level.

Passing to the limit, the map $G_{n+1} \to G_n$ corresponds to the natural homomorphism of group algebras. On the other hand we have an *injection*

$$\frac{1}{p^n}\mathbf{Z}_p/\mathbf{Z}_p[\Gamma_n] \to \frac{1}{p^{n+1}}\mathbf{Z}_p/\mathbf{Z}_p[\Gamma_{n+1}]$$

in a natural way. The coefficients are both contained in $\mathbf{Q}_p/\mathbf{Z}_p$, and there is a natural embedding

$$\Gamma_n \to \Gamma_{n+1}$$

which sends an element of Γ_n to the formal sum of elements in Γ_{n+1} lying above it under the canonical surjection $\Gamma_{n+1} \to \Gamma_n$. Passing to the limit may be represented by the arrows:

(9)
$$\begin{array}{ccccc} G_{n+1} & \times & \hat{G}_{n+1} & \longrightarrow & \boldsymbol{\mu}_{p^{n+1}} \\ \downarrow & & \uparrow & & \uparrow \\ G_n & \times & \hat{G}_n & \longrightarrow & \boldsymbol{\mu}_{p^n} \end{array}$$

In the limit, we then find a duality of compact and discrete abelian groups,

(10) $$G \times \hat{G} \to \boldsymbol{\mu}^{(p)}, \quad \text{where } \hat{G} = \varinjlim \frac{1}{p^n}\mathbf{Z}_p/\mathbf{Z}_p[\Gamma_n].$$

In any case, it is apparent from the group algebras that

(11) **G is a 1-dimensional free module over R.**

CHAPTER 5

The Cuspidal Divisor Class Group on $X(N)$

In the most classical situation, the Galois group of the field $\mathbf{Q}(\boldsymbol{\mu}_N)$ of roots of unity operates on ideal classes and units modulo cyclotomic units. Classical problems of number theory are concerned with the eigenspace decomposition of the p-primary part of these groups when $N = p$, and of the structures as Galois modules in general. Results include those of Kummer, Stickelberger, Herbrand, and more recently Iwasawa, Leopoldt, and Ribet. Also in recent times, such results have been the object of study in the case of complex multiplication of elliptic curves for the elliptic units as in Robert and Coates-Wiles.

Here we are concerned with the "generic" case of the modular function field of the modular curve $X(N)$. The divisor class group generated by the cusps can be represented as a quotient of the group ring of the Cartan group by a Stickelberger ideal. This makes use of the characterization of units in Chapter 4, and the quadratic relations of Chapter 3, which determine which units have a given level N. The connection with the geometry is made via the Fricke-Wohlfart theorem: intersecting the Stickelberger module with the integral group ring corresponds to requiring that the functions have integral orders at each cusp, and is equivalent to their having level N.

This allows us to compute the order of the cuspidal divisor class group in a way analogous to that of Iwasawa in the case of cyclotomic fields. Whereas Iwasawa meets the Bernoulli numbers $B_{1,\chi}$, we encounter here the second Bernoulli numbers $B_{2,\chi}$ (a twist). For the cyclotomic theory, cf. for instance [L 8], Chapter 2, partly reproducing Iwasawa's result, and partly also dealing with general Bernoulli numbers $B_{k,\chi}$.

After that, we analyze the eigenspace decomposition at level p on $X(p)$. Again we are dealing with a purely algebraic question in the group ring modulo the Stickelberger ideal, and the eigenspaces are generated by a product of the integralizing ideal of quadratic relations times Bernoulli numbers, on the Cartan group. These decompose as a product of Gauss sums times ordinary Bernoulli numbers on $\mathbf{Z}(p)^*$. In that case one knows their ideal factorization by Stickelberger's theorem. This allows us to determine completely the p-adic order of the eigenspace.

One phenomenon appears here which did not appear in the cyclotomic theory: in some cases, we find the existence of a special group which contributes one more piece than that indicated by the Bernoulli number. It turns out that this special group is exactly the same that was already found in Chapter 2, in connection with the Weierstrass forms.

The fact that the cuspidal divisor class group is finite, i.e. that the cusps are of finite order in the Jacobian of the modular curves, was originally proved by Manin-Drinfeld [Ma-Dr]. Indeed, let x_1, x_2 be two cusps. We denote by

$$\{x_1, x_2\}$$

the functional on the space of differentials of first kind given by

$$\{x_1, x_2\} : \omega \mapsto \int_{x_1}^{x_2} \omega.$$

A priori, $\{x_1, x_2\}$ lies in $H^1(X(N), \mathbf{R})$. Manin-Drinfeld show that $\{x_1, x_2\}$ in fact lies in $H^1(X(N), \mathbf{Q})$. They use Hecke operators. Cf. [L 7], Chapter 4. Their method suggests generalizations to higher dimensional bounded symmetric domains. We leave this method aside in the present book. For further comments in connection with diophantine analysis, see Chapter 8.

§1. The Stickelberger Ideal

We fix an integer $N > 1$. We let:

$G(N) \cong C(N)/\pm 1$, where $C(N)$ is the Cartan group of degree 2.

$R = \mathbf{Z}[G(N)] =$ group ring of $G(N)$ over $\mathbf{Z}$.

$R_0 =$ subgroup of R consisting of elements of degree 0.

$\mathscr{D} =$ group of divisors on the modular curve $X(N)$ generated by the cusps.

$\mathscr{D}_0 =$ subgroup of divisors of degree 0.

$\mathscr{F} =$ group of divisors of units in the modular function field F_N of level N.

$\mathscr{C} = \mathscr{D}_0/\mathscr{F} =$ cuspidal divisor class group.

$\mathscr{S}$ = group of divisors of Siegel functions

$$g = \prod g_a^{m(a)}$$

such that $a \in (1/N)\mathbf{Z}^2$, $a \notin \mathbf{Z}^2$, and g has level N.

We index these groups by N if we need to emphasize this reference. If $\mathfrak{S}$ = group of Siegel functions as above, then we may also write

$$\mathscr{S} = \operatorname{div} \mathfrak{S}.$$

In this chapter, we take the divisor by measuring the order using the local parameter $q^{1/N}$, *which depends on* N. *Therefore this order is* N *times the absolute order taken in Chapter* 2. We have inclusions

$$\mathscr{D}_0 \supset \mathscr{F} \supset \mathscr{S}.$$

Theorem 1.1.

(i) *If* $N = p^n$ *is a prime power, then*

$$\mathscr{F} = \mathscr{S} \quad \text{and} \quad \mathscr{C} = \mathscr{D}_0/\mathscr{S}.$$

(ii) *If* N *is arbitrary, then the factor group* $\mathscr{F}/\mathscr{S}$ *is a* 2-*group, and is finite.*

Proof. This is a reformulation of Theorem 1.3 of Chapter 4.

We now wish to describe more accurately the ideal $\mathscr{S}$. Such description relies on the following characterization, which relates the geometry with the group theory inside $\mathrm{SL}_2(\mathbf{Z})$. Let

$$\sigma = \begin{pmatrix} 1 & 1 \\ 0 & 1 \end{pmatrix}$$

and let $E(N)$ be the smallest normal subgroup of $\mathrm{SL}_2(\mathbf{Z})$ containing σ^N.

Theorem 1.2. (Fricke-Wohlfart) *Let* $N \mid N'$. *Then*

$$\Gamma(N) = \Gamma(N')E(N).$$

Proof. Let

$$\alpha = \begin{pmatrix} a & b \\ c & d \end{pmatrix} \in \Gamma(N).$$

We have to show that there exists $\beta \in E(N)$ such that

$$\alpha \equiv \beta \bmod N'.$$

Suppose first that $c \not\equiv 0 \bmod N'$. We find $x, y \in \mathbf{Z}$ such that

$$cx + dy = 1 \quad \text{and} \quad (y, N'/N) = 1.$$

Then select an integer z such that $yzN \equiv c \bmod N'$. It follows that

$$\gamma = \begin{pmatrix} 1 - xyzN & x^2zN \\ -y^2zN & 1 + xyzN \end{pmatrix} \in E(N) \quad \text{and} \quad \alpha\gamma \equiv \begin{pmatrix} * & * \\ 0 & * \end{pmatrix} \bmod N'.$$

This reduces the problem to the second case, when we may assume that $c \equiv 0 \bmod N'$. In that case, $ad - 1 \equiv 0 \bmod N'$, and therefore

$$\begin{pmatrix} 1 & 0 \\ 1-d & 1 \end{pmatrix} \begin{pmatrix} a & a-1 \\ 1-a & 2-a \end{pmatrix} \begin{pmatrix} 1 & (b+1)d-1 \\ 0 & 1 \end{pmatrix}$$
$$\equiv \begin{pmatrix} 1 & 0 \\ 1-d & 1 \end{pmatrix} \begin{pmatrix} a & b \\ 1-a & d+b(d-1) \end{pmatrix} \equiv \begin{pmatrix} a & b \\ c & d \end{pmatrix} \bmod N'.$$

Since the three matrices on the left hand side lie in $E(N)$, the theorem follows.

For the above proof, cf. Wohlfart [Wo] and Leuchtbecher [Leu].

Let:

U_N = group generated by the Siegel functions g_a with $a \in \frac{1}{N}\mathbf{Z}^2/\mathbf{Z}^2$.

For each $g \in U_N$, let

$$\operatorname{div}(g) = \sum_{\alpha \in C(N)/\pm 1} \operatorname{ord}_{q^{1/N}}(\sigma_\alpha g)\sigma_\alpha^{-1}.$$

As before, we identify the group ring and the group of divisors:

$$R = \mathscr{D},$$

and we measure the order *in terms of the parameter* $q^{1/N}$.

Theorem 1.3. *We have*

$$\operatorname{div} U_N \cap R = \mathscr{S}.$$

In other words, a Siegel function g in U_N has level N if and only if it has integral order at each cusp.

Proof. This is a reformulation of the Fricke-Wohlfart theorem.

In Chapter 2 we had used the map

$$T:(r_1, r_2) \mapsto r_1.$$

This is for $r_1, r_2 \in \mathbf{Z}(N)$. We use the same map

$$T:(a_1, a_2) \mapsto a_1$$

on $(1/N)\mathbf{Z}^2/\mathbf{Z}^2$. Then the divisor of g_a may be written in the form

$$\operatorname{div}(g_a) = (g_a) = N \sum_{\beta \in C(N)/\pm 1} \tfrac{1}{2}\mathbf{B}_2(\langle T(\alpha\beta)\rangle)\sigma_\beta^{-1}.$$

The factor N in front of the right hand side comes from the fact that we use $q^{1/N}$ as parameter instead of q. We may now use the terminology of Stickelberger elements, and write

$$(g_a) = \mathrm{St}_N(a).$$

We let:

$\mathrm{St}(N)$ = module generated over $\mathbf{Z}$ by all Stickelberger elements

$$\mathrm{St}_N(a) \quad \text{with } a \in \left(\frac{1}{N}\mathbf{Z}^2/\mathbf{Z}^2\right)\Big/ \pm 1.$$

The **Stickelberger ideal** is then defined to be

$$\mathrm{St}(N) \cap R.$$

Theorem 1.4. *We have* $\mathrm{St}(N) \cap R = \mathscr{S}$.

Proof. This is a reformulation of the Fricke-Wohlfart theorem, taking into account the divisor of the Siegel functions.

This theorem reduces the study of the cuspidal divisor class group to an algebraic situation where no modular functions appear, namely up to 2-torsion we have an isomorphism

$$\boxed{\mathscr{C} = \mathscr{C}(N) \approx R_0/(\mathrm{St}(N) \cap R).}$$

It is convenient to define the map T on 2×2 matrices, namely

$$T:\begin{pmatrix} a & b \\ c & d \end{pmatrix} \mapsto a.$$

We may index the primitive Siegel functions by elements α of the Cartan group, putting

$$g_\alpha = g_{e_1\alpha} \quad \text{where } e_1 = \left(\frac{1}{N}, 0\right).$$

Then

$$g_\alpha = g_{a/N, b/N},$$

and the divisor of g_α may be written in the form

$$\operatorname{div}(g_\alpha) = (g_\alpha) = N \sum_{\beta \in C(N)/\pm 1} \tfrac{1}{2}\mathbf{B}_2\left(\left\langle \frac{T(\alpha\beta)}{N} \right\rangle\right)\sigma_\beta^{-1}.$$

We now shift to the prime power case, where these primitive elements give us all the information we need.

§2. The Prime Power Case, $p \geqq 5$

Let us define the (primitive) **Stickelberger element**

$$\boxed{\theta = N \sum_{\beta \in C(N)/\pm 1} \tfrac{1}{2}\mathbf{B}_2\left(\left\langle \frac{T\beta}{N} \right\rangle\right)\sigma_\beta^{-1}.}$$

Then

$$(g_\alpha) = \sigma_\alpha \theta.$$

This is particularly advantageous when $N = p^n$ is a prime power, because in that case, the distribution relations show that the Stickelberger element generates the Stickelberger module:

Theorem 2.1. *Let $N = p^n$ be a prime power. Then*

$$\operatorname{St}(N) = R\theta.$$

Proof. Immediate from the distribution relations.

We now wish to determine $R\theta \cap R$. Let:

I = ideal generated by "parallelograms"

$$\pi(\alpha, \beta) = (\alpha + \beta) + (\alpha - \beta) - 2(\alpha) - 2(\beta)$$

in the group ring R. Note that the degree of a parallelogram is -2. For any positive integer d, we let:

R_d = ideal of R consisting of those elements whose degree is $\equiv 0 \bmod d$.
$I_d = I \cap R_d$.

Theorem 2.2. *Let $N = p^n$ where p is a prime $\geqq 5$. Then*

$$R\theta \cap R = I_{12}\theta.$$

Let $\mathscr{C}$ be the cuspidal divisor class group on $X(N)$. Then we have a natural isomorphism

$$\mathscr{C} \approx R_0/I_{12}\theta = R_0/\mathscr{S},$$

where $\mathscr{S}$ is the Stickelberger ideal.

Proof. It is convenient to phrase the proof in the language of modular forms. The module $R\theta$ is isomorphic to the group of divisors of products

$$g = \prod_{\alpha} g_{\alpha}^{m(\alpha)}.$$

Intersecting the group of such divisors with R corresponds to restricting g to have integral order at each cusp on $X(N)$. By the theorem of Fricke-Wohlfart, this is equivalent to requiring that g has level N. Theorem 5.2 of Chapter 3 then implies that the divisor (g) lies in $I_{12}\theta$. This proves the inclusion

$$R\theta \cap R \subset I_{12}\theta.$$

The reverse inclusion and the rest of the theorem follow from Theorems 1.3, 1.4, 2.1. This concludes the proof.

Observe that

$$\deg \theta = 0,$$

which we can see either from the distribution relations, or from the corresponding geometric fact that a function has as many zeros as it has poles, and that the Stickelberger element gives the divisor of a function. Thus

$$\theta \in \mathbf{Q}R_0.$$

Besides the primitive Stickelberger element, we shall have to consider a variant of it to take into account the trivial character. For this purpose, we define

$$\theta' = \theta'_G = \sum_{\alpha \in C(N)/\pm 1} \frac{N}{2}\left(\mathbf{B}_2\left(\left\langle \frac{T\alpha}{N} \right\rangle\right) - \frac{1}{6}\right)\sigma_\alpha^{-1}.$$

The group G is still $G \approx C(N)/\pm 1$. We let

$$s(G) = \sum_{\sigma \in G} \sigma = \text{augmentation element.}$$

For any character χ of G (identified with an even character of $C(N)$), we let

$$\boxed{B_{2,\chi} = \sum_{\alpha \in C(N)/\pm 1} \mathbf{B}_2\left(\left\langle \frac{T\alpha}{N} \right\rangle\right)\chi(\alpha).}$$

For any non-trivial character χ we have

$$\chi(\theta) = \chi(\theta') = \frac{N}{2} B_{2,\bar{\chi}}.$$

We know from Theorem 5.3 and Corollary 2 of Chapter 1 that

$$B_{2,\bar{\chi}} \neq 0$$

for such characters. If χ is the trivial character, then

$$\theta' s(G) = \chi(\theta')s(G) = \deg(\theta')s(G) \neq 0.$$

Indeed, we have already remarked that $\deg \theta = 0$, so after subtracting $\frac{1}{6}$, we get a non-zero value for $\deg \theta'$, namely

$$\deg \theta' = -\frac{N}{12}|G|.$$

Thus $\theta \in R_0$ but $\theta' \notin R_0$. Since none of the characters vanish on θ', we get:

Lemma 2.3. *The element θ' is invertible in the group algebra $\mathbf{Q}[G]$.*

Remark. Since automorphisms σ_α of the modular curve induce a corresponding automorphism of the Jacobian, the element θ' induces an isogeny of the Jacobian because under a representation, units go into units.

§3. Computation of the Order

The order of the cuspidal divisor class group $\mathcal{D}_0/\mathcal{F}$ on $X(N)$ has been reduced to a computation of the index of a certain ideal in the group ring

$$R = \mathbf{Z}[G] \quad \text{where } G \approx C(N)/\pm 1.$$

We now give this computation in the same case as that of the preceding section, and we continue with the same notation.

Theorem 3.1. *Let $N = p^n$ where p is prime $\neq 2, 3$. Then the order of the cuspidal divisor class group on $X(N)$ is given by the index*

$$(R_0 : \mathcal{S}) = \frac{6N^3}{|G|} \prod_{\chi \neq 1} \frac{N}{2} B_{2,\chi}.$$

Proof. We go through a sequence of lemmas giving indices of various ideals in each other.

Let Δ be the usual modular form of weight 12 on $\mathrm{SL}_2(\mathbf{Z})$. On $X(N)$ its divisor is

$$(\Delta) = Ns(G).$$

Indeed, Δ has a zero at each cusp of order N since $X(N)$ is ramified of order N over $j = \infty$. It is suggestive to keep this interpretation in mind when we deal with the element $Ns(G)$ in the group ring.

We recall that

$$\theta' = \theta - \frac{N}{12} s(G)$$

and

$$\xi s(G) = \deg(\xi) s(G).$$

Lemma 3.2. $R_0 \cap (I\theta' + RNs(G)) = I_{12}\theta.$

Proof. Suppose $\xi \in I$ and

$$\deg(\xi\theta' + vNs(G)) = 0$$

for some integer v. Since $\deg \theta' = -N|G|/12$, we get

$$\tfrac{1}{12} \deg \xi = v, \quad \text{and so} \quad \deg \xi \equiv 0 \bmod 12.$$

The inclusion $\subset$ is proved. The reverse inclusion is obvious.

From the lemma, we obtain an injection

$$0 \to R_0/I_{12}\theta \to R/(I\theta' + RNs(G)).$$

We have $R_0 + (I\theta' + RNs(G)) = R_d$ for some positive integer d, and

$$\begin{aligned} R_0/I_{12}\theta &\approx [R_0 + I\theta' + RNs(G)]/[I\theta' + RNs(G)] \\ &\approx R_d/(I\theta' + RNs(G)) \end{aligned}$$

where $d = \text{g.c.d.}(\deg(I\theta'), \deg(RNs(G)))$. By the degree of an ideal, we mean the degree of a positive generator for the degrees of all elements in the ideal. It is immediate that

$$d = \frac{N}{6}|G|,$$

because I has degree 2.

We have an inclusion of ideals:

$$R \supset R_d \supset I\theta' + RNs(G) \supset I\theta'.$$

We want the middle index for Theorem 3.1. It is obvious that

$$(R : R_d) = d.$$

We shall compute successively the indices

$$(R : I\theta') \quad \text{and} \quad (I\theta' + RNs(G) : I\theta')$$

to conclude the proof.

Lemma 3.3. $(I\theta' + RNs(G) : I\theta') = \frac{1}{12}|G|$.

Proof. By Noether isomorphism the left hand side is equal to

$$(RNs(G) : RNs(G) \cap I\theta').$$

We shift to modular forms. The module $I\theta'$ is the module of divisors of products of Klein forms satisfying the quadratic relations, and its intersection with $RNs(G)$ corresponds to the condition that such a product is equal to a power of Δ, that is

$$\prod_{\alpha} \mathfrak{k}_{\alpha}^{m(\alpha)} = \Delta^{v}$$

for some integer v. Comparing weights, we must have

$$v = -\frac{1}{12} \sum_{\alpha} m(\alpha).$$

Since the Siegel function g_α satisfies $g_\alpha = \mathfrak{k}_\alpha \eta^2$, it follows that

$$\prod g_{\alpha}^{m(\alpha)} = \text{constant}.$$

From the independence of the Siegel functions for prime power level (Theorem 3.2 of Chapter 2), it follows that the exponents $m(\alpha)$ have to be constant, independent of α, say equal to an integer m. Then

$$\prod_{\alpha} \mathfrak{k}_{\alpha}^{m} = \Delta^{-m|G|/12}.$$

The lemma follows at once, because the family $\{m(\alpha)\}$ with $m(\alpha) = m$ satisfies the quadratic relations.

Next we have

$$(R : I\theta') = (R : R\theta')(R\theta : I\theta').$$

The element θ' is invertible in the rational group ring $\mathbf{Q}[G]$. Since the Bernoulli polynomial is quadratic, the element $N\theta'$ lies in R. Hence

$$(R : RN\theta') = \pm \det_{\mathbf{C}[G]} N\theta',$$

where the determinant of multiplication by $N\theta'$ can be computed in the group algebra over the complex numbers, where the group algebra splits into its simple components, which are 1-dimensional, and correspond to the characters of G. Since

$$(R : R\theta') = \frac{(R : RN\theta')}{(R\theta' : RN\theta')},$$

and $(R\theta' : RN\theta') = N^{|G|}$, we find:

Lemma 3.4. $(R : R\theta') = \prod_{\chi} \chi(\theta') = (\deg \theta') \prod_{\chi \neq 1} \chi(\theta).$

Proof. We have separated the product over all characters into one term with the trivial character, which gives deg θ', and the other product, where θ' can be replaced by θ, since the sum of a non-trivial character over all group elements is 0. So the lemma is clear.

Lemma 3.5. $(R:I) = (R\theta' : I\theta') = N^3$.

Proof. We actually prove an isomorphism

$$R/I \approx \mathbf{Z}(N)^3.$$

We may first identify the elements of the Cartan group with primitive pairs (r_1, r_2) of integers mod N, and secondly by quadratic relations Chapter 3, Theorem 4.3 we may identify the ideal I with the linear combinations of pairs satisfying the quadratic relations mod N. We then claim that the elements $(0, 1)$, $(1, 0)$, $(1, 1)$ form a basis for R/I, in other words given a primitive pair (r_1, r_2) there exist unique integers x, y, z such that

$$(r_1, r_2) + x(1, 0) + y(0, 1) + z(1, 1)$$

satisfies the quadratic relations mod N. This means:

$$\begin{aligned} r_1^2 + x + z &\equiv 0 \bmod N, \\ r_2^2 + y + z &\equiv 0 \bmod N \\ r_1 r_2 + z &\equiv 0 \bmod N \end{aligned}$$

or equivalently

$$\begin{aligned} z &\equiv -r_1 r_2 \\ x &\equiv -r_1 r_2 - r_1^2 \\ y &\equiv -r_1 r_2 - r_2^2. \end{aligned}$$

This shows x, y, z exist. It is also clear that they are unique mod N, thus proving $(R:I) = N^3$.

The element θ' is invertible in the rational group ring, and so the map

$$x \mapsto x\theta'$$

is an isomorphism on R, so we have $(R:I) = (R\theta' : I\theta')$. This concludes the proof of the lemma.

The theorem is proved by putting all the lemmas together, with the value

$$d = \frac{N}{6}|G|.$$

For the case of composite N, cf. Bergelson [Be].

§4. Eigencomponents at Level p

We consider the modular curve $X(N)$ with a prime $N \geqq 7$. We omitted the primes 2, 3 before because of the technical complications arising from the quadratic relations in this case. We now omit 7 also because $X(5)$ has genus 0, so the group of divisor classes is trivial.

It will be useful to define

$$Q(x) = N\mathbf{B}_2\left(\left\langle\frac{x}{N}\right\rangle\right).$$

We don't use h because we now want to use h for class numbers.

Let p be a prime $\geqq 7$ which does not divide the order of

$$C(\pm) = C(N)/\pm 1.$$

In particular, if $N = p$, then $C(p)$ has order $p^2 - 1$, and p itself can be taken as the prime under consideration. We let q be a power of p such that the multiplicative group $\mathbf{F}_q^*$ of the field with q elements has order divisible by the order of $C(\pm)$. If $N = p$, then we take $q = p^2$. Let

$R_p = \mathbf{Z}_p[G(\pm)]$

$I_p = \mathbf{Z}_p I$, where I is the ideal of parallelograms as in §1.

$\mathfrak{o}_q =$ ring of integers in the unramified extension of $\mathbf{Z}_p$ whose residue class field has q elements.

We have an isomorphism

$$\mathscr{C}^{(p)} \approx R_{p,0}/I_p\theta,$$

where $\mathscr{C}^{(p)}$ is the p-primary part of $\mathscr{C}$ (the cuspidal divisor class group), and $R_{p,0}$ is the submodule of R_p consisting of the elements of degree 0. This follows from Theorem 2.2, because 12 is prime to p.

It will now be convenient to tensor all objects with $\mathfrak{o}_q$. Using elementary divisors, we have an isomorphism

$$\mathbf{Z}_p \otimes \mathscr{C}^{(p)} \approx (\mathbf{Z}_p \otimes \mathscr{D}_0)/(\mathbf{Z}_p \otimes \mathscr{F}),$$

and hence an isomorphism

$$\mathfrak{o}_q \otimes \mathscr{C}^{(p)} \approx (\mathfrak{o}_q \otimes \mathscr{D}_0)/(\mathfrak{o}_q \otimes \mathscr{F}).$$

If, as an abelian group, $\mathscr{C}^{(p)}$ is of type $(p^{e_1}, \ldots, p^{e_m})$, then $\mathfrak{o}_q \otimes \mathscr{C}^{(p)}$ is of the same type as a module over $\mathfrak{o}_q$. Thus to determine the structure of $\mathscr{C}^{(p)}$ it suffices to determine $\mathfrak{o}_q \otimes \mathscr{C}^{(p)}$. For any non-trivial even character χ of C

we let

$$\mathscr{C}^{(p)}(\chi)$$

be the χ-eigenspace of $\mathfrak{o}_q \otimes \mathscr{C}^{(p)}$. Since $\mathscr{D}/\mathscr{D}_0$ is the eigenspace for the trivial character, we have a direct sum decomposition

$$\mathfrak{o}_q \otimes \mathscr{C}^{(p)} = \bigoplus_{\chi \neq 1} \mathscr{C}^{(p)}(\chi).$$

Let

$$\pi(\alpha, \beta) = (\alpha + \beta) + (\alpha - \beta) - 2(\alpha) - 2(\beta).$$

Let

$$\pi_\chi(\alpha, \beta) = \chi(\alpha + \beta) + \chi(\alpha - \beta) - 2\chi(\alpha) - 2\chi(\beta).$$

We let I_χ be the ideal of $\mathfrak{o}_q$ generated by all elements $\pi_\chi(\alpha, \beta)$. Then on $\mathscr{C}^{(p)}(\chi)$ we have

$$\begin{aligned} \pi(\alpha, \beta)\theta \,|\, \mathscr{C}^{(p)}(\chi) &= \pi_\chi(\alpha, \beta) \sum_{\gamma \in C(\pm)} \frac{N}{2} \mathbf{B}_2\left(\left\langle \frac{T\gamma}{N} \right\rangle\right) \bar{\chi}(\gamma) \\ &= \pi_\chi(\alpha, \beta) \tfrac{1}{4} S_C(\bar{\chi}, Q \circ T). \end{aligned}$$

We have used our standard notation for the sum,

$$S_C(f, g) = \sum_{\alpha \in C(\pm)} f(\alpha) g(\alpha)$$

for two functions f, g on $C(\pm)$. This gives us the structure of $\mathscr{C}^{(p)}(\chi)$ in terms of the latter sum:

Theorem 4.1. *Let χ be a non-trivial even character of C. Then we have an isomorphism*

$$\mathscr{C}^{(p)}(\chi) \approx \mathfrak{o}_q / S_C(\bar{\chi}, Q \circ T) I_\chi.$$

In particular, $\mathscr{C}^{(p)}(\chi) = 0$ if and only if $S_C(\bar{\chi}, Q \circ T) I_\chi = (1)$.

The remaining task is to give the exact order of the ideal

$$S_C(\bar{\chi}, Q \circ T) I_\chi.$$

As already mentioned, we pass to the most interesting case, $N = p$.

If χ is a character on C, we let

$\chi_{\mathbf{Z}}$ = restriction of χ to $\mathbf{Z}(N)^*$.

If χ is even, then we view χ as character on $C(\pm)$, and then $\chi_{\mathbf{Z}}$ is the restriction to $\mathbf{Z}(N)^*/\pm 1$.

Theorem 4.2. *Let $N = p$ so $X(N) = X(p)$. If $\chi_{\mathbf{Z}} = 1$ then*

$$\mathscr{C}^{(p)}(\chi) = 0.$$

Next we state the result when $\chi_{\mathbf{Z}} \neq 1$. We recall that for any non-trivial character ψ on $\mathbf{Z}(p)^*$, we let

$$B_{2,\psi} = p \sum_{a=1}^{p-1} \psi(a)\mathbf{B}_2\left(\frac{a}{p}\right).$$

We define the **Teichmuller character**

$$\omega : \mathbf{F}_q^* \to \mathfrak{o}_q^*$$

to be the character such that

$$\omega(a) \equiv a \pmod{p}.$$

Then ω has values in the $(q-1)$-th roots of unity. The restriction of ω to $\mathbf{F}_p^*$ is also the Teichmuller character on $\mathbf{F}_p^*$, that is satisfies the same congruence relation.

Any non-trivial *even* character χ on $\mathbf{F}_q^*$ can be written as a power of the Teichmuller character,

$$\chi = \omega^k, \qquad 2 \leqq k \leqq q-3, \qquad k \text{ even}.$$

We define the positive integer $m(k)$ to be the order of the ideal $S_C(\bar{\chi}, Q \circ T)I_\chi$, so that we have

$$S_C(\bar{\chi}, Q \circ T)I_\chi = (p^{m(k)}).$$

To determine $m(k)$ in certain cases, we have to introduce more notation. We write

$$k = k_0 + k_1 p, \qquad 0 \leqq k_i \leqq p-1.$$

We note that $k_0 + k_1 \not\equiv 0 \pmod{p-1}$, otherwise $k \equiv 0 \pmod{p-1}$ and $\chi_{\mathbf{Z}}$ is trivial, contrary to assumption. The following table then describes $\mathscr{C}^{(p)}(k)$ and $m(k)$, where $\mathscr{C}^{(p)}(k) = \mathscr{C}^{(p)}(\omega^k) \approx \mathfrak{o}_q/p^{m(k)}\mathfrak{o}_q$.

Theorem 4.3. *Assume that $\chi_{\mathbf{Z}}$ is not trivial and again $N = p$.*

Case	k	$\mathscr{C}^{(p)}(k)$	$m(k)$
1	$k = 2, 2p, 1+p$	0	$1 + \text{ord } B_{2,\bar{\chi}_{\mathbf{Z}}} = 0$
2	$k \equiv -2, -2p, -(1+p) \bmod p^2 - 1$	$\neq 0$	$1 + \text{ord } B_{2,\bar{\chi}_{\mathbf{Z}}}$
	$k \not\equiv \pm 2, \pm 2p, \pm(1+p) \bmod p^2 - 1$		
3	$k \equiv 2 \bmod p - 1$	0	$1 + \text{ord } B_{2,\bar{\chi}_{\mathbf{Z}}} = 0$
4	$k \equiv -2 \bmod p - 1$	$\mathscr{C}^{(p)}(k) \neq 0 \Leftrightarrow p \mid B_{2,\bar{\chi}_{\mathbf{Z}}}$	$\text{ord } B_{2,\bar{\chi}_{\mathbf{Z}}}$
5	$k \not\equiv \pm 2 \bmod p - 1$	$\mathscr{C}^{(p)}(k)$ or $\mathscr{C}^{(p)}(-k) \neq 0$	$\text{ord } B_{2,\bar{\chi}_{\mathbf{Z}}}$ if $k_0 + k_1 < p - 1$
6			$1 + \text{ord } B_{2,\bar{\chi}_{\mathbf{Z}}}$ if $k_0 + k_1 > p - 1$

The cases have been numbered for ease of reference. Cases 1 and 3 together are those when $\chi_{\mathbf{Z}}$ is the square of the Teichmuller character, and $\mathscr{C}^{(p)}(\chi) = 0$ in that case. This is analogous to the classical cyclotomic case of p-th roots of unity, when the character is *equal* to the Teichmuller character.

In the table, the orders of the Bernoulli numbers are the orders at p, with respect to the unramified prime in $\mathfrak{o}_q$ from which the Teichmuller character was obtained by reduction mod p.

Part of what we shall prove are the relations

$$1 + \text{ord } B_{2,\bar{\chi}_{\mathbf{Z}}} = 0$$

under the given congruences (i.e. when $k \equiv 2 \bmod p - 1$). These amount to Von Staudt type congruences. We recall the congruence:

Theorem 4.4. *Let $2 \leqq k \leqq p - 2$. Let $\omega = \omega_{\mathbf{Z}}$ be the Teichmuller character on $\mathbf{Z}(p)^*$. Then for any integer $n \geqq 1$ we have*

$$\frac{1}{n} B_{n,\omega^{k-n}} \equiv \frac{1}{k} B_k \pmod{p}.$$

For the proof, cf. [L 8], Chapter 2, Theorem 2.5.

In the next section, we give a sequence of lemmas which describe the orders of certain exponential sums. After that, we apply these lemmas to prove the validity of the table.

Suppose last that $l \neq p$ is another prime, $l \geqq 7$.

Theorem 4.5. *Assume that l does not divide $p^2 - 1$. Let χ be a non-trivial even character of $C(p)$.*

(i) *If $\chi_{\mathbf{Z}}$ is trivial, then $\mathscr{C}^{(l)}(\chi) = 0$.*
(ii) *Suppose $\chi_{\mathbf{Z}}$ is non-trivial. Then $I_\chi = (1)$ and*

$$\operatorname{ord}_l S_C(\bar{\chi}, Q \circ T) I_\chi = \operatorname{ord}_l B_{2,\bar{\chi}_{\mathbf{Z}}}$$

The isomorphism of Theorem 4.1 yields

$$\mathscr{C}^{(l)}(\chi) \approx \mathfrak{o}_{l^n}/B_{2,\bar{\chi}_{\mathbf{Z}}}\mathfrak{o}_{l^n}.$$

The proof of Theorem 4.5 is a direct consequence of Theorem 5.1 below, and the fact that the character sums are divisible only by primes dividing p.

Theorem 4.3 contains the analogue for the modular case of the classical Kummer-Herbrand theorems, but goes further since in the present case we know a priori that the cuspidal divisor class group is cyclic over the group. In the cyclotomic case, this is the Iwasawa-Leopoldt conjecture (for the p-primary part). Here the structure should be viewed as twist of the cyclotomic case, and it is a major problem to relate the two by a direct mapping.

Similar algebraic geometric interpretations for the $B_{k,\bar{\chi}}$ when $k > 2$ are still lacking.

We also recall that Coates-Sinnott [Co-S] prove that a Stickelberger ideal formed also with $\mathbf{B}_2$ annihilates K_2 of the ring of integers of, say, a real cyclotomic field. This raises the question whether one can establish a direct connection between such K_2 and the cuspidal divisor class group on $X_1(p)$.

§5. p-Adic Orders of Character Sums

In this section we evaluate the order of $S_C(\bar{\chi}, Q \circ T)$ by means of a sequence of lemmas.

The group C contains $Z = (\mathbf{Z}/p\mathbf{Z})^*$, and if f is a function on C we let $f_{\mathbf{Z}}$ denote its restriction to Z. We may define a sum $S_{\mathbf{Z}}(f_{\mathbf{Z}}, g_{\mathbf{Z}})$ by a similar sum as $S_C(f, g)$ taken over elements of Z.

We let λ be a non-trivial character of the additive group $\mathbf{F}_q$, restricting to a non-trivial character of $\mathbf{F}_p$.

The next theorem is a special case of Chapter 1, Corollary 2 of Theorem 5.3. We reproduce a direct proof shown to us originally by Tate. At prime level, one does not need the more elaborate machinery used in Chapter 1, §5 to get the desired result.

Theorem 5.1. *Assume that χ is non-trivial, and even.*

(i) *If $\chi_{\mathbf{Z}}$ is trivial, then*

$$S_C(\bar{\chi}, Q \circ T) = \frac{p^2 - 1}{6p} S_C(\bar{\chi}, \lambda \circ T).$$

(ii) *If $\chi_{\mathbf{Z}}$ is non-trivial, then*

$$S_C(\bar{\chi}, Q \circ T) = \frac{1}{p} S_C(\bar{\chi}, \lambda \circ T) S_{\mathbf{Z}}(\chi_{\mathbf{Z}}, \bar{\lambda}_{\mathbf{Z}}) B_{2,\bar{\chi}_{\mathbf{Z}}}.$$

Proof. Let f be any function on $\mathbf{Z}(p)$. Recall the Fourier inversion:

$$f(x) = \sum_{y \in \mathbf{Z}(p)} \hat{f}(y)\lambda(yx)$$

$$\hat{f}(y) = \frac{1}{p} \sum_{x \in \mathbf{Z}(p)} f(x)\lambda(-xy).$$

For purpose of this proof, we write λ instead of $\lambda_{\mathbf{Z}}$, because no Fourier transform other than that on $\mathbf{Z}(p)$ will occur. We consider the sum

$$\begin{aligned} S_C(\chi, f \circ T) &= \sum_{\alpha \in C} \chi(\alpha) f(T\alpha). \\ &= \sum_{\alpha} \sum_{y} \chi(\alpha)\hat{f}(y)\lambda(T(y\alpha)) \\ &= \sum_{\alpha} \sum_{y \neq 0} \chi(\alpha)\hat{f}(y)\lambda(T(y\alpha)) + \sum_{\alpha} \chi(\alpha)\hat{f}(0). \end{aligned}$$

Since χ is assumed non-trivial, the sum on the right is 0, so changing the order of summation and making a change of variables $\alpha \mapsto y^{-1}\alpha$, we find:

$$\begin{aligned} &= \sum_{y \neq 0} \sum_{\alpha} \chi(\alpha)\bar{\chi}(y)\hat{f}(y)\lambda(T\alpha) \\ &= \sum_{y \neq 0} \bar{\chi}(y)\hat{f}(y) \sum_{\alpha} \chi(\alpha)\lambda(T\alpha). \end{aligned}$$

$$= S_Z(\bar{\chi}_Z, \hat{f}) S_C(\chi, \lambda \circ T). \tag{1}$$

Suppose that $\chi_{\mathbf{Z}}$ is trivial. Then

$$\sum_{y \neq 0} \bar{\chi}(y)\hat{f}(y) = \sum_{y \neq 0} \hat{f}(y) = f(0) - \hat{f}(0).$$

In particular, if $f = Q$ we obtain

$$Q(0) - \hat{Q}(0) = \frac{p}{6} - \frac{p}{6p^2} = \frac{p^2 - 1}{6p},$$

evaluating $\hat{Q}(0)$ by means of the distribution relation for Bernoulli polynomials. This proves (i).

Suppose that $\chi_{\mathbf{Z}}$ is non-trivial. Then

$$\begin{aligned} S_{\mathbf{Z}}(\bar{\chi}_{\mathbf{Z}}, \hat{f}) &= \frac{1}{p} \sum_{y \neq 0} \sum_{x} \bar{\chi}(y)\hat{f}(x)\lambda(-xy) \\ &= \frac{1}{p} \sum_{y \neq 0} \sum_{x \neq 0} \bar{\chi}(y)f(x)\lambda(-xy) + 0 \\ &= \frac{1}{p} \sum_{x \neq 0} \sum_{y \neq 0} \bar{\chi}(y)\chi(x)f(x)\lambda(-y) \\ &= \frac{1}{p} S_{\mathbf{Z}}(\bar{\chi}, \bar{\lambda})S_{\mathbf{Z}}(\chi, f). \end{aligned} \tag{2}$$

But by definition,

$$B_{2,\chi} = pS_{\mathbf{Z}}(\chi, Q).$$

This proves (ii).

Lemma 5.2.

(i) *If* $k = 2, 2p,$ *or* $1 + p$ *then* $I_\chi = (p)$.
(ii) *For other values of* k, $I_\chi = (1)$.

Proof. In both parts we may reduce the value of the character mod p to get a value for $\pi_\chi(x, y)$ in the finite field, which is

$$(x + y)^k + (x - y)^k - 2x^k - 2y^k, \qquad x, y \in \mathbf{F}_{p^2}.$$

This is a homogeneous polynomial in which the terms of degree k and $k - 1$ in x vanish. Thus the polynomial has degree $\leqq k - 2$ in x, and is equal to

$$k(k - 1)x^{k-2}y^2 + \text{lower degree terms in } x.$$

Let $y = 1$ to get a polynomial in one variable

$$(x+1)^k + (x-1)^k - 2x^k - 2 = k(k-1)x^{k-2} + \text{lower terms.}$$

If the polynomial is not identically 0, then there is some element $x \in \mathbf{F}_q$, $x \neq 0, \pm 1$, at which the polynomial does not take the value 0. Hence we must show that the polynomial is identically 0 only for the three cases $k = 2, 2p$, or $1 + p$. Write $k = p^e n$ with n prime to p. Because $k \leqq p^2 - 3$ we have $e = 0$ or 1, and the polynomial can be written in $t = x^{p^e}$ as

$$(t+1)^n + (t-1)^n - 2t^n - 2 = n(n-1)t^{n-2} + \text{lower terms.}$$

Assume that $n \neq 2$. The polynomial is identically 0 only if $n \equiv 1 \pmod{p}$, so $n = 1 + mp$, $m \geqq 1$ and m is odd since k is even. Then $e = 0$ (otherwise $n > p^2$). We write the polynomial in the form

$$(t^p + 1)^m(t+1) + (t^p - 1)^m(t-1) - 2t^{pm}t - 2.$$

We leave it to the reader to verify that this polynomial is identically 0 only for $m = 1$. This concludes the list of cases when I_χ is divisible by p.

Finally we want to show that in Case (i) the ideal is not divisible by p^2. By Theorem 1.2 of Chapter 3, for any positive integer n not divisible by p, the formal expression

$$(nx) - n^2(x)$$

can be written as linear combination with integer coefficients of "parallelograms"

$$(x+y) + (x-y) - 2(x) - 2(y),$$

hence the ideal I_χ contains values

$$\chi(nx) - n^2\chi(x).$$

Pick an integer $n = \zeta + pz$, where z is a p-adic unit. Then

$$\zeta^2 = \chi(n)$$

is a $(p-1)$-th root of unity. Then I_χ contains

$$\chi(x)(\zeta^2 - n^2)$$

which is not divisible by p^2, as desired.

In the next lemma, we use a primitive root w in $\mathbf{F}_p^*$, and we let

$$r = w^{k+2}.$$

Lemma 5.3. *Assume that* $\chi_{\mathbf{Z}}$ *is non-trivial. Then* $B_{2,\chi_{\mathbf{Z}}}$ *satisfies the following congruences* mod $\mathbf{Z}_p$.

$$B_{2,\chi_{\mathbf{Z}}} \equiv \begin{cases} \dfrac{p-1}{p} & \text{if } k+2 \equiv 0 \pmod{p-1} \\[2ex] \dfrac{1}{p}\dfrac{r^{p-1}-1}{r-1} & \text{if } k+2 \not\equiv 0 \pmod{p-1}. \end{cases}$$

The value in case $k+2 \not\equiv 0 \pmod{p-1}$ *is p-integral.*

Proof. We have

$$\begin{aligned} B_{2,\chi_{\mathbf{Z}}} &\equiv p \sum_{a=1}^{p-1} \left(\frac{a}{p}\right)^2 \chi_{\mathbf{Z}}(a) \quad \pmod{\mathbf{Z}_p} \\ &\equiv \frac{1}{p} \sum_{a=1}^{p-1} a^{k+2} \qquad\qquad \pmod{\mathbf{Z}_p}. \end{aligned}$$

When a ranges over $1, \ldots, p-1$ the residue classes range over all elements of $\mathbf{F}_p^*$, so that the sum is a geometric series which yields the expression stated in the lemma.

Lemma 5.4. *Assume that* $\chi_{\mathbf{Z}}$ *is non-trivial. Then* $B_{2,\bar{\chi}_{\mathbf{Z}}}I_{\chi}$*, which is* $B_{2,\omega^{-k}}$*, is equal to the following*:

(i) $pB_{2,\bar{\chi}_{\mathbf{Z}}}\mathfrak{o}_q = (1)$ *if* $k = 2, 2p, 1+p$.

Otherwise, i.e. if $k \neq 2, 2p, 1+p$:

(ii) $B_{2,\bar{\chi}_{\mathbf{Z}}}\mathfrak{o}_q = \dfrac{1}{p}\mathfrak{o}_q$ *if* $k \equiv 2 \bmod p-1$

(iii) $B_{2,\bar{\chi}_{\mathbf{Z}}}\mathfrak{o}_q =$ *p-integral* *if* $k \not\equiv 2 \bmod p-1$.

Proof. This is immediate from Lemma 5.2 and Lemma 5.3, distinguishing all the possible cases.

Next we deal with the divisibility property of the ordinary Gauss sums. These are well known as Stickelberger's theorem. Cf. for instance [L 8], Chapter 1, §2.

We continue to assume that $\chi_{\mathbf{Z}}$ is non-trivial. Let

$$S = S_C(\bar{\chi}, \lambda \circ T) S_{\mathbf{Z}}(\chi_{\mathbf{Z}}, \bar{\lambda}).$$

Then

$$S\bar{S} = p^3.$$

The factors other than S in Theorem 4.1 are all elements of the field of $(p-1)$-th roots of unity, and have therefore integral orders at *a* p-adic valuation because that field is unramified at p. Hence the order of S is in $\mathbf{Z}$. The relation $S\bar{S} = p^3$ shows that

$$\operatorname{ord}_p S = 0, 1, 2, \text{ or } 3.$$

Lemma 5.5. *The cases* ord $S = 0$, ord $\bar{S} = 3$ *(or vice versa) do not occur. In particular $p^{-1}S$ is p-integral.*

Proof. By Stickelberger, we know that the Gauss sums $S_C(\bar{\chi}, \lambda \circ T)$ and $S_{\mathbf{Z}}(\chi_{\mathbf{Z}}, \bar{\lambda})$ are divisible by all the primes dividing p in the field of $(q-1)$ and $(p-1)$-th roots of unity respectively. Therefore the order cannot be 0, and it cannot be 3 as one sees by replacing χ with its conjugate. This proves the lemma.

§6. Proof of the Theorems

We put together the lemmas of §5 in order to prove the theorems. We need the exact order given for Gauss sums in terms of the Teichmuller character. We write

$$k = k_0 + k_1 p, \qquad 0 \leqq k_i \leqq p - 1$$

and define

$$s_q(k) = k_0 + k_1.$$

We also have $s_p(k)$. In that case, we have to select

$$k' \equiv -k \,(\operatorname{mod} p-1) \quad \text{and} \quad 0 \leq k' < p - 1.$$

Then

$$s_p(-k) = k'.$$

By Stickelberger's theorem,

$$\text{ord } S_C(\omega^{-k}, \lambda \circ T) = \frac{1}{p-1}(k_0 + k_1)$$

$$\text{ord } S_{\mathbf{Z}}(\omega_{\mathbf{Z}}^k, \bar{\lambda}) = \frac{1}{p-1} k'.$$

Let us start with the case $\chi_{\mathbf{Z}}$ trivial. Then $k \equiv 0 \pmod{p-1}$, and then $k_0 + k_1 \equiv 0 \pmod{p-1}$. But since $0 \leqq k_i \leqq p-1$, we get

$$k_0 + k_1 = p - 1 \quad \text{or} \quad 2(p-1).$$

Hence

$$\text{ord } S_C(\bar{\chi}, \lambda \circ T) = 1.$$

By Lemma 5.2 in the present case we have $I_\chi = (1)$. Using Theorem 4.1 and Theorem 5.1(i) concludes the proof of Theorem 4.2.

Assume that $\chi_{\mathbf{Z}}$ is non-trivial. As the technique in all cases is the same, we shall work out in detail the first case and the last case in the table of Theorem 4.3, and leave the others to the reader. By Theorem 4.1 and Theorem 5.1(ii) we have to determine the order of

$$S_C(\bar{\chi}, Q \circ T) I_\chi = \frac{1}{p} S_C(\bar{\chi}, \lambda \circ T) S_{\mathbf{Z}}(\chi_{\mathbf{Z}}, \bar{\lambda}_{\mathbf{Z}}) B_{2,\bar{\chi}_{\mathbf{Z}}} I_\chi.$$

Let $k = 2$, or $2p$, or $1 + p$. Then $k_0 + k_1 = 2$. Furthermore,

$$\text{ord } S_C(\omega^{-k}) = \text{ord } S_C(\bar{\chi}) = \frac{1}{p-1}(k_0 + k_1) = \frac{2}{p-1}.$$

$$\text{ord } S_{\mathbf{Z}}(\omega^k) = \text{ord } S_{\mathbf{Z}}(\chi) = \frac{1}{p-1} s_p(-2) = \frac{p-3}{p-1}.$$

Hence ord $S = 1$. Finally Lemma 5.3 shows that

$$\text{ord } B_{2,\chi} = -1,$$

and Lemma 5.2 shows that $I_\chi = (p)$. This concludes the proof in the present case.

Next, suppose that $k \not\equiv \pm 2 \pmod{p-1}$, and $k_0 + k_1 < p - 1$. Then

$$k' = p - 1 - (k_0 + k_1).$$

Hence ord $S = 1$. Lemma 5.3 shows that $B_{2,\bar{\chi}\mathbf{z}}$ is p-integral, and Lemma 5.2 shows that $I_\chi = (1)$. This gives the entry in the table when $k_0 + k_1 < p - 1$.

On the other hand, suppose that $k_0 + k_1 > p - 1$. Then

$$0 < k_0 + k_1 - (p - 1) < p - 1.$$

Since $k \equiv k_0 + k_1 \pmod{p - 1}$ we get

$$-k \equiv k' = 2(p - 1) - (k_0 + k_1).$$

Hence ord $S = 2$. Again, Lemma 5.3 shows that $B_{2,\bar{\chi}\mathbf{z}}$ is p-integral, and Lemma 5.2 shows that $I_{\bar{\chi}} = (1)$. This gives the entry in the table when

$$k_0 + k_1 > p - 1,$$

and concludes the proof of the theorem.

§7. The Special Group

In the table of Theorem 4.3 we note that the divisor class group is not a priori zero in Cases 2, 4, 5, 6. In Cases 2 and 6, the order is

$$1 + \text{ord } B_{2,\bar{\chi}\mathbf{z}}.$$

In this section, we describe a special component of order 1 occurring precisely in those cases. We shall use the Weierstrass $\wp$-function, as in Chapter 2, §6.

We essentially use the notation of that chapter, on the curve $X(p)$ with p *prime* $\geqq 5$. We repeat some facts for convenience. We express modular points in terms of an elliptic curve A, rather than in terms of lattices. As before, we let $\mathfrak{C}$ denote a cyclic subgroup of order p in A, and let A_p be the subgroup of all points of order p. A pair of points $a, b \in A_p$ is said to be **admissible** for $\mathfrak{C}$ if it satisfies the condition:

> *The cyclic groups (a), (b) generated by a, b are equal. The points a, b, $a \pm b$, are not in $\mathfrak{C}$.*

Since $p \geqq 5$, admissible pairs obviously exist. As before, we define

$$\wp[a, b; \mathfrak{C}] = \wp(a, A/\mathfrak{C}) - \wp(b, A/\mathfrak{C}).$$

Write $b = ra$ for some integer r (prime to p, by the hypotheses). The association

$$(a, \mathfrak{C}, A) \mapsto \wp[a, ra; \mathfrak{C}]$$

defines a modular form on $X(p)$ whose divisor is the p-th multiple of a divisor on $X(p)$ by Theorem 6.3 of Chapter 2.

Theorem 7.1. *Let (a_1, b_1) be admissible for $\mathfrak{C}_1$ and (a_2, b_2) be admissible for $\mathfrak{C}_2$. Then*

$$u = \frac{\wp[a_1, b_1; \mathfrak{C}_1]}{\wp[a_2, b_2; \mathfrak{C}_2]}$$

defines a modular function on $X(p)$ whose divisor is the p-th multiple of a divisor on $X(p)$. If $p > 5$, then the p-th root of this function is not modular of any level.

Proof. The first statement merely repeats Theorem 6.3 of Chapter 2, for the convenience of the reader. The second statement about the p-th root can be proved easily from the main result of Chapter 4, and will be left to the reader.

From the definition of the Klein forms (which are equal to the sigma function times an exponential factor) and the Siegel functions (having an extra factor η^2), we have the factorization

$$\wp[a, b; \mathfrak{C}] = \frac{\mathfrak{k}(a + b, A/\mathfrak{C})\mathfrak{k}(a - b, A/\mathfrak{C})}{\mathfrak{k}^2(a, A/\mathfrak{C})\mathfrak{k}^2(b, A/\mathfrak{C})}$$

$$= \eta^4(A/\mathfrak{C}) \frac{g(a + b, A/\mathfrak{C})g(a - b, A/\mathfrak{C})}{g^2(a, A/\mathfrak{C})g^2(b, A/\mathfrak{C})} \tag{1}$$

$$= \lambda_1 \eta^4(A/\mathfrak{C}) \prod_{c \in \mathfrak{C}} \frac{g(a + b + c, A)g(a - b + c, A)}{g^2(a + c, A)g^2(b + c, A)} \tag{2}$$

for some constant λ_1.

For convenience, we repeat the distribution relations of Chapter 2, §4 in the present notation.

Dist 1.
$$\prod_{c \in \mathfrak{C}} g(a + c, A) = g(a, A/\mathfrak{C}).$$

Dist 2.
$$\prod_{\substack{c \in \mathfrak{C}(\pm) \\ c \neq 0}} g(c, A) = \lambda_2 \frac{\eta(A/\mathfrak{C})}{\eta(A)}$$

for some constant λ_2. As usual, $\mathfrak{C}(\pm)$ denotes $\mathfrak{C}/\pm 1$.

For $p = 5$, because of the obvious restriction for an admissible pair, one has the additional relation

$$\prod_{c \in \mathfrak{C}} g(a+b+c, A)g(a-b+c, A) = \lambda_3 \prod_{\substack{c \in \mathfrak{C}(\pm) \\ c \neq 0}} \frac{1}{g(c, A)}.$$

So we obtain:

Theorem 7.2. *For $p = 5$,*

$$\left(\frac{\eta(A/\mathfrak{C}_1)}{\eta(A/\mathfrak{C}_2)}\right)^5 = \frac{\wp[a_1, b_1; \mathfrak{C}_1]}{\wp[a_2, b_2; \mathfrak{C}_2]} \lambda_4.$$

The cuspidal divisor class group is trivial when $p = 5$, and one expects a priori to have the divisor of the function in Theorem 7.1 expressible as a fifth power, of a modular function of some level. We gave Theorem 7.2 to show explicitly what this function is.

In the sequel we assume that $p \geqq 7$.

We define the **special divisor group** to be the group generated by the divisors of all the functions as in Theorem 7.1, for all choices of $(a_1, b_1; \mathfrak{C}_1)$ and $(a_2, b_2; \mathfrak{C}_2)$, and denote it by

$$\mathscr{S}\not{p} \subset p\mathscr{D}_0.$$

Factoring out the special divisor class group from the full cuspidal divisor class group amounts to considering the factor group

$$\mathscr{C}_{\mathscr{S}\not{p}} = \mathscr{D}_0/(\mathscr{F} + p^{-1}\mathscr{S}\not{p}).$$

Again we tensor with $\mathfrak{o}_q$ (where $q = p^2$), and denote by $\mathscr{C}^{(p)}_{\mathscr{S}\not{p}}(\chi)$ the χ-eigenspace. We want to know when $\mathscr{C}^{(p)}_{\mathscr{S}\not{p}}(\chi)$ is different from $\mathscr{C}^{(p)}(\chi)$. This again amounts to determining the order of the associated $\mathfrak{o}_q$-ideal, which can differ by at most 1 in the two groups. A priori, we can omit from consideration those cases when $\mathscr{C}^{(p)}(\chi) = 0$, so when $\chi_{\mathbf{Z}} = 1$, or Cases 1 and 3 of the table in Theorem 4.3.

Theorem 7.3. *Assume that $\chi_{\mathbf{Z}} \neq 1$. Among the Cases 2, 4, 5, 6 of Theorem 4.3, the group $\mathscr{C}^{(p)}_{\mathscr{S}\not{p}}(\chi)$ differs (necessarily by order 1) from $\mathscr{C}^{(p)}(\chi)$ precisely in Cases 2 and 6.*

The proof will be quite similar to the preceding proofs. In the rest of this section, we describe more closely the defining Stickelberger ideal for the new group $\mathscr{C}^{(p)}_{\mathscr{S}p}(\chi)$, and then we analyse its order to finish the proof.

We denote by Γ a cyclic subgroup of $\mathbf{F}_q = \mathbf{F}_{p^2}$. Its elements are of the form

$$r\gamma_0, \qquad r = 0, \ldots, p-1$$

for some single element γ_0.

From formula (2), we see that there exist two cyclic subgroups Γ_1 and Γ_2 such that the divisor of the Weierstrass function of Theorem 7.1 has the following form. We identify cuspidal divisors with elements of the group ring $\mathbf{Q}[C]$.

$$(3)\quad \frac{1}{p}\operatorname{div}\frac{\wp[a_1,b_1;\mathfrak{C}_1]}{\wp[a_2,b_2;\mathfrak{C}_2]} = \frac{1}{2}\sum_{\gamma\in C}\left(\mathbf{B}_2\left(\left\langle\frac{T\gamma}{p}\right\rangle\right) - \frac{1}{6}\right)\gamma^{-1}$$
$$+ \pi(\alpha_1,\beta_1;\Gamma_1) - s(\Gamma_1) - \pi(\alpha_2,\beta_2;\Gamma_2) + s(\Gamma_2),$$

where we use the notation:

$$\pi(\alpha,\beta;\Gamma) = \sum_{\gamma\in\Gamma}\{(\alpha+\beta+\gamma) + (\alpha-\beta+\gamma) - 2(\alpha+\gamma) - 2(\beta+\gamma)\}$$

and

$$s(\Gamma) = \sum_{\substack{\gamma\in\Gamma\\ \gamma\neq 0}}(\gamma).$$

The expression $-\frac{1}{6}$ occurs because there is no factor of the Dedekind η-function in the quotient of Weierstrass forms, so the usual term with $\mathbf{B}_2$ has to be corrected by the constant term. It will disappear again when we evaluate with a non-trivial character.

The elements $\alpha, \beta \in C$ correspond to the "vectors" a, b respectively, and form an admissible pair for Γ, namely:

The cyclic groups generated by α and β are equal.
The elements α, β, $\alpha \pm \beta$ are not in Γ.

For any character χ on the Cartan group C, we let

$$\pi_\chi(\alpha,\beta;\Gamma) = \sum_{\gamma\in\Gamma}\{\chi(\alpha+\beta+\gamma) + \chi(\alpha-\beta+\gamma) - 2\chi(\alpha+\gamma) - 2\chi(\beta+\gamma)\}.$$

We assume that $\chi_{\mathbf{Z}}$ is a non-trivial even character, in which case

$$\chi(s(\Gamma)) = \sum \chi(\gamma) = 0$$

because $\chi(s(\Gamma)) = \sum_{r} \chi_{\mathbf{Z}}(r)\chi(\gamma_0) = 0.$

We let $I_{\chi,\mathscr{S}_p}$ be the ideal generated by all expressions

$$\pi_\chi(\alpha_1, \beta_1; \Gamma_1) - \pi_\chi(\alpha_2, \beta_2; \Gamma_2)$$

with arbitrary groups of order p and admissible pairs for them. In this manner we obtain the analogue of Theorem 4.1.

Theorem 7.4. *Let χ be an even character on C such that $\chi_{\mathbf{Z}}$ is non-trivial. Then we have an isomorphism*

$$\mathscr{C}^{(p)}_{\mathscr{S}_p}(\chi) \approx \mathfrak{o}_q/S_C(\bar{\chi}, Q \circ T)(I_\chi + p^{-1}I_{\chi,\mathscr{S}_p}).$$

Our problem is therefore to determine the order of $I_{\chi,\mathscr{S}_p}$, and in particular to find those cases when $p^{-1}I_{\chi,\mathscr{S}_p}$ is strictly larger than I_χ. The arguments will simply refine those of §5, Lemma 5.2 and §6, but will follow the same pattern.

To begin with, we simplify the generators for $I_{\chi,\mathscr{S}_p}$.

Lemma 7.5. *The ideal $I_{\chi,\mathscr{S}_p}$ is generated by all elements*

$$\pi_\chi(\alpha, \beta; \Gamma).$$

In other words, we don't have to take differences of such elements in the definition of the ideal.

Proof. Let $r \in \mathbf{Z}(p)^*$ be such that $\chi_{\mathbf{Z}}(r) \neq 1$. Then

$$\pi_\chi(\alpha, \beta; \Gamma) - \pi_\chi(r\alpha, r\beta; \Gamma) = (1 - \chi(r))\pi_\chi(\alpha, \beta; \Gamma)$$

has the same p-order as $\pi_\chi(\alpha, \beta; \Gamma)$. The lemma is then obvious.

Lemma 7.6. *Let $\chi = \omega^k$ with $2 < k < p^2 - 1$. Assume $\chi_{\mathbf{Z}}$ non-trivial, and $k \not\equiv 2 \pmod{p-1}$. Then $I_{\chi,\mathscr{S}_p} = (1)$ if and only if $k_0 + k_1 > p - 1$. If $k_0 + k_1 < p - 1$ then $p \mid I_{\chi,\mathscr{S}_p}$.*

Proof. Fix Γ and omit it from the notation. As we are only interested in whether $I_{\chi,\mathscr{S}_p} = (1)$ or not, we work mod p, so $\pi_\chi(x, y)$ is now interpreted mod p, lying in the finite field. Then

$$\pi_\chi(x, y) = \sum_{z \in \Gamma} \sum_{j=0}^{k} [(x + y)^j z^{k-j} + (x - y)^j z^{k-j} - 2x^j z^{k-j} - 2y^j z^{k-j}].$$

If $j = k$ the sum over z is 0 because of a factor p. The elements of Γ are of the form rz_0 with r in $\mathbf{F}_p$. If

$$k - j \not\equiv 0 \pmod{p - 1}$$

then

$$\sum_z z^{k-j} = 0.$$

Interchanging the order of summation shows that

$$\pi_\chi(x, y) = - \sum_{\substack{0 \leqq j < k \\ j \equiv k \ (\mathrm{mod}\ p-1)}} \binom{k}{j} z_0^{k-j} [(x + y)^j + (x - y)^j - 2x^j - 2y^j].$$

The expression on the right is homogeneous of degree k in z_0, x, y. We divide by x^k. Since by assumption of admissibility, the elements x, y generate the same group, putting $x = 1$ then implies that y is in the prime field, and so $y^j = y^k$. Hence

$$\pi_\chi(1, y) = -\sum \binom{k}{j} z_0^{k-j} [(1 + y)^k + (1 - y)^k - 2 - 2y^k],$$

and the sum is taken over j such that

$$0 \leqq j < k \quad \text{and} \quad j \equiv k \pmod{p - 1}.$$

By Lemma 5.2, the expression in y is not identically zero as function on the prime field. We are therefore reduced to considering the polynomial

$$P(z) = \sum \binom{k}{j} z^{k-j}$$

summed over the above values of j, and determining when this polynomial is identically zero. Observe that P has degree $\leqq p^2 - p$. Since $P(0) = 0$, if P is not identically zero, then there exists an element z_0 not in $\mathbf{F}_p$ such that $P(z_0) \neq 0$, and this element generates a group Γ, from which we see that the ideal $I_{\chi,\mathscr{S}_p}$ is the unit ideal. Thus to prove Lemma 7.6, it suffices to prove:

P is identically zero if and only if $k_0 + k_1 < p - 1$.

We obviously have ord $k! = k_1$. Suppose first that $k_0 + k_1 < p - 1$. Then the possible values for j are of the form

$$j = k_0 + k_1 + r(p-1) \quad \text{with } r \leqq k_1 - 1,$$

because $j < k$. Hence ord $j! = r$, and one verifies at once that

$$\operatorname{ord}(k-j)! = k_1 - r - 1.$$

Hence p divides the binomial coefficient, and P is identically zero. On the other hand, suppose that $k_0 + k_1 > p - 1$. Let

$$j = k_0 + k_1 - (p-1).$$

For this value of j, the binomial coefficient is not divisible by p and hence the polynomial P is not identically zero. This concludes the proof of Lemma 7.6.

Lemma 7.7. *Among Cases* 2, 4, 5, 6 *of the Table in Theorem* 4.3, *we have* $I_{\chi,\mathscr{S}_p} = (1)$ *exactly in Cases* 2 *and* 6. *In the other cases, the ideal* $I_{\chi,\mathscr{S}_p}$ *is divisible by* p.

Proof. In each case one has to determine k_0 and k_1, and verify that

$$k_0 + k_1 > p - 1$$

precisely in Cases 2 and 6. The matter is trivial. For example, in Case 2 we have

$$k \equiv -2, -2p, -(1+p) \pmod{p^2 - 1},$$

so $k = p^2 - 3,\ p^2 - 2p - 1,\ p^2 - p - 2$. The determination of k_0, k_1 is immediate and shows that their sum is $> p - 1$. We leave the rest to the reader.

By Lemma 5.2, we know that $I_\chi = (1)$ in Cases 2 and 6. Hence

$$I_\chi + p^{-1} I_{\chi,\mathscr{S}_p} \neq I_\chi$$

precisely in Cases 2 and 6. This concludes the proof of Theorem 7.3.

§8. The Special Group Disappears on $X_1(p)$

We deal with the general theory on the modular curve $X_1(N)$ (N prime power) in the next chapter, but it is convenient here to point out that the direct image of the special group from $X(p)$ to $X_1(p)$ vanishes. We assume that the reader is acquainted with the basic definitions for $X_1(N)$, which is the modular curve corresponding to the subgroup $\Gamma_1(N)$ of matrices α such that

$$\alpha \equiv \begin{pmatrix} 1 & * \\ 0 & 1 \end{pmatrix} \bmod N.$$

Theorem 8.1. *The direct image of the special divisor class group*

$$(\mathcal{F} + p^{-1}\mathcal{S}\wp)/\mathcal{F}$$

into the cuspidal divisor class group on $X_1(p)$ is trivial.

Proof. We have to analyse the norm of the forms

$$\wp(a, A/\mathfrak{C}) - \wp(b, A/\mathfrak{C}) = \wp[a, b; \mathfrak{C}]$$

from $X(p)$ to $X_1(p)$, and show that they are p-th powers (up to constant factors). We distinguish two cases.

Case 1. Under the identification $A = A_\tau \approx \mathbf{C}/[\tau, 1]$, the group $\mathfrak{C}$ is $\mathfrak{C}_\tau$, generated by $1/p$. Then $\wp[a, b; \mathfrak{C}_\tau]$ is a form on $X_1(p)$, and on $X(p)$ by pull-back from the projection

$$X(p) \to X_1(p).$$

Consequently the norm is the p-th power,

$$\text{Norm of } \wp[a, b; \mathfrak{C}] = \wp[a, b; \mathfrak{C}]^p.$$

Case 2. The group $\mathfrak{C}$ is not $\mathfrak{C}_\tau$. Then for some constant λ,

$$\text{Norm } \wp[a, b; \mathfrak{C}] = \lambda(\wp(a, A) - \wp(b, A))^p.$$

Before giving the proof we recall other distribution relations. We let $\mathfrak{D}$ also range over cyclic subgroups of order p in A.

Dist 3. Let $a \in \mathfrak{C}$. Then

$$\prod_{\mathfrak{D} \neq \mathfrak{C}} \prod_{d \in \mathfrak{D}} g(a + d, A) \prod_{\substack{c \in \mathfrak{C} \\ c \neq 0}} g(c, A) = \lambda_5 g^p(a, A).$$

This is obvious by separating the terms with $d \in \mathfrak{D}$, $d \neq 0$, and $d = 0$. The product over the former yields a constant, and the product over the latter yields $g^p(a, A)$.

Dist 4.
$$\prod_{\mathfrak{D}} \eta^4(A/\mathfrak{D}) = \eta^{4(p+1)}(A).$$

This is obvious by looking at the q-expansions.

We now go to the proof of Case 2. We let $\mathfrak{C} = \mathfrak{C}_\tau$. By the periodicity of the Weierstrass function, we can then choose $a, b \in \mathfrak{C}$. Then the desired norm is equal to:

$$\prod_{\mathfrak{D} \neq \mathfrak{C}} (\wp(a, A/\mathfrak{D}) - \wp(b, A/\mathfrak{D}))$$
$$= \prod_{\mathfrak{D} \neq \mathfrak{C}} \frac{g(a+b, A/\mathfrak{D})g(a-b, A/\mathfrak{D})}{g^2(a, A/\mathfrak{D})g^2(b, A/\mathfrak{D})} \eta^4(A/\mathfrak{D})$$
$$= \prod_{\mathfrak{D} \neq \mathfrak{C}} \prod_{d \in \mathfrak{D}} \frac{g(a+b+d, A/\mathfrak{D})g(a-b+d, A/\mathfrak{D})}{g^2(a+d, A/\mathfrak{D})g^2(b+d, A/\mathfrak{D})} \prod_{\mathfrak{D} \neq \mathfrak{C}} \eta^4(A/\mathfrak{D})$$
$$= \lambda_6 \frac{g^p(a+b, A)g^p(a-b, A)}{g^{2p}(a, A)g^{2p}(b, A)} \prod_{\substack{c \in \mathfrak{C} \\ c \neq 0}} g^2(c, A) \prod_{\mathfrak{D} \neq \mathfrak{C}} \eta^4(A/\mathfrak{D}) \quad \text{(by \textbf{Dist 3})}$$
$$= \lambda_7 \frac{g^p(a+b, A)g^p(a-b, A)}{g^{2p}(a, A)g^{2p}(b, A)} \prod_{\mathfrak{D}} \eta^4(A/\mathfrak{D})\eta^4(A)^{-1} \quad \text{(by \textbf{Dist 2})}$$
$$= \lambda_7 \frac{g^p(a+b, A)g^p(a-b, A)}{g^{2p}(a, A)g^{2p}(b, A)} \frac{\eta^{4(p+1)}(A)}{\eta^4(A)} \quad \text{(by \textbf{Dist 4})}$$
$$= \lambda_7 \frac{\mathfrak{k}^p(a+b, A)\mathfrak{k}^p(a-b, A)}{\mathfrak{k}^{2p}(a, A)\mathfrak{k}^{2p}(b, A)}$$
$$= \lambda_7(\wp(a, A) - \wp(b, A))^p$$

thereby proving the theorem.

§9. Projective Limits

Let p be a prime $\geqq 5$. We shall use the following notation.

$X(p^n) = X_n =$ modular curve of level p^n.

$G_{n,n-1} = \mathrm{Gal}(X_n/X_{n-1}) = \{\sigma \in \mathrm{SL}_2(p^n) \text{ such that } \sigma \equiv id \bmod p^{n-1}\}$.

$C_n =$ set of cusps on X_n.

$\mathscr{C}_n = p$-primary part of the cuspidal divisor class group on X_n.

$$\begin{aligned} \mathscr{D}(n) &= \text{free abelian group generated by the cusps on } X_n. \\ \mathscr{D}_0(n) &= \text{subgroup of divisors of degree 0.} \\ \operatorname{div} \mathscr{F}(n) &= \text{subgroup of divisors of functions having zeros and poles only} \\ &\quad\ \text{at the cusps.} \end{aligned}$$

Then

$$\mathscr{C}_n \approx \mathscr{D}_0(n)/\operatorname{div} \mathscr{F}(n) \otimes \mathbf{Z}_p.$$

We also have

$$\mathscr{D}_0(n) \otimes \mathbf{Z}_p = \mathbf{Z}_p[C_n]_0,$$

and if $\pi_n : \mathscr{D}(n) \to \mathscr{D}(n-1)$ is the natural homomorphism (direct image), then we have a natural commutative diagram

$$\begin{array}{ccc} \mathbf{Z}_p[C_n]_0 & \longrightarrow & \mathscr{C}_n \\ \downarrow & & \downarrow \pi_{n*} \\ \mathbf{Z}_p[C_{n-1}]_0 & \longrightarrow & \mathscr{C}_{n-1} \end{array}$$

We shall write more simply π_* instead of π_{n*}. We wish to determine the projective limit of the cuspidal divisor class group. Let

$$R_0 = \text{proj. lim } \mathbf{Z}_p[C_n]_0.$$

We shall write lim instead of proj. lim to denote projective limit. From the above commutative diagram, we have a homomorphism

$$R_0 \to \lim \mathscr{C}_n.$$

Theorem 9.1. *This homomorphism is an isomorphism.*

Proof. First note that R_0 is compact, being a limit of compact groups (finitely generated free modules over $\mathbf{Z}_p$). Since the image of R_0 in $\lim \mathscr{C}_n$ is obviously dense (being surjective on each finite level), it follows that the homomorphism is surjective. We have to show that it is injective.

Lemma 9.2. *For $n \geqq 2$ we have $\pi_* \operatorname{div} \mathscr{F}(n) \subset p \cdot \operatorname{div} \mathscr{F}(n-1)$.*

Before proving the lemma, we indicate how the theorem follows from it. Let $\{\alpha_n\}$ be a projective system of divisors of degree 0 whose image in $\lim \mathscr{C}_n$

is 0. This means that for each n, α_n is the divisor of a function f_n on X_n. Fix an integer n_0. Then the lemma implies that

$$\alpha_{n_0} \in p^r \operatorname{div} \mathscr{F}(n_0)$$

for all positive integers r, so $\alpha_{n_0} = 0$, thus proving the desired injectivity.

We shall now prove Lemma 9.2.

A unit in the modular function field of X_n has a product representation up to a constant multiple,

$$g = \prod_a g_a^{m(a)}$$

where a ranges over the primitive elements of order p^n in $p^{-n}\mathbf{Z}^2/\mathbf{Z}^2 \bmod \pm 1$, and the family $\{m(a)\}$ satisfies the quadratic relations, as well as

$$\sum m(a) \equiv 0 \bmod 12.$$

See Theorem 1.3 of Chapter 4. The quadratic relations can be written in the form

$$\sum m(a)a_1^2 \equiv \sum m(a)a_2^2 \equiv \sum m(a)a_1a_2 \equiv 0 \bmod \frac{1}{N}\mathbf{Z},$$

where $N = p^n$ and we are dealing with the case N odd. The combination of these two conditions on $\{m(a)\}$ is in fact necessary and sufficient that such a product be of level p^n. The direct image of the divisor of g to level p^{n-1} is the divisor of the norm $N_{n,n-1}g$. It will therefore suffice to prove that this norm is a p-th power of a function on X_{n-1}, up to a constant multiple.

Let $\sigma \in G_{n,n-1}$ and let $\bar{\sigma}$ be any lifting of σ in $\mathrm{SL}_2(\mathbf{Z}_p)$. Then

$$g_a^{\bar{\sigma}} = \lambda(\bar{\sigma}, a) g_{a\sigma}$$

for some constant $\lambda(\bar{\sigma}, a)$. Hence

$$g^\sigma \doteq \prod_a g_{a\sigma}^{m(a)},$$

where the sign $\doteq$ means that the expressions on the left and right differ by a constant multiple.

Lemma 9.3. $\displaystyle\prod_{\sigma \in G_{n,n-1}} g_{a\sigma} \doteq g_{pa}^p.$

Proof. Let us prove this first when

$$a = e_n = \left(\frac{1}{p^n}, 0\right).$$

The isotropy group of e_n in $G_{n,n-1}$ consists of the matrices

$$\begin{pmatrix} 1 & 0 \\ xp^{n-1} & 1 \end{pmatrix}.$$

A system of coset representatives is given by matrices

$$\begin{pmatrix} 1 + rp^{n-1} & sp^{n-1} \\ * & ** \end{pmatrix}$$

where $*$, $**$ consist of one choice of the second row for each pair (r, s), and r, s range over integers mod p. Then formally,

$$\sum_{\sigma \in G_{n,n-1}} (e_n \sigma) = p \sum_{r,s} \left(\frac{1}{p^n} + \frac{r}{p}, \frac{s}{p}\right).$$

Since the functions g_a satisfy the distribution relations we see that Lemma 9.3 is proved in case $a = e_n$.

For arbitrary a, there exists $\varphi \in \mathrm{SL}_2(\mathbf{Z}_p)$ such that $a = e_n \varphi$. Write e instead of e_n for typographical reasons. Let

$$\sigma' = \varphi \sigma \varphi^{-1}.$$

Then

$$\begin{aligned} \prod_\sigma g_{a\sigma} = \prod_\sigma g_{e\varphi\sigma} &= \prod_\sigma g_{e\sigma'\varphi} \doteq \prod_\sigma (g_{e\sigma'})^\varphi \\ &\doteq g_{pe}^{p\varphi} \doteq g_{pe\varphi}^p \doteq g_{pa}^p. \end{aligned}$$

This proves Lemma 9.3.

We then find the norm:

$$\begin{aligned} N_{n,n-1}(g) &\doteq \prod_a \prod_\sigma g_{a\sigma}^{m(a)} \\ &\doteq \prod_a g_{pa}^{pm(a)} \\ &\doteq \left(\prod_b g_b^{n(b)}\right)^p \end{aligned} \tag{9.4}$$

where the product over b is taken for b primitive of level p^{n-1}, and

$$n(b) = \sum_{pa=b} m(a).$$

It is then immediate that the family $\{n(b)\}$ satisfies the quadratic relations at level p^{n-1}, and that

$$\sum n(b) \equiv 0 \bmod 12.$$

This proves Lemma 9.2.

CHAPTER 6

The Cuspidal Divisor Class Group on $X_1(N)$

In this chapter we assume that the reader is acquainted with the basic definitions concerning the modular curve $X_1(N)$. We wish to analyze its cuspidal divisor class group. It turns out we find an analogous theory to the case of $X(N)$, provided we restrict ourselves to part of this divisor class group, as described in §2 and §3. Then the same Cartan group that occurs in cyclotomic theory also occurs here, namely the Cartan group of degree 1,

$$\mathbf{Z}(N)^*/\pm 1.$$

The part of the cuspidal divisor class group which we analyze is isomorphic to a group ring modulo an ideal, which is the Stickelberger ideal at prime level ($p \geqq 5$). For prime power level, a "special" group occurs again.

The index of Stickelberger ideals for prime power level was carried out in [K-L 7], [K-L 8]. See also [L 8], Chapter 2. For the convenience of the reader, we reproduce the computation of the index in the case which concerns us here, to determine the appropriate cuspidal class number, which is related to the Birch-Tate-Lichtenbaum conjecture. This is done in §1.

Assuming that the cuspidal group under consideration is isomorphic to the group ring modulo the Stickelberger ideal at level p, we then compute the structure of each χ-component, in a way entirely analogous to the cyclotomic case, but with $B_{2,\chi}$ appearing instead of $B_{1,\chi}$. The integralizing ideal is the ideal of quadratic relations. Such ideals had already appeared in Coates-Sinnott [Co-Si 2] and [Co-Si 3].

In the third section, we derive the fact for prime power level that the group under consideration is isomorphic to the group ring modulo a suitable ideal, which may be smaller than the Stickelberger ideal if the level is p^n with $n \geqq 2$. This requires a combinatorial analysis of the cusps and the Siegel functions, as usual, but slightly more complicated than on the full modular curve $X(p^n)$. The more technical parts of the proof are finished in the last section of the chapter. In any case we get the appropriate class number formula for our divisor class group. For $X_1(p)$, this class number was also obtained by Klimek (Kl].

For Stickelberger ideals of order 1 (formed with the first Bernoulli number) in the prime power case, this was done by Iwasawa [Iw 1], and by Sinnott [Sin] in the composite case, which we do not treat here, but refer to [Yu].

The methods work for $p = 2$ and 3, but Theorem 5.3 of Chapter 3 (quadratic relations in these cases) shows that the situation needs to be split into several tedious cases, and we don't do it.

The Stickelberger ideal in §1 also occurs in the context of K-theory, specifically $K_2\mathfrak{o}$ where $\mathfrak{o}$ is the ring of integers in the real cyclotomic fields. Coates [Co 2] shows that from work of Tate, $K_2\mathfrak{o}$ is the first twist of the ideal class group. Coates and Sinnott [C-S 1], [C-S 2] show that the Stickelberger ideal annihilates this twist. The Birch-Tate-Lichtenbaum conjecture predicts the order of $K_2\mathfrak{o}$. Our computation of the index coincides with this order, thus giving evidence to conjecture that there is an isomorphism of $K_2\mathfrak{o}$ with $R_0/\mathcal{S}$ in this case (for the p-primary part).

In that connection it is possible to develop some general statements concerning the twistings of group rings modulo Stickelberger ideals. This is done in [KL 7], see also [L 8], Chapter 2, §7.

§1. Index of the Stickelberger Ideal

We let $N = p^n$ with p prime $\neq 2, 3$, and we let

$$G \approx \mathbf{Z}(N)^*/\pm 1$$

under the correspondence $a \mapsto \sigma_a$. As before, we consider the group ring:

$R = \mathbf{Z}[G]$

$R_d =$ ideal of elements of degree $\equiv 0 \bmod d$.

$R_0 =$ augmentation ideal $=$ ideal of elements of degree 0.

If I is an ideal of R, we let $I_d = I \cap R_d$.

$$s(G) = \sum_{\sigma \in G} \sigma$$

For any $\xi \in R$ we have

$$\xi s(G) = (\deg \xi) s(G).$$

If J is an ideal of R, we write $d = \deg J$ to mean that d is the smallest integer $\geqq 0$ which generates the $\mathbf{Z}$-ideal of elements $\deg \xi$, with $\xi \in R$. We let:

$I =$ ideal of R generated by the elements $\sigma_c - c^2$ with c prime to N, $c \in \mathbf{Z}$.

$$\theta = \sum_{b \in \mathbf{Z}(N)^*/\pm 1} \frac{N}{2} \mathbf{B}_2\left(\left\langle \frac{b}{N} \right\rangle\right) \sigma_b^{-1}.$$

For any character χ on G (identified with a character on $\mathbf{Z}(N)^*/\pm 1$, or an even character of $\mathbf{Z}(N)^*$), we have the Bernoulli number

$$B_{2,\chi} = \sum_{b \in \mathbf{Z}(N)^*/\pm 1} N\mathbf{B}_2\left(\left\langle \frac{b}{N} \right\rangle\right) \chi(b).$$

Note here the presence of the factor N in $B_{2,\chi}$ because we are working with the Cartan group of degree 1. Then

$$\chi(\theta) = \tfrac{1}{2} B_{2,\bar{\chi}}.$$

Let

$$\theta' = N \sum_{b \in G} \frac{1}{2}\left(\mathbf{B}_2\left(\left\langle \frac{b}{N} \right\rangle\right) - \frac{1}{6}\right) \sigma_b^{-1} = \theta - \frac{N}{12} s(G)$$

The next lemma is a very special case of Lemma 12 in [C-S 3].

Lemma 1.1. *Let Q be the ideal in R generated by all elements satisfying the quadratic relations* mod N, *that is*

$$\sum m(c)\sigma_c \quad \textit{such that} \quad \sum m(c)c^2 \equiv 0 \bmod N.$$

Then $Q = I$, and N itself lies in I.

Proof. It is clear that $N\sigma_1$ lies in Q. As for I, we consider the two elements

$$\sigma_{1+N} - (1 + N)^2 \quad \text{and} \quad \sigma_{1-N} - (1 - N)^2.$$

Their difference is equal to $4N$. On the other hand, the first of these two elements is also equal to the odd number $-(N^2 + 2N)$, so N lies in I also.

Now we write

$$\sum m(c)\sigma_c = \sum m(c)(\sigma_c - c^2) + \sum m(c)c^2.$$

It is then clear that $Q \subset I$, and the reverse inclusion is immediate, thus proving the lemma.

The next lemma shows there is no ambiguity in the possible definition of the Stickelberger ideal.

Lemma 1.2. *$R\theta' \cap R = I\theta'$. In fact, if $\xi \in R$ is such that $\xi\theta' \in R$ then $\xi \in Q$. Hence $R_0\theta' \cap R = I_0\theta'$.*

Proof. The element θ' has only a possible N in its denominator, because of the quadratic term. Write

$$\xi = \sum z(b)\sigma_b.$$

Multiplying with θ' and assume that $\xi\theta' \in R$, we conclude that

$$\sum z(b)b^2 \equiv 0 \bmod N,$$

so $\xi \in Q$. The converse is obvious.

For our purposes, we define the **Stickelberger ideal**

$$\mathscr{S} = I_0\theta' = I_0\theta.$$

Theorem 1.3. *We have* $(R_0 : I_0\theta') = N \prod_{\chi \neq 1} \frac{1}{2}B_{2,\chi}$.

The proof will result from computing a sequence of other indices.

Note: Here and elsewhere, the right hand side is to be interpreted only up to ± 1, taking whichever sign makes it positive. Using other arguments, one could in fact determine the precise sign, but it is not necessary to go into this here.

Since $I\theta' \cap R_0 = I_0\theta'$ (for instance by Lemma 1.2), we have an injection

$$0 \to R_0/I_0\theta' \to R/I\theta'.$$

The image consists of $(R_0 + I\theta')/I\theta'$, and $R_0 + I\theta' = R_d$ where

$$d = \deg I\theta'.$$

Since $\deg I = 1$ (because $p \neq 2, 3$), we also get $d = \pm\deg \theta'$. It turns out that d will cancel later, so we don't need to know it explicitly, but an easy computation shows that

$$2 \deg \theta = \frac{1-p}{12} \quad \text{and} \quad 2 \deg \theta' = \frac{1-p}{12} - \frac{N}{6}|G|$$

The degree of θ is computed using the distribution relation of the Bernoulli polynomials. Be that as it may, we have

$$(R_0 : I_0\theta') = (R_d : I\theta').$$

Furthermore, we have inclusions

$$R \supset R_d \supset I\theta',$$

and obviously $(R : R_d) = d$. On the other hand, we have formally

$$(R : I\theta') = (R : R\theta')(R\theta' : I\theta').$$

Of course, $R\theta'$ need not be contained in R, but the index $(R : R\theta')$ can be easily interpreted as follows. The element $N\theta'$ lies in R. Hence

$$(R : R\theta') = \frac{(R : RN\theta')}{(R\theta' : RN\theta')}.$$

Lemma 1.4. $(R : R\theta') = (\deg \theta') \prod_{\chi \neq 1} \chi(\theta).$

Proof. The index $(R : RN\theta')$ is equal to the absolute value of the determinant of $N\theta'$ acting on R by multiplication. This determinant can be computed on the vector space $\mathbf{C}[G]$, which splits into 1-dimensional eigenspaces corresponding to the characters. The trivial character gives the degree. The formula is then obvious since

$$(R\theta' : RN\theta') = N^{|G|},$$

so the extra power of N cancels.

Lemma 1.5. $(R : I) = N.$

Proof. This is easy, and one can show in fact that

$$R/I \approx \mathbf{Z}/N\mathbf{Z}.$$

We leave the details to the reader. A harder case with a more complicated Cartan group was treated in full in Lemma 3.5 of Chapter 5.

Putting all the indices together yields the formula as stated in the theorem.

Let $\mathfrak{o}$ be the ring of integers in the real cyclotomic field

$$K = \mathbf{Q}(\boldsymbol{\mu}_N)^+.$$

The Birch-Tate-Lichtenbaum conjecture predicts that the order of $K_2\mathfrak{o}$ is given by

$$w_2(N)\zeta_K(-1),$$

where $w_2(N)$ is the order of the group of roots of unity in the composite of all quadratic extensions of K. As we took $N = p^n$ with $p \geqq 5$, it follows that

$$w_2(N) = 12N.$$

On the other hand, decomposing the zeta function into L-series, we find

$$\begin{aligned} \zeta_K(-1) &= \zeta_{\mathbf{Q}}(-1) \prod_{\chi \neq 1} L(1-2,\chi) \\ &= \tfrac{1}{12} \prod_{\chi \neq 1} -\tfrac{1}{2}B_{2,\chi}. \end{aligned}$$

(Here we determine the sign by the analysis.) Thus the computation of Theorem 1.3 yields the same value.

§2. The p-Primary Part at Level p

In the next section, we shall identify a subgroup of the cuspidal divisor class group with a certain factor group of the group ring,

$$\mathscr{C}_1^0 \approx R_0/\mathscr{S},$$

where $\mathscr{S}$ is the Stickelberger ideal of §1, for $N = p$. In this section, let us take this factor group as the definition of $\mathscr{C}_1^0$. We put an upper index

$$\mathscr{C}_1^{0,(p)}$$

for the p-primary part of this group, i.e. that part obtained by tensoring with $\mathbf{Z}_p$. We let:

$$R_p = \mathbf{Z}_p[G] \quad \text{and} \quad I_p = R_p I.$$

Theorem 2.1. *Let $N = p$ be prime $\geqq 5$. Let χ be an even character of* $\mathbf{Z}(N)^*$. *There is an isomorphism*

$$\mathscr{C}_1^{0,(p)}(\chi) \approx \mathbf{Z}_p/B_{2,\bar{\chi}}I_\chi$$

where $I_\chi = \chi(I_p)$.

(i) *If χ is trivial, then $\mathscr{C}_1^{0,(p)}(\chi) = 0$.*
(ii) *If $\chi = \omega^2$ where ω is the Teichmuller character, then again*

$$\mathscr{C}_1^{0,(p)}(\chi) = 0.$$

In fact, $I_\chi = (p)$ and $B_{2,\bar{\chi}}$ has p-order -1.
(iii) *If χ is non-trivial and $\neq \omega^2$, then $I_\chi = (1)$ and*

$$\mathscr{C}_1^{0,(p)}(\chi) \approx \mathbf{Z}_p/B_{2,\bar{\chi}}\mathbf{Z}_p,$$

so this group is cyclic of order ord $B_{2,\bar{\chi}}$.

Proof. The orders of the Bernoulli numbers are standard in the theory of such numbers. Cf. for instance [L 8], Chapter 2, Theorem 2.5, combined with the Kummer and von Staudt congruences. It is an easy exercise to determine the order of I_χ, easier on the present Cartan group of degree 1 than it was on the corresponding Cartan group of degree 2 treated in the last chapter. Thus we leave that determination as an exercise. The theorem is then clear.

§3. Part of the Cuspidal Divisor Class Group on $X_1(N)$

As in §1, we take $N = p^n$ where p is prime $\geqq 5$.

We already know the cusps (primes at infinity on the modular curve $X(N)$) which could be represented by vectors

$$\pm\begin{bmatrix} x \\ y \end{bmatrix}$$

where $x,\ y \in \mathbf{Z}(N) = \mathbf{Z}/N\mathbf{Z}$, and $(x, y, N) = 1$. The cusps on $X_1(N)$ are orbit classes under the action of

$$\begin{pmatrix} 1 & 1 \\ 0 & 1 \end{pmatrix},$$

so under the equivalence

$$\pm\begin{bmatrix} x \\ y \end{bmatrix} \sim \pm\begin{bmatrix} x + ry \\ y \end{bmatrix} \quad \text{with } r \in \mathbf{Z}.$$

We can take a normalized representative in such a class with

$$0 \leqq y < N \quad \text{and} \quad 0 \leqq x < (N, y).$$

We then distinguish three kinds of cusps.

C 1. The cusps $\pm\begin{bmatrix} 0 \\ y \end{bmatrix}$. In the upper half plane representation these are the cusps lying above 0, and are characterized by $(N, y) = 1$.

C 2. The cusps $\pm\begin{bmatrix} x \\ 0 \end{bmatrix}$. These are characterized by $(x, N) = 1$.

C 3. All others.

We let:

$\mathscr{F}_1(N)$ = group of function on $X_1(N)$ whose divisors have support in the cusps.

$\mathscr{D}_1^0(N)$ = group of divisors of degree 0 on $X_1(N)$ whose support lies among the cusps of first kind.

$\mathscr{F}_1^0(N)$ = group of functions on $X_1(N)$ whose divisors have support in the cusps of first kind.

$\mathscr{C}_1^0(\mathrm{N})$ = corresponding divisor class group,

$$\mathscr{C}_1^0(N) = \mathscr{D}_1^0(N)/\mathrm{div}\, \mathscr{F}_1^0(N).$$

We let again $G \cong \mathbf{Z}(N)^*/\pm 1$. The cusps of first kind form a principal homogeneous set over G, so the divisor group $\mathscr{D}_1^0(N)$ can be identified with the augmentation ideal in the group ring $\mathbf{Z}[G]$. We are concerned with determining the subgroup div $\mathscr{F}_1^0(N)$, and we shall describe the extent to which it differs from the Stickelberger ideal, see Theorem 3.3.

We wish to give a characterization of $\mathscr{F}_1^0(N)$. Usually we index the Siegel functions g_a with

$$a = (a_1, a_2) \in \mathbf{Q}^2, \qquad a \notin \mathbf{Z}^2.$$

Here we shall deal here only with a such that $a_1 = 0$. Consequently we shall abbreviate the notation, and write

$$g_a = g_{(0,a)} \quad \text{for } a \in \mathbf{Q}, \quad a \notin \mathbf{Z}, \quad Na \in \mathbf{Z}.$$

We let

$$T_N = (\mathbf{Q}/\mathbf{Z})_N/\pm 1,$$

where $(\mathbf{Q}/\mathbf{Z})_N$ denotes the subgroup of $\mathbf{Q}/\mathbf{Z}$ consisting of the elements of order N. We let T_N^* be the subset of T_N consisting of primitive elements (those with exact period N).

We shall consider three conditions on a divisor

$$\sum m(a)(a).$$

D 0. $\sum m(a) = 0$ (i.e. the divisor is of degree 0).

To state the next condition, observe that $(1/p)\mathbf{Z}/\mathbf{Z}$ operates on T_N by addition. The next condition then reads:

D 1. For every orbit O of $\frac{1}{p}\mathbf{Z}/\mathbf{Z}$, we have

$$\sum_{a \in O} m(a) = 0.$$

Note that **D 1** implies **D 0**. If $N = p$, the sum is to be interpreted as taken over T_p^*.

D 2. The quadratic relations mod N, that is:

$$\sum m(a)(Na)^2 \equiv 0 \bmod N.$$

To analyze $\mathscr{F}_1^0$, we shall first analyze a subgroup. Let

$\mathscr{F}'_1(N)$ = group of products of Siegel functions

$$g = \prod g_a^{m(a)}$$

with $a \in T_N$, $a \neq 0$, and integer exponents $m(a)$, satisfying the relations **D 0** and **D 2**.

Since $g_a = \mathfrak{k}_a \eta^2$ where $\mathfrak{k}_a$ is the Klein form, we see from condition **D 0** that we could have written $\mathfrak{k}_a$ instead of g_a in the product. By Theorem 5.2 of Chapter 3, the quadratic relations are necessary to make g have level N, and **D 0**, **D 2** are sufficient. (Actually, the quadratic relations and the condition "12 divides $\sum m(a)$" are necessary and sufficient.)

The divisor group of $\mathscr{F}'_1(N)$ is easily determined since we know the order of the Klein form. Indeed, from the q-expansion:

$$\begin{aligned}\operatorname{ord}_{\left[\begin{smallmatrix} x \\ y \end{smallmatrix}\right]} \mathfrak{k}_a &= \tfrac{1}{2}(\mathbf{B}_2(\langle ay \rangle) - \tfrac{1}{6}) \\ &= \tfrac{1}{2}\langle ay \rangle(\langle ay \rangle - 1).\end{aligned}$$

The order is taken with respect to the parameter $q = e^{2\pi i\tau}$, even though there may be ramification over q. So the order is 0 for any cusp of type **C 2**:

$$\operatorname{ord}_{\left[\begin{smallmatrix} x \\ 0 \end{smallmatrix}\right]} \mathfrak{k}_a = 0.$$

Let $\mathscr{F}''_1$ = subgroup of functions $f \in \mathscr{F}_1$ such that:

$\operatorname{ord}_{\left[\begin{smallmatrix} x \\ y \end{smallmatrix}\right]} f = \operatorname{ord}_{\left[\begin{smallmatrix} x' \\ y \end{smallmatrix}\right]} f$ for any pair x, x';

$\operatorname{ord}_{\left[\begin{smallmatrix} x \\ 0 \end{smallmatrix}\right]} f = 0.$

Then $\mathscr{F}'_1 \subset \mathscr{F}''_1$.

Theorem 3.1. *We have* $\mathscr{F}'_1 = \mathscr{F}''_1$.

Proof. It is clear that $\mathscr{F}''_1$ has no cotorsion in the group $\mathscr{F}_1(N)$ of units on $X_1(N)$, i.e. $\mathscr{F}_1(N)/\mathscr{F}''_1(N)$ has no torsion. We contend that $\mathscr{F}'_1$ also has no cotorsion. To see this, suppose that

$$g = \prod g_a^{m(a)}$$

is in $\mathscr{F}'_1$ and $g = h^l$, where h is modular and l is prime. Let $m(a) \neq 0$ in the product for some a of maximal level. By the beginning of the proof of Theorem 4.3 of Chapter 4, we must have

$$m(a) \equiv 0 \bmod l.$$

Indeed, we can use an element $b = (1/p, a_2)$ so b has the same level as $a_2 = a$. Then $m(b) = 0$ and $m(a) \equiv m(b) \bmod l$. In this way we can absord in h those factors of the product with a of maximal level, and continue by induction to see that h also has a product expression satisfying **D 0**. If h is on $X_1(N)$ then **D 2** is also satisfied, as desired.

To prove the theorem, it now suffices to prove that $\mathscr{F}'_1$ and $\mathscr{F}''_1$ have the same rank. The functions g_a with $a \in T_N$, $a \neq 0$, are independent, modulo constants. (The argument in Theorem 3.1 yields a proof, selecting infinitely many l for instance.) Hence

$$\operatorname{rank} \mathscr{F}'_1(N) = \frac{N-1}{2} - 1$$

because there are $(N-1)/2$ elements $\pm a$ with $a \neq 0$, and we subtract 1 for the relation $\sum m(a) = 0$.

On the other hand, the group $\mathscr{F}_1''$ is also defined by linear conditions on the divisors, whose orders depend only on y. There are $(N-1)/2$ elements $\pm y$ (other than 0), and we again subtract 1 to account for the fact that the sum of the orders of zeros of a function is 0. This proves Theorem 3.1.

It is clear that

$$\mathscr{F}_1^0 \subset \mathscr{F}_1'.$$

We wish to describe the position of $\mathscr{F}_1^0$ in $\mathscr{F}_1'$.

Theorem 3.2. *The group $\mathscr{F}_1^0$ is the subgroup of $\mathscr{F}_1'$ consisting of those functions*

$$g = \prod g_a^{m(a)}$$

satisfying **D 1** *and* **D 2**.

Proof. Denote by $\mathscr{F}_1'(\mathbf{D\,1})$ the group of elements in $\mathscr{F}_1'$ satisfying **D 1**. The same argument as in Theorem 3.1 shows that $\mathscr{F}_1^0$ and $\mathscr{F}_1'(\mathbf{D\,1})$ have no cotorsion. Again we are reduced to showing that they have the same rank. The rank of $\mathscr{F}_1^0$ is

$$\operatorname{rank} \mathscr{F}_1^0 = \frac{\phi(N)}{2} - 1.$$

The rank of $\mathscr{F}_1'(\mathbf{D\,1})$ is determined by a linear condition for each orbit, and the number of orbits of $(1/p)\mathbf{Z}/\mathbf{Z}$ in T_N is equal to

$$\frac{p^{n-1}-1}{2} + 1.$$

On the other hand, the linear condition **D 0** is implied by **D 1**. Hence

$$\operatorname{rank} \mathscr{F}_1'(\mathbf{D}) = \frac{N-1}{2} - \frac{p^{n-1}-1}{3} - 1 = \frac{\phi(N)}{2} - 1,$$

as was to be shown.

Let pr_0 be the projection operator, which projects a divisor on the subgroup of divisors having support in the cusps of first type **C 1**. Then

$$\mathrm{pr}_0(g_a) = N \sum_{b \in G} \tfrac{1}{2}\mathbf{B}_2(\langle ab \rangle)\sigma_b^{-1}.$$

If $g \in \mathscr{F}_1^0$ and

$$g = \prod g_a^{m(a)}$$

where the family $\{m(a)\}$ satisfies **D 1**, **D 2**, then

$$\begin{aligned}(g) = \mathrm{pr}_0(g) &= N \sum_a m(a) \sum_b \tfrac{1}{2}\mathbf{B}_2(\langle ab \rangle)\sigma_b^{-1} \\ &= N \sum_b \sum_a m(a)\tfrac{1}{2}\mathbf{B}_2(\langle ab \rangle)\sigma_b^{-1}.\end{aligned}$$

But by the distribution relation, lifting non-primitive elements a to primitive ones, we have

$$\sum_a m(a)\mathbf{B}_2(\langle ab \rangle) = \sum_{r=0}^{n-1} p^r \sum_z m(p^r z)\mathbf{B}_2(\langle zb \rangle),$$

where the sum over z is taken for

$$z \in T_N^*,$$

i.e. for primitive elements z. Let

$$Z_N = \mathbf{Z}(N)/\pm 1 \quad \text{and} \quad Z_N^* = \mathbf{Z}(N)^*/\pm 1.$$

Changing variables, letting $b \mapsto c^{-1}b$ where $c = Nz \in Z_N^*$, we find

$$(g) = \sum_{c \in Z_N^*} \sum_{r=0}^{n-1} p^r m(p^r c/N)\sigma_c \cdot \theta$$

where θ is the Stickelberger element discussed in §1.

Let J_1 = ideal in $R = \mathbf{Z}[G]$ generated by all elements

$$\sum_{c \in Z_N^*} \sum_{r=0}^{n-1} p^r m(p^r c/N)\sigma_c$$

where m ranges over all functions satisfying **D 1**, **D 2**. Then we have shown

Theorem 3.3. *The group of divisors of $\mathscr{F}_1^0(N)$ is given by*

$$\operatorname{div} \mathscr{F}_1^0(N) = J_1\theta,$$

and thus we have a natural isomorphism

$$\mathscr{C}_1^0(N) \approx R_0/J_1\theta.$$

We shall now analyse the ideal J_1 further combinatorially. This does not involve modular forms, only combinatorial juggling with ideals in the group ring. The ideal J_1 is easily seen to be contained in Q_0, cf. §1, Lemma 1.1, and also Lemma 4.1 of §4, so

$$J_1\theta \subset Q_0\theta = Q_0\theta' = \mathcal{S},$$

in other words $J_1\theta$ is contained in the Stickelberger ideal, and $R_0/\mathcal{S}$ appears as a natural factor group of $\mathcal{C}_0^1(N)$. We shall determine the index

$$(R_0 : J_1\theta).$$

Since we already know $(R_0 : \mathcal{S})$, it suffices to determine $(\mathcal{S} : J_1\theta)$ (carried out in the next section). We shall find:

Theorem 3.4. $(\mathcal{S} : J_1\theta) = p^{p^{n-1}}/Np^{2n-2}$, *and therefore*

$$\text{order of } \mathcal{C}_1^0(N) = \frac{p^{p^{n-1}}}{p^{2n-2}} \prod_{\chi \neq 1} \tfrac{1}{2}B_{2,\chi}.$$

If $n = 1$, $N = p$ *then* $\mathcal{S} = J_1\theta$ *and*

$$\mathcal{C}_1^0(p) \approx R_0/\mathcal{S}.$$

To analyse $\mathcal{S}/J_1\theta$ geometrically, it is necessary to enter into considerations of special divisor classes, analogous to those already discussed in Chapter 5, §7, which can be exhibited explicitly and have to be factored out to get a divisor class group precisely isomorphic to $R_0/\mathcal{S}$.

Note that the result for $n = 1$, $N = p$ that $\mathcal{S} = J_1\theta$ and

$$\mathcal{C}_1^0(p) \approx R_0/\mathcal{S},$$

is compatible with Theorem 8.1 of Chapter 5, where there was no special group at level p.

Since θ is invertible in R, the map multiplication by θ is injective on R. Consequently

$$(\mathcal{S} : J_1\theta) = (Q_0\theta) : J_1\theta) = (Q_0 : J_1).$$

In the next section, we analyse the factor group

$$Q_0/J_1 \approx \mathcal{S}/J_1\theta.$$

§4. Computation of a Class Number

Recall that Q, defined in §1, is the ideal of elements in the group ring R satisfying the quadratic relations.

Let J_Q be the ideal in R generated by the elements

$$\sum_{c\in Z_N^*}\sum_{r=0}^{n-1} p^r m\left(p^r \frac{c}{N}\right)\sigma_c$$

with families $\{m(a)\}$ satisfying the quadratic relations **D 2**. Then the coefficient $m'(c)$ of σ_c is given by the expression

$$m'(c) = \sum_{r=0}^{n-1} p^r m\left(p^r \frac{c}{N}\right).$$

Lemma 4.1. *$J_Q \subset Q$, and in fact we have the formula*

$$\sum_{a\in T_N} m(a)(Na)^2 = \sum_{c\in Z_N^*} m'(c)c^2.$$

Proof. Let us write

$$\sum_{a\in T_N} m(a)(Na)^2 = \sum_{r=0}^{n-1}\sum_{a\in T_{p^{n-r}}^*} m(a)(Na)^2.$$

Each $a \in T_{p^{n-r}}^*$ can be written $a = p^r c/N$ with $c \in Z_N^*$, and there are p^r such possible values of c. Hence

$$\begin{aligned}\sum_{a\in T_{p^{n-r}}^*} m(a)(Na)^2 &= \frac{1}{p^r}\sum_{c\in Z_N^*} m\left(p^r\frac{c}{N}\right)\left(Np^r\frac{c}{N}\right)^2\\ &= \sum_{a\in Z_N^*} p^r m\left(p^r\frac{c}{N}\right)c^2.\end{aligned}$$

Summing over r yields the desired formula and shows that $\{m(a)\}$ satisfies **D 2** if and only if $\{m'(c)\}$ satisfies the quadratic relations, and proves the lemma.

Let $J_{T_p} = J_{T_p}(N)$ be the ideal in R_0 consisting of elements

$$\sum_{a\in T_N^*}\sum_{r=0}^{n-1} p^r m(p^r a)\sigma_{Na}$$

such that the family $\{m(a)\}$ satisfies the orbit condition **D 1**. Then by Lemma 4.1,

$$J_{T_p} \cap Q_0 = J_1.$$

Lemma 4.2. *We have an injection*

$$Q_0/J_1 \to R_0/J_{T_p},$$

and the index of the image is $N/p = p^{n-1}$.

Proof. The natural map induced by inclusion is injective by Lemma 4.1. We have to analyse its image. Let

$$\psi : R_0 \to \mathbf{Z}/N\mathbf{Z}$$

be the map

$$\sum m'\left(\frac{b}{N}\right)\sigma_b \longmapsto \sum m'\left(\frac{b}{N}\right)b^2.$$

The kernel of ψ is precisely Q_0. It will suffice to show that the image $\psi(J_{T_p})$ is equal to $p^{n-1}\mathbf{Z}/N\mathbf{Z}$, as is clear for instance from the following exact diagram.

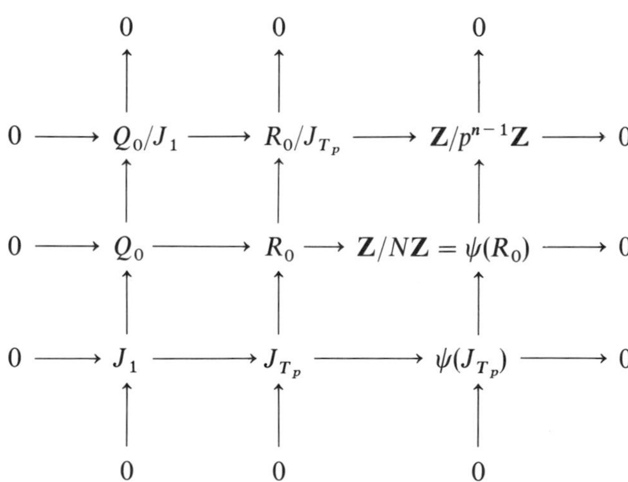

$$\begin{array}{ccccccccc}
 & & 0 & & 0 & & 0 & & \\
 & & \uparrow & & \uparrow & & \uparrow & & \\
0 & \longrightarrow & Q_0/J_1 & \longrightarrow & R_0/J_{T_p} & \longrightarrow & \mathbf{Z}/p^{n-1}\mathbf{Z} & \longrightarrow & 0 \\
 & & \uparrow & & \uparrow & & \uparrow & & \\
0 & \longrightarrow & Q_0 & \longrightarrow & R_0 & \longrightarrow & \mathbf{Z}/N\mathbf{Z} = \psi(R_0) & \longrightarrow & 0 \\
 & & \uparrow & & \uparrow & & \uparrow & & \\
0 & \longrightarrow & J_1 & \longrightarrow & J_{T_p} & \longrightarrow & \psi(J_{T_p}) & \longrightarrow & 0 \\
 & & \uparrow & & \uparrow & & \uparrow & & \\
 & & 0 & & 0 & & 0 & &
\end{array}$$

Consider first an orbit of maximal level, and a function m' on this orbit such that

$$\sum_{x \in \text{orbit}} m'(x) = 0,$$

and m' is 0 outside this orbit. Write $x = b/N$, and select some element $x_0 = b_0/N$ in the orbit. Then the image under ψ of

$$\sum m'\left(\frac{b}{N}\right)\sigma_b$$

is

$$\sum_{b \neq b_0} m'\left(\frac{b}{N}\right)(b^2 - b_0^2).$$

We have $b \equiv b_0 \bmod p^{n-1}$. The function m' can be chosen arbitrarily on $b \neq b_0$, so we can choose it so that one term in the sum is divisible exactly by p^{n-1} and the other terms are 0. It is then clear that the image of elements in J_{T_p} formed with functions m which are 0 outside cosets of maximal level is precisely $p^{n-1}\mathbf{Z}/N\mathbf{Z}$. A similar argument applied to elements in J_{T_p} formed with functions m which are 0 outside cosets of intermediate level shows that the image of such elements is contained in $p^{n-1}\mathbf{Z}/N\mathbf{Z}$. This proves Lemma 4.2.

It is to a certain extent inconvenient to go back and forth between T_N and Z_N (isomorphic under multiplication with N). Hence we now eliminate the group ring notation, and use functional notation, viewing elements of the group ring as functions on T_N^*. Thus we let:

$F_0(T_N^*)$ = additive group of $\mathbf{Z}$-valued functions on T_N^*, of degree 0, i.e. satisfying

$$\sum_{x \in T_N^*} f(x) = 0.$$

Let $M | N$, $M = p^{n-s}$. Let $\pi_s : T_N \to T_M$ be multiplication by p^s. Let φ be a function on T_M^*. We define its **lifting**

$$\pi_s^*\varphi(y) = p^s\varphi(p^s y) \quad \text{for } y \in T_N^*.$$

Thus π_s^* is a homomorphism from $F_0(T_M^*)$ into $F_0(T_N^*)$.

With this functional notation, we see that the ideal J_{T_p} is generated by functions of the following type. We select an orbit C of $(1/p)\mathbf{Z}/\mathbf{Z}$ in $T_{p^{n-s}}^*$, and we let $m = m_C$, be any function such that

$$m(x) = 0 \quad \text{if } x \notin C \quad \text{and} \quad \sum_{a \in C} m(a) = 0.$$

We take the lifted function $\pi_s^* m$. Such lifted functions generate J_{T_p}.

We shall characterize J_{T_p} by another condition which allows us to perform an inductive procedure, ultimately leading to the desired computation of the index. Let:

$F_0^{(1)}(T_N^*)$ = subgroup of functions f such that for every r with $1 \leqq r < n$ and every orbit K of $(1/p^r)\mathbf{Z}/\mathbf{Z}$ in T_N^* we have

$$\sum_{x \in K} f(x) \equiv 0 \bmod p^{2r},$$

and also

$$\sum_{x \in T_N^*} f(x) = 0.$$

Lemma 4.3. $J_{T_p} = F_0^{(1)}(T_N^*)$.

Proof. We first prove the inclusion $\subset$. Let $m = m_{C,s}$ be one of the functions described above, such that the lifted function

$$\pi_s^* m$$

is one of the generators of J_{T_p}. Let K be an orbit of $(1/p^r)\mathbf{Z}/\mathbf{Z}$ in T_N^*. We have to verify that for $r < n$,

$$\sum_{z \in K} p^s m(p^s z) \equiv 0 \bmod p^{2r}.$$

Let $K' = \pi_s^{-1}(C)$. Three cases arise:

Case 1. $K \cap K'$ is empty. The congruence condition is clear.

Case 2. $K \supset K'$. The condition is again clear, since

$$\sum_{x \in C} m_{C,s}(x) = 0.$$

Case 3. $K \subset K'$ and $K \neq K'$. Then $r < s + 1$, so $r \leqq s$ and $m(p^s z)$ is constant for $z \in K$, so we get the congruence

$$\sum_{z \in K} p^s m(p^s z) \equiv 0 \bmod p^{r+s},$$

whence mod p^{2r} as desired.

Conversely, let $f \in F_0^{(1)}(T_N^*)$. We follow an inductive descending procedure, subtracting an appropriate function from J_{T_p} to reduce the level. For each orbit C of $(1/p)\mathbf{Z}/\mathbf{Z}$ we define

$$g_C(x) = f(x) - \frac{1}{p}\sum_{y \in C} f(y) \quad \text{if } x \in C$$

$$g_C(x) = 0 \quad \text{if } x \notin C.$$

Let

$$g(x) = \sum_C g_C(x).$$

Then it is immediately verified that $g \in J_{T_p}$ and we can define a function φ by

$$(f - g)(x) = p \sum_{pz=x} \varphi(px),$$

since

$$\sum_{y \in C} f(y) \equiv 0 \bmod p^2.$$

Then $\varphi \in F_0^{(1)}(T_{N/p}^*)$. Since

$$\pi_1^* J_{T_p}(N/p) \subset J_{T_p}(N),$$

the converse inclusion follows by induction.

Let $F^{(1)}(T_N^*)$ be defined by the same congruence relations as $F_0^{(1)}(T_N^*)$, but omitting the degree 0 condition, i.e. only by the congruences for every r with $1 \leqq r < n$ and every orbit K of $(1/p^r)\mathbf{Z}/\mathbf{Z}$ in T_N^*,

$$\sum_{x \in K} f(x) \equiv 0 \bmod p^{2r}.$$

It is immediately verified that the augmentation homomorphism applied to $F^{(1)}(T_N^*)$ yields the degree

$$\deg F^{(1)}(T_N^*) = (p^{2n-2}).$$

We have an injection

$$R_0/F_0^{(1)} \to R/F^{(1)},$$

from which we already see that Theorem 3.4 follows when $n = 1$, i.e. $R_0/F_0^{(1)}$ is trivial, and Lemmas 4.2 and 4.3 conclude the proof.

The degree stated above shows that the image of the injection is

$$p^{2n-2}R/F^{(1)}.$$

Suppose that $n \geqq 2$. Then from the value of augmentation we find:

Lemma 4.4. $(R/F^{(1)} : R_0/F_0^{(1)}) = p^{2n-2}$.

We define a **filtration**

$$F^{(1)} \subset F^{(2)} \subset \cdots \subset F^{(n)} = R$$

by letting:

$F^{(s)}$ = group of functions f on T_N^* such that for every r with

$$0 \leqq r \leqq n - s$$

and every orbit K of $(1/p^r)\mathbf{Z}/\mathbf{Z}$ we have

$$\sum_{x \in K} f(x) \equiv 0 \bmod p^{2r}.$$

We shall determine the factor group $F^{(s)}$ mod $F^{(s-1)}$. We let:

$$\tau(s) = \text{number of orbits of } \frac{1}{p^{n-s}}\mathbf{Z}/\mathbf{Z} \quad \text{in} \quad T_N^* = \frac{p-1}{2}p^{s-1}.$$

Lemma 4.5. $(F^{(s)}:F^{(s-1)}) = p^{2\tau(s-1)}$, for $s \geqq 2$.

Proof. To each function $f \in F^{(s)}$ we associate a vector

$$f \mapsto (\ldots, \sum_{x \in K} f(x), \ldots) \bmod p^{2(n-s+1)}$$

whose components are indexed by orbits K of $(1/p^{n-s+1})\mathbf{Z}/\mathbf{Z}$ in T_N^*. This gives rise to a homomorphism

$$\psi_s : F^{(s)} \to (p^{2(n-s)}\mathbf{Z}/p^{2(n-s+1)}\mathbf{Z})^{\tau(s-1)} \cong \mathbf{Z}(p^2)^{\tau(s-1)}$$

into the product of $\mathbf{Z}(p^2)$ taken $\tau(s-1)$ times. The kernel is clearly $F^{(s-1)}$, and it is easy to verify that the map is surjective. Indeed, we just consider functions divisible by $p^{2(n-s)}$, which thus belong to $F^{(s)}$, and otherwise arbitrary. This proves the lemma.

Taking the product over s in Lemma 4.5 yields the index

$$(F^{(n)} : F^{(1)}),$$

and $F^{(n)} = R$. Combining this with Lemma 4.4 yields the value stated in Theorem 3.4.

§5. Projective Limits

We now wish to prove the theorem analogous to that of Chapter 5, §9, on the curve $X_1(p^n)$. This deals with a certain subgroup of the cuspidal divisor class group whose definition we recall from §3.

The cusps on $X_1(p^n)$ are identifiable with orbit classes of column vectors

$$\pm\begin{bmatrix} x \\ y \end{bmatrix} \quad \text{under the action of} \quad \begin{pmatrix} 1 & 1 \\ 0 & 1 \end{pmatrix}.$$

The components x, y lie in $\mathbf{Z}/N\mathbf{Z}$, with $(x, y, N) = 1$. We can take normalized representatives in an orbit class with

$$0 \leqq y < p^n \quad \text{and} \quad 0 \leqq x < (y, p^n).$$

We then have the three kinds of cusps:

C 1. The cusps $\pm\begin{bmatrix} 0 \\ y \end{bmatrix}$.

C 2. The cusps $\pm\begin{bmatrix} x \\ 0 \end{bmatrix}$.

C 3. All others.

We let:

C_n^0 = set of cusps of first type on $X_1(p^n)$.

$\mathscr{D}_1^0(n)$ = divisor group of degree 0 with support in C_n^0.

div $\mathscr{F}_1^0(n)$ = divisor group of functions on $X_1(p^n)$ having support in C_n^0.

$\mathscr{C}_1^0(n) = \mathscr{D}_1^0(n)/\text{div}\,\mathscr{F}_1^0(n)$ = corresponding divisor class group.

$\mathscr{C}_n$ = p-primary part of $\mathscr{C}_1^0(n) = \mathscr{C}_1^0(n) \otimes \mathbf{Z}_p$.

As before we have a natural commutative diagram

$$\begin{array}{ccc} \mathbf{Z}_p[C_n^0]_0 & \longrightarrow & \mathscr{C}_n \\ \downarrow & & \downarrow \\ \mathbf{Z}_p[C_{n-1}^0]_0 & \longrightarrow & \mathscr{C}_{n-1}. \end{array}$$

We let

$$R_0 = \lim \mathbf{Z}_p[C_n]_0.$$

Again we have a homomorphism $R_0 \to \lim \mathscr{C}_n$.

Theorem 5.1. *This homomorphism is an isomorphism.*

Proof. As before, the proof depends on the following lemma.

Lemma 5.2. *For $n \geqq 2$, we have $\pi_* \operatorname{div} \mathscr{F}_1^0(n) \subset p \cdot \operatorname{div} \mathscr{F}_1^0(n-1)$.*

For the proof, we have to recall some more explicit facts concerning the functions on $X_1(p^n)$ having divisors with support in the cusps of first kind.

We now write $a \in \mathbf{Q}/\mathbf{Z}$, $a \neq 0$, $p^n a = 0$. We write

$$g_a = g_{0,a},$$

well defined up to a constant multiple. We let

$$T_N = (\mathbf{Q}/\mathbf{Z})_N/\pm 1 \quad \text{with } N = p^n.$$

We note that $(1/p)\mathbf{Z}/\mathbf{Z}$ operates on T_N by addition. By Theorem 3.2 of §3, we know that:

$\mathscr{F}_1^0(n)$ = group of functions (up to constant multiples)

$$\prod_{\substack{a \in T_N \\ a \neq 0}} g_a^{m(a)}$$

such that the family $\{m(a)\}$ satisfies the two conditions:

D 1. For every orbit O we have

$$\sum_{a \in O} m(a) = 0.$$

If $N = p$ so $n = 1$, this sum is to be interpreted as taken over primitive elements of T_p.

D 2. The quadratic relations

$$\sum m(a)a^2 \equiv 0 \bmod \frac{1}{p^n}\mathbf{Z}.$$

To take the norm $N_{n,n-1}$ from $X_1(p^n)$ down to $X_1(p^{n-1})$, we use the groups:

H_n = group of $\sigma \in \mathrm{SL}_2(p^n)$ such that

$$\sigma \equiv \begin{pmatrix} 1 & b \\ 0 & 1 \end{pmatrix} \bmod p^n.$$

H_{n-1} = group of elements σ satisfying the same congruence but mod p^{n-1}, or equivalently,

$$\sigma \equiv \begin{pmatrix} 1 + tp^{n-1} & b' \\ rp^{n-1} & 1 + sp^{n-1} \end{pmatrix} \bmod p^n;$$

The integers r, s range over residue class mod p, and for each pair (r, s) the integer t is uniquely determined mod p. Then

$$(0, a)\sigma = (arp^{n-1}, a(1 + sp^{n-1})).$$

Lemma 5.3.

$$N_{n,n-1}(g_a) \doteq \begin{cases} g_a^{p^2} & \text{if } a \text{ is not primitive} \\ g_{pa} & \text{if } a \text{ is primitive.} \end{cases}$$

Proof. The norm is given by taking the product of $g_{a\sigma}$ for $\sigma \in H_n \backslash H_{n-1}$. By the distribution relation, this is equal to

$$\prod_{r,s} g_{(r/p,\, a+s/p)} \doteq g_{(0,a)}.$$

Now taking the norm $N_{n,n-1}g$, we decompose the product over those a which are primitive, and those a which are not primitive. If $a = c/p^n$ is not primitive, write $a = b/p^{n-1}$. Then

$$(0, a)\sigma = (rb, b/p^{n-1} + sb) \equiv (0, a).$$

Taking the norm essentially gives the product of the same function taken p^2 times. If a is primitive, then the set

$$\{(rc/p,\, c/p^n + sc/p)\}$$

is the set of the p^2 elements (u, v) of $(1/p^n)\mathbf{Z}/\mathbf{Z}$ such that

$$p(u, v) = (0, c/p^{n-1}) \doteq (0, pa).$$

Hence by the distribution relation

$$N_{n,n-1}(g_a) = g_{pa}.$$

This proves the lemma.

We now conclude:

$$N_{n,n-1}(g) = \Big(\prod_{a \text{ not prim.}} g_a^{pm(a)}\Big)^p, \tag{5.4}$$

and it suffices to prove that each partial product taken for a in an orbit has level p^{n-1}. For this it suffices to prove:

The family $\{pm(a)\}$ for a in an orbit satisfies the quadratic relations of level p^{n-1}. In other words,

$$\sum_{\substack{a \in 0 \\ a \text{ not prim.}}} pm(a)a^2 \equiv 0 \bmod \frac{1}{p^{n-1}} \mathbf{Z}.$$

Equivalently,

$$\sum_{a \in 0} m(a)a^2 \equiv 0 \bmod \frac{1}{p^n} \mathbf{Z}.$$

Write $a = a_0 + r/p$ with $r \in \mathbf{Z}/p\mathbf{Z}$, with one value of r omitted if the orbit has level p. Then

$$\begin{aligned}\sum_{a \in 0} m(a)a^2 &= \sum_r m\Big(a_0 + \frac{r}{p}\Big)\Big(a_0 + \frac{r}{p}\Big)^2 \\ &= \sum_r m\Big(a_0 + \frac{r}{p}\Big)a_0^2 + 2\sum_r m\Big(a_0 + \frac{r}{p}\Big)a_0\frac{r}{p} + \sum_r m\Big(a_0 + \frac{r}{p}\Big)\frac{r^2}{p^2}.\end{aligned}$$

The first term on the right is 0 by the orbit condition **D 1**. The second term satisfies the desired congruence because a_0 is not primitive. The third term satisfies the desired congruence because $n \geqq 2$. This concludes the proof.

§6. Projective Limit of the Trivial Group

When we computed the order of $\mathscr{C}_1^0(n)$, we saw that div $\mathscr{F}_1^0(n)$ was smaller than the Stickelberger ideal, and was in fact equal to $J_1\theta$, where J_1 was defined in §3. The Stickelberger ideal is $Q_0\theta$, where Q is the ideal in the group ring satisfying the quadratic relations, and Q_0 is the subideal of elements of degree 0. Thus we may consider the factor group

$$\mathscr{T}(n) = Q_0(n)\theta/J_1(n)\theta \approx Q_0(n)/J_1(n)$$

as a "**trivial**" **part of the cuspidal divisor class group** under consideration.

As we are interested here in the p-primary part of the divisor class group, we take the group ring to have coefficients in $\mathbf{Z}_p$, rather than $\mathbf{Z}$. Thus the group ring is

$$R(n) = \mathbf{Z}_p[G(n)], \quad \text{where } G(n) \approx \mathbf{Z}(p^n)^*/\pm 1.$$

In §4 we had established a sequence of injections

$$Q_0/J_1 \to R_0/F_0^{(1)} \to R/F^{(1)} = F^{(n)}/F^{(1)},$$

where the $F^{(s)}$ were an appropriate filtration of $R = F^{(n)}$. Furthermore, $F^{(n)}/F^{(1)}$ is a finite p-group.

We are now interested in the projective limit of $\mathscr{T}(n)$. The group $G(1)$ is a direct factor of $G(n)$ for each n, and has order $(p-1)/2$. It operates on $R(n)$, so on the augmentation ideal $R_0(n)$, and finally on the projective limit $\lim R_0(n)$. For each character χ of $G(1)$, we let

$$e(\chi) = \frac{1}{|G(1)|} \sum_{\sigma} \chi(\sigma)\sigma^{-1}$$

be the idempotent corresponding to χ. Then $e(\chi)$ lies in $R(n)$. If $R(n)$ operates on a module A, then we let $A(\chi) = e(\chi)A$ be the χ-eigenspace of A. We abbreviate for instance

$$R(n)(\chi) = R(n, \chi).$$

Lemma 6.1. *For $\chi \neq 1, \omega^2$ we have $e(\chi) \in Q_0(n)$ and*

$$e(\chi)R(n) = Q_0(n, \chi).$$

Proof. Write

$$e(\chi) = \frac{1}{|G(1)|} \sum_{c} \chi(\sigma_c)\sigma_c^{-1},$$

and let $\chi = \omega^k$. Then the relation to be satisfied for the quadratic relations is that

$$\sum_c \zeta_c^k \zeta_c^{-2} \equiv 0 \bmod p^n,$$

where ζ_c is the Teichmuller representative of c mod p. The sum on the left is actually equal to 0 when $\chi \neq \omega^2$, so in that case the quadratic relations are satisfied. The idempotent has degree 0 when $\chi \neq 1$, so the lemma follows.

Theorem 6.2. *For $\chi \neq 1, \omega^2$ we have*

$$\lim \mathscr{T}(n)(\chi) = \lim R(n)(\chi) = R_0 \cdot \lim \theta(n)(\chi).$$

Proof. We have a commutative diagram

$$\begin{array}{ccc} Q(n, \chi) & \longrightarrow & \mathscr{T}(n, \chi) \\ \downarrow & & \downarrow \\ Q(n-1, \chi) & \longrightarrow & \mathscr{T}(n-1, \chi). \end{array}$$

As in the preceding section,

$$\pi_* J_1(n, \chi) \subset p J_1(n-1, \chi),$$

so the projective limit of the $J_1(n, \chi)$ is 0. The theorem follows.

The limit

$$\lim \theta(n, \chi)$$

is the usual power series in $\mathbf{Z}_p[[X]]$ associated with the Stickelberger elements, as in Iwasawa theory.

The projective limit is thus seen to be rather simple, and the more interesting part is the remaining part due to the Stickelberger ideal. For $\chi = \omega^2$ the situation is different.

Theorem 6.3. *If $\chi = \omega^2$ then*

$$e(\chi) Q(n) \theta(n) = R(n, \chi).$$

Proof. We can write

$$G(n) = G(1) \times \Gamma_n,$$

where Γ_n = group of units in $\mathbf{Z}(p^{n-1})^*$ which are $\equiv 1 \bmod p$. Thus $R(n, \chi)$ is isomorphic to the usual truncated polynomial ring

$$R(n, \chi) \approx \mathbf{Z}_p[X]/h_n(X),$$

where $h_n(X) = (X + 1)^{p^n} - 1$. We need the trivial fact:

An element

$$z_0 + z_1 X + \cdots + z_{N-1} X^{N-1}, \quad \text{with } N = p^n,\ z_i \in \mathbf{Z}_p,$$

is a unit in $\mathbf{Z}_p[X]/h_n(X)$ *if and only if* $z_0 \in \mathbf{Z}_p^*$.

The proof is obvious, as usual. First if z_0 is not a unit, then it is clear that the element is not invertible. If z_0 is a unit, then we may assume that $z_0 = 1$. In that case, we invert $1 - Xg(X)$ by the usual geometric series. Since

$$X^{p^n} \equiv 0 \bmod p, h_n(X)$$

it follows that this series converges in $\mathbf{Z}_p[X]/h_n(X)$, because

$$x^{mp^n + r} \equiv 0 \bmod p^m.$$

Now we note that the canonical map

$$e(\chi)Q(n)\theta(n) \to e(\chi)Q(n-1)\theta(n-1)$$

is surjective, for $n \geqq 2$. If we can show that $e(\chi)Q(n-1)\theta(n-1)$ contains a unit in $R(n-1, \chi)$, then the above remark shows that $e(\chi)Q(n)\theta(n)$ must contain a unit in $R(n, \chi)$, namely any polynomial which reduces to a unit in $R(n-1, \chi)$ mod $h_{n-1}(X)$. This reduces the proof of the theorem to the case when $n = 1$.

In that case, however, the theorem is a consequence of the standard computation of p-adic orders of the second Bernoulli numbers, already used in §2. In this case, one verifies trivially that $Q(\chi) = (p)$ and $B_{2,\bar{\chi}}$ has order -1 (von Staudt type theorem, which can be seen directly from the definition).

CHAPTER 7

Modular Units on Tate Curves

The generic units (which are algebraic functions of j) can be specialized whenever j is specialized, say into a number field. Three cases arise: when j is not integral, when j is integral without complex multiplication, and when j is integral with complex multiplication. The first case will be discussed in this chapter. The third case is in some sense the oldest and will be discussed later. The middle case is the one about which the least is known. A recent result of Harris [Har] gives an asymptotic estimate for the rank of the specialized units.

The first case corresponds to a Tate curve. We assume that the reader is acquainted with the fundamental facts about these curves, as given for instance in [L 5], Chapter 15. According to Serre [Se], see also [L 5], Chapter 16, the Galois group of torsion points is essentially as large as possible, so that in this case, the analysis of independence of the specialized units can be carried out in a manner which is similar to that of the generic situation, although some more delicate points arise when dealing with the intervening Gauss sums.

In §1 we give a way of specializing functions and cuspidal divisor classes from the generic case to the special case on a Tate curve over a number field. Not much is known about the non-degeneracy of this homomorphism. Geometrically, it corresponds to intersecting, or pulling back, divisor classes from the modular scheme to the base scheme.

We state the main theorem in §2. It gives a sufficient condition under which the units remain independent in terms of the non-degeneracy of the Galois group of torsion points. The proof relies on the non-vanishing of a certain character sum, which is taken care of in §3.

§1. Specializations of Divisors and Functions at Infinity

In this section we analyse what happens to a function in the modular function field when specialized at a point of the modular curve where the j-invariant is not integral.

We return to an arbitrary positive integer N. We let R be the integral closure of $\mathbf{Z}[j]$ in F_N, and R^∞ the integral closure of $\mathbf{Z}[1/j]$ in F_N. We let *in this section* $V(N)$ be the scheme whose function field is F_N, given by

$$V(N) = \operatorname{spec}(R) \cup \operatorname{spec}(R^\infty).$$

We let

$$\mathscr{D}^\infty$$

be the free abelian group on $V(N)$ generated by the minimal prime ideals in R^∞ lying above the ideal $(1/j)$ in $\mathbf{Z}[1/j]$. We call $\mathscr{D}^\infty$ the group of **divisors at infinity** on $V(N)$.

Remark. *An element f of F_N has a $V(N)$-divisor in $\mathscr{D}^\infty$ if and only if f is a unit in R_N.*

Proof. Suppose that the $V(N)$-divisor of f is at infinity. The ring R, which is Noetherian integrally closed, is equal to the intersection of the local (discrete valuation) rings

$$R = \bigcap_{\mathfrak{p}} R_{\mathfrak{p}}$$

taken over all minimal prime ideals $\mathfrak{p}$ in R. By assumption, f is a unit in each $R_{\mathfrak{p}}$, whence f lies in R. The same argument applied to $1/f$ shows that f is a unit in R. Conversely, if f is a unit in R, then its $V(N)$-divisor carries only minimal primes in R^∞ which do not contain any prime number [otherwise they would be represented also on spec(R)]. Hence these minimal primes lie over $(1/j)$ in $\mathbf{Z}[1/j]$ and are at infinity, thereby proving the remark.

The Tate curve over a complete local ring, rather than discrete valuation ring, was first considered in [L 3], cf. [L 5], Chapter 15. The question of regularity of the modular schemes is studied deeply in Deligne-Rapoport [De-Ra]. For the convenience of the reader, we describe the regularity at infinity, where we can give a very short and simple proof just from what is done in [L 5], Chapter 15.

Theorem 1.1. *Let R^∞ be the integral closure of $\mathbf{Z}[1/j]$ in the modular function field F_N, for any positive integer N. Let $\mathfrak{M}$ be a maximal ideal in*

R^∞ containing a prime number p and $1/j$. Then the local ring $R^\infty_{\mathfrak{M}}$ is regular. In fact, let $\mathfrak{o}$ be the ring of algebraic integers in the cyclotomic field $\mathbf{Q}(\zeta_N)$ and let $\mathfrak{p} = \mathfrak{M} \cap \mathfrak{o}$. Let ^ denote completion. Then there is a natural isomorphism

$$\hat{R}^\infty_{\mathfrak{M}} \cong \hat{\mathfrak{o}}_{\mathfrak{p}}[[q^{1/N}]].$$

Proof. The completion $\hat{R}^\infty_M$ is integrally closed by commutative algebra (e.g. EGA, Chapter IV, 7.8.3), and contains

$$\hat{\mathfrak{o}}_{\mathfrak{p}}[[1/j]] = \hat{\mathfrak{o}}_{\mathfrak{p}}[[q]].$$

The field of modular functions F_N can be identified with the field of "normalized" x-coordinates of N-th division points on the generic elliptic curve A with invariant j, as in Shimura [Sh] or [L 5], p. 67. We let h_A be the normalized Weber function. Let K be the quotient field of $\hat{\mathfrak{o}}_{\mathfrak{p}}[[q]]$. Let B be the Tate curve as in [L 5], Chapter 15, defined over K, also having invariant j. Then there exists an isomorphism

$$\lambda : A \to B$$

over some finite extension of K, uniquely determined up to ± 1 since $\mathrm{End}(A) = \mathbf{Z}$. We have

$$h_B(\lambda a) = h_A(a), \quad \text{all } a \in A_N.$$

Furthermore, $K(B_N) = K(q^{1/N})$. The integral closure of $\hat{\mathfrak{o}}_{\mathfrak{p}}[[q]]$ in this field of N-th division points $K(B_N)$ is therefore obviously equal to the power series ring

$$\hat{\mathfrak{o}}_{\mathfrak{p}}[[q^{1/N}]]$$

(which is integrally closed, and integral over $\hat{\mathfrak{o}}_{\mathfrak{p}}[[q]]$).

We conclude that the isomorphism λ induces an embedding of $\hat{R}^\infty_{\mathfrak{M}}$ into $\hat{\mathfrak{o}}_{\mathfrak{p}}[[q^{1/N}]]$. The only subfields of $K(q^{1/N})$ containing K are of type $K(q^{1/d})$ with $d \mid N$ by Kummer theory. Since the modular function field F_N is ramified of order N at infinity, it follows that $\hat{R}^\infty_{\mathfrak{M}}$ must be equal to $\hat{\mathfrak{o}}_{\mathfrak{p}}[[q^{1/N}]]$, thus proving the last assertion of the theorem.

In particular, a local ring is regular if and only if its completion is regular, so the first assertion also follows.

Let $V(N)^\infty$ be the subset of spec R^∞ consisting of those prime ideals containing $1/j$. We view the maximal ideal $\mathfrak{M}$ in Theorem 1.1 as the maximal ideal of a closed point on $V(N)^\infty$, and we have the inclusion

$$\operatorname{spec} R^\infty_{\mathfrak{M}} \to V(N)^\infty.$$

A prime divisor on $V(N)^\infty$ passing through $\mathfrak{M}$ is identified with a minimal prime of $R^\infty_{\mathfrak{M}}$ containing $(1/j)$.

Corollary. *Given a point $\mathfrak{M}$ as in the theorem, there exists a unique prime divisor on $V(N)^\infty$ passing through $\mathfrak{M}$,* i.e. *$\mathfrak{M}$ contains a unique minimal prime containing $1/j$.*

Proof. It suffices to prove the assertion for the completion

$$\operatorname{spec} \hat{R}^\infty_{\mathfrak{M}} = \operatorname{spec} \hat{\mathfrak{o}}_{\mathfrak{p}}[[q^{1/N}]].$$

In this case, it is clear that the ideal generated by $q^{1/N}$ is the unique minimal prime ideal of $\hat{\mathfrak{o}}_{\mathfrak{p}}[[q^{1/N}]]$ containing q (or $1/j$).

Just as in algebraic number theory, or the abstract theory of extensions of discrete valuations to an extension field, we can describe the (geometric) prime divisors at infinity also in the following manner. Let $\mathbf{Q}_N = \mathbf{Q}(\zeta_N)$ be the cyclotomic field of N-th roots of unity over $\mathbf{Q}$. The rational function field $\mathbf{Q}_N(j)$ is embedded canonically as a subfield of $\mathbf{Q}_N((1/j)) = \mathbf{Q}_N((q))$. The primes at infinity of F_N are in bijection with the extensions of this embedding to embeddings of F_N in $\mathbf{Q}_N((q^{1/N}))$, regarding two such extended embeddings as equivalent if they differ by an automorphism of $\mathbf{Q}_N((q^{1/N}))$ over $\mathbf{Q}_N((q))$. Such automorphisms are obviously of Kummer type, multiplying $q^{1/N}$ by an N-th root of unity. The (equivalence class) of embeddings inducing the same prime P will be called a **P-embedding**.

The closed point $\mathfrak{M}$ on $V(N)^\infty$ will often be called an **arithmetic point**. If P is a prime divisor at infinity passing through $\mathfrak{M}$, then the ideal

$$PR^\infty_{\mathfrak{M}}$$

is principal because the local ring $R^\infty_{\mathfrak{M}}$ is regular, and is generated by a prime element t which is called a local parameter at P. In geometric language, P is defined locally in a neighborhood of $\mathfrak{M}$ by the equation

$$t = 0.$$

The maximal ideal $\mathfrak{M}$ is generated by two elements,

$$\mathfrak{M} = (\pi, t),$$

and the point $\mathfrak{M}$ is locally the intersection of the two hypersurfaces

$$\pi = 0 \quad \text{and} \quad t = 0,$$

which intersect transversally at $\mathfrak{M}$. A P-embedding of F_N in the power series field $\mathbf{Q}_\zeta((q^{1/N}))$ determines a power series expansion for t, called the **q-expansion at P,**

$$t = c_1 q^{1/N} + \cdots.$$

The coefficients c_i lie in $\mathbf{Q}(\zeta_N)$, $i = 1, 2, \ldots$.

Theorem 1.2. *Given a prime P at infinity, there exists a local parameter t in F_N whose q-expansion at P has coefficients c_i in $\mathbf{Z}[\zeta_N]$,* i.e. *algebraic integers, and such that c_1 is a unit in $\mathbf{Z}[\zeta_N]$.*

Proof. The modular function field F_N has a "standard" prime at infinity, determined by the complex q-expansions, and denoted by P_∞. Given any other prime P at infinity, there exists an element $\alpha \in SL_2(\mathbf{Z})$ such that $P = \alpha P_\infty$. In other words, two primes at infinity are conjugate under the action of the Galois group of F_N over F_1. It suffices to prove that there exists a function $f \in F_N$ whose q-expansion at P_∞ has the desired form,

$$f_{P_\infty}(q) = \sum c_i q^{i/N},$$

with c_i integral, and c_1 equal to a unit, for then we can take

$$t = f^{\alpha^{-1}}$$

as the desired parameter at P.

The construction of f is then obvious. The function

$$u = \frac{\wp_{1,1} - \wp_{1,0}}{\wp_{0,1} - \wp_{1,0}}$$

is immediately seen to have a q-expansion of the form

$$\zeta^{-1}(\zeta - 1)q^{1/N} + \text{higher terms},$$

and the coefficients of the higher terms are all integral, divisible by differences $\zeta^n - 1$ or $\zeta^{-n} - 1$, which in turn are divisible by $\zeta - 1$. Thus the function

$$f = (\zeta - 1)^{-1} u$$

has the desired properties.

Remark. For level $N = 2$, this is the significance of the factor 2 which appears in the classical formula, reproduced as $\mathbf{E}_{31}^{1/4}$ of [L 5], Chapter 18, §4, p. 251.

Let $k \supset \mathbf{Q}(\zeta_N)$. A point of $V(N)$ into a number field k, not at infinity, is identified with a homomorphism

$$\varphi : R = R_N \to k.$$

At a given prime $\mathfrak{p}$ in k we have the local ring $\mathfrak{o}_\mathfrak{p}$, whence a map

$$\operatorname{spec}(\mathfrak{o}_\mathfrak{p}) \to X(N).$$

By pull back we can define a homomorphism from the divisors on $X(N)$ into the group of divisors of $\operatorname{spec}(\mathfrak{o}_\mathfrak{p})$, essentially the cyclic group generated by $\mathfrak{p}$. We describe this pull back completely in the following elementary manner.

For w in the local ring of φ, denote $\varphi(w)$ by $\bar{w}$. Let P be a prime divisor at infinity. Let $\mathfrak{p}$ be a prime ideal in k. If $\bar{j}$ is integral at $\mathfrak{p}$, we let

$$m(\varphi, \mathfrak{p}, P) = 0.$$

If $\bar{j}$ is not integral at $\mathfrak{p}$, we consider the composite homomorphism

$$R^\infty \to \bar{R}^\infty \to \bar{R}^\infty \bmod \mathfrak{p},$$

and let $\mathfrak{M}$ be its kernel. Then by Theorem 1.1

$$PR_\mathfrak{M}^\infty = (t)$$

is a principal ideal, whose generator t is determined up to a unit in $R_\mathfrak{M}^\infty$. Thus the order

$$m(\varphi, \mathfrak{p}, P) = \operatorname{ord}_\mathfrak{p} \bar{t}$$

is well defined, and we let

$$(\varphi P)(\mathfrak{p}) = \bar{P}(\mathfrak{p}) = (\bar{t})_\mathfrak{p} = \mathfrak{p}^{m(\varphi,\mathfrak{p},P)}.$$

We may view $m(\varphi, \mathfrak{p}, P)$ as the multiplicity of intersection of the geometric prime divisor P and the arithmetic prime divisor

$$\operatorname{spec}\, \mathfrak{o}_\mathfrak{p} \to V(N)$$

at M.

Finally, to each prime divisor P at infinity we associate the ideal

$$\varphi P = \bar{P} = \prod_{\mathfrak{p}} (\varphi P)(\mathfrak{p}),$$

and extend this to divisors $D = \sum v_P P$ by linearity, so that we obtain a fractional ideal

$$\varphi D = \bar{D} = \prod_{P} \varphi(P)^{v_P}.$$

The association

$$D \mapsto \varphi D = \bar{D}$$

is a homomorphism of $\mathscr{D}^\infty$ into the group of fractional ideals in k.

We let $\mathscr{C}^\infty$ be the factor group of $\mathscr{D}^\infty$ by the subgroup of divisors of units in R_N. Then the preceding association gives rise to a homomorphism of the divisor classes $\mathscr{C}^\infty$ into the group of ideal classes in k. Nothing is known about the non-degeneracy of this homomorphism, which relates the existence of rational points on $V(N)$ with the group of ideal classes.

Theorem 1.3. *If a unit u in R_N has a divisor at infinity which is the m-th multiple of some divisor, then for any point φ as above, the ideal factorization of $\bar{u} = \varphi u$ is an m-th power.*

The proof is obvious, from the preceding discussion.

The theorem is a formulation in terms of commutative algebra of the following simple geometric idea. We view u as a function on the "variety" $V(N)$ viewed as scheme over $\mathbf{Z}$. If V is a variety, and W is a subvariety, and u is a function on V, then under appropriate conditions of completeness and non-singularity, the divisors of u on V, and the induced function

$$\bar{u} = u|W$$

on W are related by the relation

$$(\bar{u})_W = (u)_V \cdot W,$$

where the dot is the intersection product. In particular, if $(u)_V$ is the N-th multiple of a divisor on V, then $(\bar{u})_W$ is the N-th multiple of a divisor on W. The corollary (which obviously holds in a general context) formalizes this argument in the arithmetic case, when W is the "subvariety" associated with the ring of integers in a number field.

If the function u has a q-expansion whose first coefficient is a unit, and such that all other coefficients are p-integral, then the order of the specialized value $\bar{u}$ at $\mathfrak{p}$ can also be described in terms of the order of the q-expansion, as follows. Let A be a complete discrete valuation ring with maximal ideal $\mathfrak{m}$. Given an element $\bar{t} \in \mathfrak{m}$, and the power series ring $A[[t]]$ in one variable t, there exists a unique continuous homomorphism

$$A[[t]] \to A$$

such that $t \mapsto \bar{t}$. This will be applied when $t = q^{1/N}$.

In addition, let $k = \mathbf{Q}(\zeta_N)$ be the cyclotomic field, let $\mathfrak{o} = \mathfrak{o}_k$ be its ring of integers. Then any homomorphism

$$\varphi : R^{\infty} \to A$$

induces a homomorphism

$$\mathfrak{o} \to A.$$

Let us assume that $A/\mathfrak{m}$ has characteristic p.

Let $\mathfrak{p}$ be the kernel in $\mathfrak{o}$ of the composite homomorphism

$$\mathfrak{o} \to A \to A/\mathfrak{m}.$$

Then we get an induced continuous homomorphism of the completion

$$\hat{\mathfrak{o}}_{\mathfrak{p}} \to A.$$

Viewing $q^{1/N} = t$ as a variable, there is a corresponding homomorphism of power series rings

$$\hat{\mathfrak{o}}_{\mathfrak{p}}[[q^{1/N}]] \to A[[q^{1/N}]].$$

Let $\bar{q}^{1/N}$ be an element of $\mathfrak{m}$. We see that there exists a unique continuous homomorphism

$$\hat{\mathfrak{o}}_{\mathfrak{p}}[[q^{1/N}]] \to A$$

which is equal to φ on $\mathfrak{o}_{\mathfrak{p}}$, and maps $q^{1/N}$ on $\bar{q}^{1/N}$.

The above notation will be preserved in the next theorem.

Theorem 1.4. *Let A be a complete discrete valuation ring with maximal ideal $\mathfrak{m}$ such that $A/\mathfrak{m}$ has characteristic p. Let*

$$\varphi : R^{\infty} \to A$$

be a homomorphism such that $1/\bar{j} = \varphi(1/j)$ *lies in* $\mathfrak{m}$. *Let* $\mathfrak{M}$ *be the kernel of the composite homomorphism*

$$R^{\infty} \to A \to A/\mathfrak{m},$$

and $R^{\infty}_{\mathfrak{M}}$ *the corresponding local ring. Let* $\mathfrak{p} = \mathfrak{M} \cap \mathfrak{o}$. *Then we have a commutative diagram*

$$\begin{array}{ccc} R^{\infty}_{M} & \xrightarrow{\varphi} & A \\ \downarrow & & \uparrow \\ \hat{R}^{\infty}_{M} & \xrightarrow{\approx} & \hat{\mathfrak{o}}_{\mathfrak{p}}[[q^{1/N}]]. \end{array}$$

Proof. The homomorphism $R^{\infty}_{\mathfrak{M}} \to A$ is continuous, since the maximal ideal $\mathfrak{M}$ is the inverse image of $\mathfrak{m}$ in R^{∞}. Therefore it extends uniquely to a continuous homomorphism of $\hat{R}^{\infty}_{\mathfrak{M}}$ into A. The isomorphism (identification) of Theorem 1.1 determines a special value $\bar{q}^{1/N}$ in A, so that Theorem 1.4 is obvious from the preceding discussion.

Corollary. *Let in addition u be in* $\hat{R}^{\infty}_{\mathfrak{M}}$, *admitting a power series expansion determined by the bottom isomorphism*

$$u = \sum_{n \geqq r} a_n q^{n/N}$$

with a_r *a unit in* $\hat{\mathfrak{o}}_{\mathfrak{p}}$ *and* $a_n \in \hat{\mathfrak{o}}_{\mathfrak{p}}$ *for all n. Then*

$$\operatorname{ord}_{\mathfrak{m}} \bar{u} = \frac{r}{N} \operatorname{ord}_{\mathfrak{m}} \bar{q}.$$

Proof. Obvious.

Theorem 1.5. *Let* $\varphi : R = R_N \to k$ *be a point of* $V(N)$ *in a number field k, not at infinity. Let* $\mathfrak{p}$ *be a prime of k dividing the denominator of* $\varphi j = \bar{j}$. *There exists only one prime at infinity P on* $V(N)^{\infty}$ *such that* $m(\varphi, \mathfrak{p}, P) \neq 0$, *and for that prime P we have*

$$m(\varphi, \mathfrak{p}, P) = -\frac{1}{N} \operatorname{ord}_{\mathfrak{p}} \varphi j.$$

Proof. The corollary of Theorem 1.1 already tells us that there is only one prime at infinity passing through $\mathfrak{M}$, so for all but one prime P the ideal

$PR_{\mathfrak{M}}^{\infty}$ is the unit ideal, and the multiplicity of intersection is 0. For the special prime P, we know from Theorem 1.2 that there is a local parameter having its corresponding q-expansion equal to

$$t = \sum c_i q^{i/N} = c_1 q^{1/N} + \cdots,$$

where c_1 is a unit in $\mathbf{Z}[\zeta_N]$ and all c_i are integral. The point φ is induced by an evaluation of the power series corresponding to

$$q^{1/N} \mapsto \bar{q}^{1/N}.$$

The order at $\mathfrak{p}$ of the special value $\varphi t = \bar{t}$ is therefore the order at $\mathfrak{p}$ of $\bar{q}^{1/N}$. But $\operatorname{ord} \bar{q} = -\operatorname{ord} j$, so our theorem is proved.

§2. Non-Degeneracy of the Units

Let k be a number field. Let A be an elliptic curve defined over k, and let

$$k(A_N/\pm)$$

be the field extension of x-coordinates of N-torsion points. Let $j_A = j(z)$ for some complex number z in the upper half plane. Then:

$k(A_N/\pm) =$ field generated by all values $f(z)$ of all modular functions f of level N defined at z.

By definition, the modular functions are taken from the function field having the cyclotomic field $\mathbf{Q}(\boldsymbol{\mu}_N)$ as constant field.

Let $\mathfrak{p}$ be a prime of k lying above the prime number p.

Let V be the multiplicative subgroup of elements $\alpha \in k(A_N/\pm)$, $\alpha \neq 0$, such that $\operatorname{ord}_{\mathfrak{P}} \alpha$ is independent of the prime $\mathfrak{P} | \mathfrak{p}$ in $k(A_N/\pm)$. We observe that V contains the units of $k(A_N/\pm)$, and also contains the cyclotomic numbers $1 - \zeta$ where $\zeta \in \boldsymbol{\mu}_N$.

For each prime $l | N$, let $\mathfrak{o}_l$ be the unramified extension of degree 2 of $\mathbf{Z}_l$. Let

$$C(l^n) = (\mathfrak{o}_l / l^n \mathfrak{o}_l)^*$$

be the Cartan group which has been considered all along in the study of modular units. If

$$N = \prod l^{n(l)}$$

we let

$$C(N) = \prod C(l^{n(l)}).$$

We often write $C(N)/\pm 1 = C(N)(\pm)$ or $C(N)/\pm$. This Cartan group acts simply transitively on the cusps of the modular curve $X(N)$. Once a basis of A_N over $\mathbf{Z}(N)$ is chosen, this Cartan group has a matrix representation with which we usually identify it. Thus we view $C(N)$ as a subgroup of $\mathrm{GL}_2(N) = \mathrm{GL}_2(\mathbf{Z}/N\mathbf{Z})$. By a theorem of Serre [Se], it is known that the Galois group of $k(A_N)$ over k is "usually" $\mathrm{GL}_2(N)$. For our purposes, we do not need the full GL_2, but will relate the independence of the specialized modular units to the size of the Cartan group in the Galois group. For simplicity, we restrict ourselves to the case when the whole Cartan group is contained in the Galois group, as follows.

We let g_a be the Siegel units, with

$$a \in \left(\frac{1}{N}\mathbf{Z}^2/\mathbf{Z}^2\right)\Big/ \pm 1.$$

Then g_a^{12N} is a "geometric unit" in the modular function field and has level N. Let U be the group generated by these Siegel functions. Let $U(z)$ be the group generated by the values

$$g_a^{12N}(z).$$

Then $U(z)$ is a subgroup of $k(A_N/\pm)^*$.

In Chapter 2, Theorem 3.1, we showed that the rank of U (modulo constants) is equal to

$$|C(N)(\pm)| - 1.$$

Our purpose here is to determine the rank of the specialized group $U(z)$ modulo V when A is a Tate curve.

Theorem 2.1. *Assume*:

(i) $\mathrm{Gal}(k(A_N/\pm)/k)$ *contains* $C(N)(\pm)$.
(ii) *The prime* $\mathfrak{p}$ *divides the denominator of* j_A. *Then*

$$\operatorname{rank} U(z)V/V = |C(N)(\pm)| - 1.$$

Proof. Let $\mathfrak{P}$ be a prime of $k(A_N/\pm)$ lying above $\mathfrak{p}$. Let $\mathrm{inj}_{\mathfrak{P}}$ be the injection of $k(A_N/\pm)$ in its completion at $\mathfrak{P}$. The Tate curve has a corresponding element

$$q_A = j_A^{-1} + \cdots \quad \text{with } j_A = \frac{1}{q_A} + 744 + 196884q_A + \cdots.$$

Cf. for instance [L 5], Chapter 15. In the analytic parametrization of the Tate curve, the N-th torsion points have a basis consisting of some chosen N-th root of q_A and a primitive N-th root of unity ζ. If r is an integer, we write

$$q_1^{r/N} = \zeta^r.$$

If α, α' are two non-zero elements of the $\mathfrak{P}$-adic field, we write

$$\alpha \sim \alpha'$$

to mean that α/α' is a $\mathfrak{P}$-adic unit.

A Siegel unit g_a has the q-expansion at the standard prime at infinity given up to a root of unity ζ by

$$g_a(z) = -q_\tau^{(1/2)\mathbf{B}_2(a_1)}\zeta(1 - q_z)\prod(1 - q_\tau^n q_z)(1 - q_\tau^n/q_z)$$

where $z = a_1\tau + a_2$, by Chapter 2, §1, formula **K 4**. All the primes at infinity of the modular function field are conjugate under GL_2. By Theorem 1.5, there exists $\sigma \in \mathrm{GL}_2(N)$ such that

O 1. $$\mathrm{inj}_{\mathfrak{P}}\, g_a^{12N}(z) \sim q_A^{(12N/2)\mathbf{B}_2(\langle(a\sigma)_1\rangle)}(1 - q_A^{(a\sigma)_1} q_1^{(a\sigma)_2})^{12N}$$

Lemma 2.2. *We can select an* N-th *root* $q^{1/N}$ *and the root of unity* ζ *such that the element* σ *above lies in the Cartan group* $C(N)$.

Proof. As remarked already in Chapter 2, §3, we have

$$\mathrm{GL}_2(N) = C(N)G_\infty,$$

where G_∞ is the isotropy group of $\begin{pmatrix}1\\0\end{pmatrix}$, consisting of the matrices

$$\begin{pmatrix}1 & b\\0 & d\end{pmatrix},$$

and every element of $\mathrm{GL}_2(N)$ has a unique decomposition as a product of an element in G_∞ and an element in $C(N)$. Let $\{e_1, e_2\}$ be a basis of $((1/N)\mathbf{Z}/\mathbf{Z})^2$.

Let T be the transformation associated with σ, and let $\{e'_1, e'_2\}$ be a basis for the image. Then

$$Te_1 = ae'_1 + be'_2.$$
$$Te_2 = ce'_1 + de'_2.$$

Let $t \in \mathbf{Z}(N)^*$ and $x \in \mathbf{Z}(N)$. Write

$$e'_2 = te''_2$$
$$e'_1 = e''_1 + xe''_2.$$

Then

$$Te_1 = ae''_1 + (bt + ax)e''_2$$
$$Te_2 = ce_1 + (cx + dt)e''_2.$$

Corresponding to the change of basis, the matrix for T is

$$\begin{pmatrix} a & b \\ c & d \end{pmatrix}\begin{pmatrix} 1 & x \\ 0 & t \end{pmatrix}.$$

We can choose t, x such that this matrix lies in $C(N)$. This proves the lemma.

For any element $\alpha \in k(A_N/\pm)$ we have for $\gamma \in C(N)/\pm 1$:

O 2. $$\operatorname{ord}_{\gamma^{-1}\mathfrak{P}} \alpha = \operatorname{ord}_{\mathfrak{P}} \gamma\alpha.$$

Furthermore,

O 3. $$\operatorname{inj}_{\mathfrak{P}} \gamma g_a^{12N}(z) = \operatorname{inj}_{\mathfrak{P}} g_{a\gamma}^{12N}(z).$$

This and the order formula **O 1** give us the order at $\gamma^{-1}\mathfrak{P}$ for the special values of the Siegel functions. We now consider the usual "logarithm map" concentrated at the orbit of $\mathfrak{P}$ under the Cartan group, namely for

$$\alpha \in k(A_N/\pm)^*$$

we let

$$L: \alpha \mapsto \sum_{\gamma \in C(N)/\pm 1} (\operatorname{ord}_{\gamma^{-1}\mathfrak{P}} \alpha)\sigma_\gamma^{-1}.$$

Then by assumption, we see that

$$L: V \to \mathbf{Z} \sum_\gamma \sigma_\gamma,$$

i.e., L maps V into the space in the group algebra corresponding to the trivial character. On the other hand,

$$L : g_a^{12N} \mapsto \sum_{\gamma} r(\gamma)\sigma_\gamma^{-1},$$

where

$$r(\gamma) = 12N\, \tfrac{1}{2}\mathbf{B}_2(\langle(a\gamma\sigma)_1\rangle)\, \mathrm{ord}_{\mathfrak{P}}\, q_A + 12N\, \mathrm{ord}_{\mathfrak{P}}(1 - q_A^{(a\gamma\sigma)_1} q_1^{(a\gamma\sigma)_2})$$

We now analyse the image of the group generated by the Siegel units by means of its position in the various eigenspaces for the characters $\chi \neq 1$ on $C(N)(\pm)$. As in our previous work, in the matrix representation

$$\gamma = \begin{pmatrix} a_1 & a_2 \\ c & d \end{pmatrix}$$

of the Cartan group, we let $T\gamma = a_1$. Let M be the conductor of χ. Let e_χ be the usual idempotent on $C(M)(\pm)$. Then

$$\mathbf{C} \otimes L(U)e_\chi$$

has dimension 0 or 1, and is spanned by

$$\frac{1}{2} \sum_{\gamma \in C(M)(\pm)} \mathbf{B}_2\left(\left\langle \frac{T\gamma}{M} \right\rangle\right)\bar{\chi}(\gamma)\, \mathrm{ord}_{\mathfrak{P}}\, q_A + \sum_{\substack{T\gamma = 0 \\ \gamma \in C(M)(\pm)}} \mathrm{ord}_{\mathfrak{P}}(1 - \zeta_M)\bar{\chi}(\gamma),$$

where ζ_M is a primitive M-th root of unity; so our assertion is clear from the expression given above for $r(\gamma)$.

We wish to prove that each such eigenspace for a non-trivial character has in fact dimension 1, so we have to prove that each generating element is $\neq 0$.

The first term is a positive multiple of the Cartan-Bernoulli number, which we know is $\neq 0$ since it is expressed as a non-zero multiple of the ordinary Bernoulli-Leopoldt number, as in Chapter 1, Theorem 5.3 and its corollaries.

The second term

$$\mathrm{ord}_{\mathfrak{P}}(1 - \zeta_M) \sum_{\substack{T\gamma = 0 \\ \gamma \in C(M)(\pm)}} \bar{\chi}(\gamma)$$

is equal to 0 if M is not a prime power, and also is equal to 0 if $\chi_{\mathbf{Z}}$ is non-trivial, because if $T\gamma = 0$, then $T(c\gamma) = 0$ for all $c \in \mathbf{Z}(M)(\pm)^*$, and the sum over $T\gamma = 0$ can be decomposed over a sum over cosets of $\mathbf{Z}(M)(\pm)^*$. In case $\chi_{\mathbf{Z}}$ is non-trivial, this shows that the generating element is $\neq 0$.

We are left with the case when $\chi_{\mathbf{Z}}$ is trivial, and M is a prime power, which we now suppose is the case. Then again the corollaries of Theorem 5.3, of Chapter 1 give us the value for

$$B_{2,\bar{\chi},T}$$

on the Cartan group, with $m = k = 2$. Since

$$B_{2,\bar{\chi}_{\mathbf{Z}}} = \tfrac{1}{6},$$

we see that

$$B_{2,\chi,T} = S_C(\bar{\chi}, T)\rho, \quad \text{where } \rho = N^{1-2}\left(1 - \frac{1}{p^2}\right)\frac{1}{6} > 0.$$

But

$$\sum_{\substack{T\gamma = 0 \\ \gamma \in C(M)(\pm)}} \bar{\chi}(\gamma) = \sum_{T(\gamma)=0} \bar{\chi}(\gamma).$$

We can therefore apply Proposition 3.2 proved in the next section to conclude the proof of the theorem.

§3. The Value of a Gauss Sum

In this section we give the value of the Gauss sum needed to complete the arguments giving the rank of the modular units on a Tate curve. The arguments are self contained.

Let k be an integer $\geqq 1$, p prime. Let $\mathfrak{o}$ be the unramified (ring) extension of $\mathbf{Z}_p$ of degree k, and let

$$\mathfrak{o}(p^n) = \mathfrak{o}/p^n\mathfrak{o}.$$

We call $C = C(p^n) = \mathfrak{o}(p^n)^*$ the **Cartan group**. We let

$$\lambda : \mathfrak{o}(p^n) \to \boldsymbol{\mu}_{p^n}$$

be a character giving rise to the usual self duality. For instance,

$$\lambda(x) = \exp(2\pi i \operatorname{Tr}(x)/p^n)$$

is such a character, where the trace is taken to $\mathbf{Z}_p$.

Let χ be a character on C. We may form the usual **Gauss sum**

$$S_C(\chi, \lambda) = S(\chi, \lambda) = \sum \chi(x)\lambda(x),$$

where the sum is taken over $x \in C$, or even $x \in \mathfrak{o}(p^n)$ since χ is defined to be 0 outside C. If χ is primitive, it is standard that the absolute value of the Gauss sum and all its conjugates satisfies

$$|S(\chi, \lambda)| = p^{nk/2}.$$

We shall first prove the proposition which is used in determining the rank of the modular units in the preceding section.

The group $\mathbf{Z}(p^n)^*$ is contained in the Cartan group. As usual, we let $\chi_{\mathbf{Z}}$ denote the restriction of χ to that group. Similarly, we let $\lambda_{\mathbf{Z}}$ denote the restriction of λ to $\mathbf{Z}(p^n)$.

Suppose that $k = 2$. If $\lambda_{\mathbf{Z}}$ is trivial, then

$$\operatorname{Ker} \lambda = \mathbf{Z}(p^n)$$

because $\mathfrak{o}(p^n)$ is free of dimension 2 over $\mathbf{Z}(p^n)$. We are of course especially interested in the case $k = 2$ for applications to the modular units.

Proposition 3.1. *Let χ be primitive on the Cartan group. Assume that $\chi_{\mathbf{Z}}$ and $\lambda_{\mathbf{Z}}$ are trivial. Let $k = 2$. Then*

$$S_C(\chi, \lambda) = p^n.$$

Proof. We give the proof when $n > 1$ (it is even easier when $n = 1$). We have

$$S_C(\chi, \lambda) = \sum_{x \in \mathbf{Z}(p^n)^*} 1 + \sum_{x \notin \mathbf{Z}(p^n)} \chi(x)\lambda(x).$$

The first sum on the right of course gives the Euler function $\phi(p^n)$. As to the second sum, it can be decomposed into a sum over non-trivial cosets of $\mathbf{Z}(p^n)^*$ in the Cartan group. Each coset consists of elements

$$\{xa\}, \quad \text{with } a \in \mathbf{Z}(p^n)^*,$$

and $\lambda(xa) = \lambda(x)^a$. Furthermore, $\chi(xa) = \chi(x)$, and

$$\sum_a \lambda(x)^a = \operatorname{Tr} \lambda(x),$$

where Tr is the trace from $\mathbf{Q}(\boldsymbol{\mu}_{p^n})$ to $\mathbf{Q}$. Since $\lambda(x) \in \boldsymbol{\mu}_{p^n}$, we see that the trace is $\neq 0$ if and only if $\lambda(x) \in \boldsymbol{\mu}_p$. Let us rewrite the second sum as

$$\sum_{x \notin \mathbf{Z}(p^n)} \chi(x)\lambda(x) = \sum \chi(\bar{x}) \operatorname{Tr} \lambda(\bar{x})$$

where $\bar{x}$ ranges over representatives of $C/\mathbf{Z}(p^n)^*$ other than the unit coset. Furthermore, since we can limit ourselves to x such that $\lambda(x) \in \boldsymbol{\mu}_p$, since $\operatorname{Ker} \lambda = \mathbf{Z}(p^n)$, and since λ is surjective, it follows that

$$x \equiv a + p^{n-1}y \bmod p^n,$$

with some element $a \in \mathbf{Z}$ and $y \in \mathfrak{o}(p)/\mathbf{Z}(p) - \{0\}$. We may therefore select the representative $\bar{x}$ to be of the form

$$\bar{x} = 1 + p^{n-1}y.$$

For the trace, we find

$$\operatorname{Tr} \lambda(x) = -\frac{\phi(p^n)}{p-1}$$

because the trace of a primitive p-th root of unity to $\mathbf{Q}$ is -1, and we are here taking the trace from the field of p^n-th roots of unity, so -1 has to be multiplied by the appropriate degree. Consequently, our second sum is equal to

$$\sum_y \chi(1 + p^{n-1}y)\left(-\frac{\phi(p^n)}{p-1}\right)$$

with $y \in \mathfrak{o}(p^n)/\mathbf{Z}(p^n)$. Since χ is assumed primitive, the sum over y is equal to -1. Hence finally

$$S_C(\chi, \lambda) = \phi(p^n) + \frac{\phi(p^n)}{p-1} = p^n.$$

This proves the proposition.

Proposition 3.2. *Again let $k = 2$. Let t be a real number, $t > 0$. Let χ be a primitive character on $C(p^n)$. Then*

$$S(\chi, \lambda) + t \sum_{\lambda(x)=0} \chi(x) \neq 0.$$

Proof. If $\chi_{\mathbf{Z}}$ is not trivial, then

$$\sum_{\lambda(x)=0} \chi(x) = 0$$

since the sum can be decomposed over sums on cosets of $\mathbf{Z}(p^n)^*$. The proposition follows because $S(\chi, \lambda) \neq 0$.

Suppose that $\chi_{\mathbf{Z}}$ is trivial. Without loss of generality, we may assume that $\lambda_{\mathbf{Z}}$ is trivial. Indeed, suppose we replace λ by $\lambda \circ \gamma$ for some element $\gamma \in C$. Then the sum $S(\chi, \lambda)$ changes by a factor $\overline{\chi}(\gamma)$, and so does the other sum over $\lambda(x) = 0$. We can then find some $\gamma \in C$ such that $\lambda \circ \gamma$ is trivial on $\mathbf{Z}(p^n)$, thus reducing the problem to the case when $\lambda_{\mathbf{Z}}$ is trivial.

Suppose this is the case. Then the desired expression is simply equal to

$$p^n + t(p - 1) > 0$$

by Proposition 3.1. This concludes the proof.

CHAPTER 8

Diophantine Applications

This brief chapter gives some diophantine applications of the existence of certain modular units. It could be read immediately at the end of Chapter 2, following the construction of Weierstrass units.

The Fermat-Thue-Siegel (Fermat for short) curve has played a basic role in the theory of integral points, see for instance [Sie 1], Lang in similar connections [L 1], and following Gelfond [G], Baker [Ba], Baker-Coates [Ba-Co], Coates [Co 1]. In these papers, it arises through the Siegel identity

$$\frac{x_3 - x_1}{x_2 - x_1}\frac{t - x_2}{t - x_3} + \frac{x_2 - x_3}{x_2 - x_1}\frac{t - x_1}{t - x_3} = 1.$$

Following [K-L 1], we shall use a simpler identity to get diophantine results on the modular curves, namely

$$\frac{x_3 - x_1}{x_2 - x_1} + \frac{x_2 - x_3}{x_2 - x_1} = 1,$$

which is for instance satisfied by the λ-function in the theory of elliptic functions.

Our construction joins a train of thought which appeared in the work of Demyanenko [Dem] and Kubert [Ku 1] in their studies of rational points on modular curves, with that of constructing units in the modular function fields [L 5], Chapter 18, §6.

§1. Integral Points

In Chapter 2, Theorem 6.2, we showed how to construct units in the affine ring R_N of the modular function field F_N. We shall now consider special cases of these units, and their applications to diophantine problems.

We shall construct units u, u' such that

$$u + u' = 1.$$

These units will be of the form

$$u = \frac{x_1 - x_2}{x_1 - x_3} \quad \text{and} \quad u' = \frac{x_2 - x_3}{x_1 - x_3}.$$

As before, we let N be an integer $\geqq 2$. Let

$$a \in \frac{1}{N}(\mathbf{Z}\tau + \mathbf{Z})$$

have exact period N mod $\mathbf{Z}\tau + \mathbf{Z}$ (we say that a is **primitive of level** N). Let r, s be integers. We say that the pair (r, s) is **admissible** (or N-admissible) if r, s, $r + s$, $r - s$ are prime to N.

Theorem 1.1. *Let r, s, m be integers such that the pairs (r, s), (r, m), (s, m) are N-admissible. Let*

$$u = \frac{\wp(ra) - \wp(sa)}{\wp(ra) - \wp(ma)} \quad \text{and} \quad u' = \frac{\wp(sa) - \wp(ma)}{\wp(ra) - \wp(ma)}.$$

Then u, u' are units in R_N and $u + u' = 1$. For such units, we have two cases. (We put $\text{ord} = \text{ord}_q$.)

Case 1. $\langle a_1 \rangle = 0$. *Then* $\text{ord}\, u = \text{ord}\, u' = 0$ *for all choices of r, s, m.*
Case 2. $\langle a_1 \rangle \neq 0$. *Then*

$$\frac{1}{N} \leqq |\text{ord}(uu')| < 1.$$

Proof. The fact that u, u' are units is a special case of Theorem 6.2 of Chapter 2, and results from the table in Lemma 2 preceding that theorem, because the quotient of two factors $1 - \zeta$ is a cyclotomic unit. Note that only Cases 1 and 2 occur in that table, in the present situation. The fact that

$$\text{ord}\, uu' \neq 0 \quad \text{and hence} \quad |\text{ord}\, uu'| \geqq \frac{1}{N}$$

comes from the relation

$$\text{ord}\, uu' = e + e'' - 2e'$$

where

$$\begin{aligned} e &= \min(\langle \pm ra_1 \rangle, \langle \pm sa_1 \rangle) \\ e' &= \min(\langle \pm ra_1 \rangle, \langle \pm ma_1 \rangle) \\ e'' &= \min(\langle \pm sa_1 \rangle, \langle \pm ma_1 \rangle). \end{aligned}$$

The three numbers

$$\alpha_1 = \langle \pm ra_1 \rangle, \qquad \alpha_2 = \langle \pm sa_1 \rangle, \qquad \alpha_3 = \langle \pm ma_1 \rangle$$

are distinct. Say $e'' = \min(\alpha_2, \alpha_3) = \alpha_2$. We distinguish two cases, according as $\alpha_1 < \alpha_2$ or $\alpha_1 > \alpha_2$. If $\alpha_1 < \alpha_2$ then

$$\operatorname{ord} uu' = \alpha_1 + \alpha_2 - 2\alpha_1 = \alpha_2 - \alpha_1 < 0.$$

In the other case, we get similarly $\operatorname{ord} uu' > 0$. In any case, $\operatorname{ord} uu' \neq 0$. It is then immediate that $\operatorname{ord} uu'$ satisfies the inequalities stated in the theorem.

Theorem 1.2. *Let $N \geqq 7$. Let R_N be the integral closure of $\mathbf{Z}[j]$ in the modular function field F_N, and let*

$$V(N)^{\text{aff}} = \operatorname{spec} R_N.$$

Let $\mathfrak{o}$ be a finitely generated subring in a number field. Then $V(N)^{\text{aff}}$ has only a finite number of points in $\mathfrak{o}$.

Proof. For each point of $V(N)^{\text{aff}}$ in $\mathfrak{o}$, we have a specialization $\bar{u}, \bar{v}$ of the units u, v in $\mathfrak{o}$ satisfying

$$\bar{u} + \bar{v} = 1.$$

Furthermore, $\bar{u}, \bar{v}$ are units in $\mathfrak{o}$ since the property of an element being a unit is preserved under homomorphisms. One can then use Gelfond's idea combined with the Baker inequalities for logarithms of algebraic numbers to deduce that the above equation has only a finite number of solutions in units, as desired. For an exposition, cf. [L 9], Chapter 6, §1.

Remark. If one knows the "modular" interpretation for $V(N)^{\text{aff}}$ then Theorem 1.2 can be proved by other means. Indeed, an integral point on $V(N)^{\text{aff}}$ corresponds to a triple (A, P, Q) where A is an elliptic curve with integral j-invariant, and P, Q is a basis for the points of order N on A, rational over the quotient field of $\mathfrak{o}$. One can then use the fact that A has potentially good reduction at all primes of $\mathfrak{o}$ to get the finiteness. However, the method used in the proof of Theorem 1.2 can be used to give a bound.

§2. Correspondence with the Fermat Curve

On the other hand, one can also use units satisfying

$$u + v = 1$$

in order to establish a correspondence with the Fermat curve.

Let $x^n = u$ and $y^n = v$. Then x, y satisfy the Fermat equation

$$x^n + y^n = 1,$$

defining the Fermat curve Φ.

Let R be a finitely generated ring of transcendence degree 1 over $\mathbf{Z}$ and let K be its quotient field. Let $V = \operatorname{spec} R$, and let k be the algebraic closure of $\mathbf{Q}$ in K. We may view V as an affine variety, defined over k, and $K = k(V)$ as the function field of V over k. We shall also assume that R is integrally closed. If x_1, x_2, x_3 above are elements of K, then

$$k(u) = k(v) = k(u, v)$$

is a subfield of K, defining a rational map of the curve V into the projective line $\mathbf{P}^1$, that is the u-line. Let

$$W = \operatorname{spec} R[x, y].$$

Then we have a natural map

$$f: W \to V,$$

and W gives a correspondence between V and the Fermat curve,

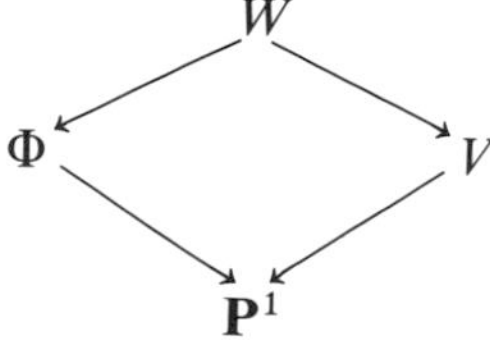

This correspondence is especially important in cases when we have additional information of an arithmetic nature about the functions u and v.

For instance, let us assume known that the Fermat curve for some $n \geqq 3$ has only a finite number of points in any subring of a number field finitely generated over $\mathbf{Z}$ (a special case of Siegel's theorem, or rather its generalization by Mahler-Lang, considerably easier to prove than the general case).

If there exist units u, v in R such that $u + v = 1$, then the affine curve V has only a finite number of points in any subring $\mathfrak{o}$ of k finitely generated over $\mathbf{Z}$.

We prove this from the corresponding property of the Fermat curve as follows. We take an integer $n \geqq 3$. For any homomorphism

$$\varphi : z \mapsto \bar{z}$$

of R into $\mathfrak{o}$, that is, any point of V in $\mathfrak{o}$, the images $\bar{u}, \bar{v}$ are units in $\mathfrak{o}$. Consequently for any extension of φ to a homomorphism of $R[x, y]$, the images $\bar{x}, \bar{y}$ are units in the integral closure of $\mathfrak{o}$. Let $E = k(\bar{x}, \bar{y})$. Then the degree

$$[k(\bar{x}, \bar{y}) : k]$$

is bounded by the generic degree $[W : V]$. Furthermore, E is unramified over k except at the primes dividing n, and the primes $\mathfrak{p}$ of k at which the elements of $\mathfrak{o}$ are not $\mathfrak{p}$-integral. This is a finite set of primes. We use an elementary lemma of algebraic number theory (for a proof, see [L 4], Chapter 17, §1).

Lemma 2.1. *Let k be a number field, let S be a finite set of primes in k, and let d be a positive integer. There is only a finite number of extensions of k of degree $\leqq d$, unramified outside S.*

We conclude that the points $(\bar{x}, \bar{y})$ can lie only in a finite number of extensions of k. Therefore, if V has infinitely many points in $\mathfrak{o}$, then the Fermat curve has infinitely many points in the integral closure of $\mathfrak{o}$ in one of these extensions, a contradiction.

In view of the preceding result, a ring with units u, v such that $u + v = 1$ should have a special name, say a **Fermat ring**.

One essential feature in the above arguments was the bound on the ramification obtained by extracting n-th roots. In the context of rational points (rather than integral points), we can also use Weil's result [We], proved jointly with Chevalley for the case of curves.

Theorem on Unramified Extensions. *Let $f : W \to V$ be an unramified covering of a projective non-singular variety V by a variety W, defined over a number field k. There exists a positive integer c having the following* *property. For any point x of V in the algebraic closure of k, the relative discriminant of $k(f^{-1}(x))$ over $k(x)$ divides c.*

In particular, for any rational point x of V in k, the extension $k(f^{-1}(x))$ ramifies only in a fixed finite set of primes of k, and its degree is bounded. Thus the lemma applies, to show that there is only a finite number of such extensions, and we conclude:

Corollary 1. *If V has infinitely many rational points in k, then W has infinitely many rational points in some finite extension of k.*

Let us say that V is **Mordellic** if it has only a finite number of rational points in any number field.

Corollary 2. *V is Mordellic if and only if W is Mordellic.*

Remark 1. In the Chevalley-Weil Theorem, if $\mathfrak{p}$ is a prime of k such that the unramified covering $f: W \to V$ has non-degenerate reduction mod $\mathfrak{p}$, to an unramified covering

$$f_{\mathfrak{p}}: W_{\mathfrak{p}} \to V_{\mathfrak{p}},$$

then for any point $x \in V_k$ the extension $k(f^{-1}(x))$ is unramified above $\mathfrak{p}$. Thus the set of primes dividing c can be determined explicitly if one knows the covering explicitly, and so can the finite number of extensions unramified outside the given finite set.

Remark 2. We recall that in any finitely generated multiplicative group of complex numbers, there exists only a finite number of elements u, v such that $u + v = 1$, cf. [L 1]. Thus the existence of such units in a natural way is always a remarkable event.

The simplest case of level 2 has especially interesting features. Let $k = \mathbf{Q}(e^{2\pi i/n})$ and again let

$$u = x^n, \qquad v = y^n,$$

so that $k(u) = k(u, v)$ is the fixed field of the function field $k(x, y)$ under the group of automorphisms

$$(x, y) \mapsto (\zeta x, \zeta' y),$$

where ζ, ζ' are n-th roots of unity. We now put

$$u = \lambda = \frac{e_2 - e_3}{e_1 - e_3},$$

where λ is the generator of the modular function field of level 2 in the theory of elliptic functions, and e_i $(i = 1, 2, 3)$ has the usual meaning, cf. [L 5], Chapter 18, §6. Let

$$\mathbf{P}^1_u = \mathbf{P}^1_\lambda$$

be the projective u-line (i.e. λ-line), so that we have rational maps

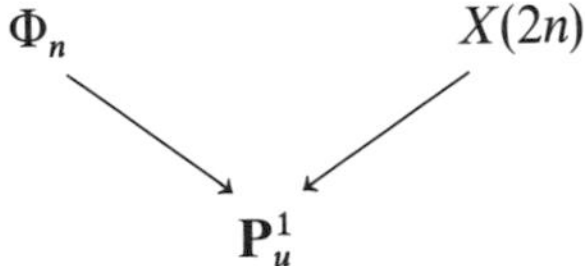

from the Fermat curve and the modular curve $X(2n)$ onto this line. Note that the points $u = 0, 1, \infty$ lie above $j = \infty$. It is obvious that the Fermat curve is ramified of order n above $u = 0, 1, \infty$, and that this is its only ramification over the projective line. It is known from the ramification theory of elliptic functions that $X(2n)$ has exactly the same ramification properties. Therefore we conclude that

the pull back of the Fermat curve Φ_n over $X(2n)$ is unramified over each of these curves.

The group belonging to $u^{1/n}$ has been considered classically, cf. [K-F], p. 658. Even though Klein and Fricke seem to use the phrase "congruence subgroup" in a sense different from the current one, it seems nevertheless that they prove that the group of $\lambda^{1/n}$ is not a congruence subgroup in the modern sense.

The Fermat conjecture here appears in a natural context ("the only rational points over $\mathbf{Q}$ should be at cusps"). Furthermore, the fact the pull backs above are unramified shows that:

The Fermat curve Φ_n is Mordellic if and only if the modular curve $X(2n)$ is Mordellic.

The proof consists of the lemma already mentioned together with the theorem on unramified extensions.

Thus we see that the diophantine properties of the Fermat curve and the modular curve affect each other. It is however not clear if ultimately the

Mordell conjecture can be proved by knowing enough about the rational points of modular curves, or if conversely, diophantine properties of the modular curve can be reduced to those of the Fermat curve as in Demyanenko and Kubert. Note that an effective solution of the Mordell conjecture implies that the Fermat conjecture over $\mathbf{Q}$ can then be verified for each n, but a proof for all n simultaneously involves additional difficulties, having to do with uniform irreducibility.

§3. Torsion Points

The modular curves parametrize elliptic curves and torsion points on them. Diophantine knowledge concerning modular curves can therefore be interpreted in terms of the above objects. We shall prove below a uniformity theorem concerning torsion points, as in [Ku 1] and [Ku 8]. We begin by discussing some conjectures. Let

$$\pi : X(N) \to \mathbf{P}^1$$

be the natural rational map of $X(N)$ onto the j-line. Assume that every subfield of F_N properly containing $\mathbf{Q}(j)$ has genus $\geqq 2$. This condition is satisfied for example when N is a sufficiently large prime. Indeed, one knows that all subgroups of $PSL_2(\mathbf{Z}/p\mathbf{Z})$ are either of Lie type, the largest being Borel, i.e. consisting of matrices

$$\begin{pmatrix} a & b \\ 0 & a^{-1} \end{pmatrix}$$

or the normalizers of a non-split Cartan subgroup, or small groups, namely A_4, S_4, A_5 (symmetric, alternating groups of the stated order). The genus of F_N and of the subfields fixed under the Borel subgroup is known, cf. for instance Shimura [Sh], and tends to infinity with N. It is clear that the genus of the fields fixed under the small groups also goes to infinity with N, by the Hurwitz genus formula.

Note that we are not asserting that all subfields of F_N for N prime tending to infinity have genus $\geqq 2$. There may exist many such subfields of genus 1, but they do not contain $\mathbf{Q}(j)$.

Assume now the Mordell conjecture for subfields of F_N properly containing $\mathbf{Q}(j)$, and of genus $\geqq 2$. Let k be a number field containing the N-th roots of unity. Assume that N is such that every subfield of kF_N properly containing $k(j)$ has genus $\geqq 2$. Then for all but a finite number of values $\bar{j}$ of j in k, we have

$$[k(\pi^{-1}(\bar{j})) : k] = [kF_N : k(j)].$$

Proof. For each value $j = \bar{j}$ in k, the point of $X(N)$ lying above $\bar{j}$ has a decomposition group, which is a subgroup H of the Galois group G of $X(N)$ over $X(1) = \mathbf{P}^1$. Suppose that for infinitely many such values of $\bar{j}$ this decomposition group is not trivial, so that we can take it equal to the same subgroup H. Its fixed field is not $k(j)$, and has genus $\geqq 2$. It has a rational point in k corresponding to such values of $\bar{j}$, contradicting the Mordell conjecture.

The above statement is an irreducibility property, and will be called the **irreducibility conjecture** for the modular curves. The hypothesis concerning the genus of intermediate fields is satisfied, for instance, when N is a sufficiently large prime number. Thus the Mordell conjecture gives a uniform bound for any degeneracy of the group of p-points on elliptic curves defined over k, with variable invariant in k. On the other hand, Serre's theorem gives a uniform bound for a fixed invariant and variable primes. One expects a theorem containing both of these aspects, i.e. uniformity with respect to pairs $(p, \bar{j})$ for all primes p and $\bar{j}$ in k, for instance a statement such as this one:

For all but a finite number of $j_0 \in k$, and all but a finite number of primes p, the Galois group of the p-primary torsion of an elliptic curve A over k with invariant j_0 is equal to $\mathrm{GL}_2(\mathbf{Z}_p)$.

Next we consider another aspect of torsion points.

The Manin-Mumford conjecture asserts that on a curve of genus $\geqq 2$ there are only a finite number of points of finite order in the Jacobian. The cusps on the modular curves provide significant examples of such points.

The question can be raised whether the cusps are also of finite order on curves which are quotients of non-congruence subgroups. This is true for the standard representation of the Fermat curve, as shown by Rohrlich [Roh], who determines completely the structure of the divisor class group generated by the cusps. On the other hand, Rohrlich has observed that the answer is negative in general. The argument goes as follows.

In [L 2], Lang reduces the Manin-Mumford conjecture to a Galois property of the field of torsion points on the Jacobian, namely that the index of the subgroup of the Galois group of the N-th torsion points over the given number field generated by the homotheties (that is, inducing multiplication by an integer prime to N on the N-th torsion points) should be bounded in $(\mathbf{Z}/N\mathbf{Z})^*$. Shimura has pointed out that this property is a simple consequence of the theory of complex multiplication, and therefore:

The Manin-Mumford conjecture is true in the case of complex multiplication.

In particular, it is true for appropriate rational images of the Fermat curve, which have complex multiplication.

By choosing infinitely many suitable non-standard correspondences of the Fermat curve with modular curves, i.e. representations as quotient of the upper half plane by non-congruence subgroups associated with units satisfying $u + v = 1$, Rohrlich shows that one would get infinitely many points on the curve of finite order in the Jacobian if the Manin-Drinfeld theorem were true in the non-congruence case, a contradiction.

For the convenience of the reader, we sketch Shimura's argument. If A is an abelian variety with complex multiplication, one may assume that it is defined over a number field k, stable under complex conjugate ρ, and such that if $\mathfrak{p}$ is a prime with good reduction and $\alpha(\mathfrak{p})$ is the endomorphism representing Frobenius at $\mathfrak{p}$, then $\alpha(\mathfrak{p}^\rho) = \alpha(\mathfrak{p})^\rho$. Then $\alpha(\mathfrak{p})\alpha(\mathfrak{p})^\rho = \mathbf{N}\mathfrak{p}$, and multiplication by $\mathbf{N}\mathfrak{p}$ on A_N (points of order N) is induced by an element of the Galois group, for N prime to $\mathfrak{p}$. But such integers $\mathbf{N}\mathfrak{p}$ in $(\mathbf{Z}/N\mathbf{Z})^*$ form a subgroup of bounded index, as desired.

We shall now deal with the uniform boundedness of torsion points of elliptic curves over a fixed number field K.

Let A be an elliptic curve defined over the number field K, in Weierstrass form

$$y^2 = 4x^3 - g_2 x - g_3.$$

Let j_A be the j-invariant of A. Let l be a positive integer. We shall say that the elliptic curve A over K is **l-deficient** if the denominator in the ideal factorization of j_A in K is the l-th power of an ideal in K.

Remark. *Suppose that l is a prime $\geqq 3$. If the order of the Galois group of the l-th torsion points of A over K is not divisible by l, then A is l-deficient.*

Proof. Let $q = q_A$ be the value corresponding to j_A on the Tate curve at a prime ideal $\mathfrak{p}$ dividing the denominator of J_A. Then A is isomorphic to the Tate curve over at most a quadratic extension E of the completion $K_\mathfrak{p}$, and

$$E(A_l) = E(\boldsymbol{\mu}_l, q_A^{1/l})$$

by the elementary theory of Tate curves, cf. [L 5], Chapter 15. Since l does not divide the order of the Galois group, it follows that some root $q_A^{1/l}$ lies in $E(\boldsymbol{\mu}_l)$. Since $\boldsymbol{\mu}_l$ has degree prime to l over $K_\mathfrak{p}$, this implies that l divides $\operatorname{ord}_\mathfrak{p} q_A$, so also divides $\operatorname{ord}_\mathfrak{p} j_A$, as desired.

It is a standard conjecture that over a number field K, the orders of K-rational torsion points of elliptic curves defined over K are uniformly bounded by a constant depending only on K. Manin has proved that for any fixed prime p, the orders of p-primary points of elliptic curves over K are uniformly bounded, cf. [Man 2]. By Manin's theorem, any uniformity theorem is reduced to one concerning sufficiently large primes.

Note that Mazur [Maz] has proved a strong version of the boundedness conjecture over the rational numbers:

Let A be an elliptic curve defined over $\mathbf{Q}$. *If t is a point of order N on A, rational over* $\mathbf{Q}$, *then the modular curve $X_1(N)$ has genus* 0, *which means that $N \leqq 10$ or $N = 12$.*

We shall prove the following weak form of the general conjecture, which is the main result of [Ku 1] and [Ku 8].

Theorem 3.1. *Let K be a number field and l a prime number $\geqq 5$. There exists a constant $C(K, l)$ such that, if N is a prime and A is an elliptic curve defined over K, which is l-deficient and has a point of order N rational over K then $N \leqq C(K, l)$.*

The techniques used in the proof are derived from ideas of Demyanenko [Dem 1] and [Dem 2] and Hellegouarch [He 1] through [He 4]. Hellegouarch showed how to associate to points on modular curves points in other algebraic varieties. Demyanenko conceived of the idea of using height arguments to prove the boundedness conjecture. Also using height arguments, Manin proved his theorem on the p-primary part for fixed p.

The rest of the section is devoted to the proof of Theorem 3.1. We shall first give a discussion concerning N-torsion points, making no assumption on A (other than it is defined over K), and on N other than it is a positive integer $\geqq 3$.

Let $t \in A_N$ be a torsion point of order N, rational over K. Any point b in the group generated by t can be written in the form $b = rt$ for some integer r. We say that b is t-**admissible** if b, $b \pm t$, $b \pm 2t$ are primitive of order N. This condition is equivalent to r, $r \pm 1$, $r \pm 2$ prime to N. We then also say that r is N-**admissible**. We let

$$u_b = \frac{x(2t) - x(t)}{x(2t) - x(rt)} \quad \text{and} \quad u'_b = \frac{x(t) - x(rt)}{x(2t) - x(rt)}.$$

Then

$$u_b + u'_b = 1.$$

We note that the two functions u_b and u'_b do not depend on the model in an isomorphism class for A.

Lemma 3.2. *Let l be a prime number $\geqq 3$, and A an elliptic curve defined over K.*

(i) *For any prime ideal* $\mathfrak{p}$ *of* K *where* j_A *is* $\mathfrak{p}$*-integral, both* u_b *and* u'_b *are* $\mathfrak{p}$*-units.*
(ii) *If* A *is* l*-deficient, and* N *is prime to* l*, then the fractional ideals* (u_b) *and* (u'_b) *are* l-th *powers in* K.

Proof. If j_A is $\mathfrak{p}$-integral, then u_b and u'_b are $\mathfrak{p}$-units since they are specializations of units over $\mathbf{Z}$ in the modular function field. Cf. Theorem 1.1, or Theorem 6.4 of Chapter 2. If j_A is not $\mathfrak{p}$-integral, then u_b, u'_b are either $\mathfrak{p}$-units, or $\sim q_A^{e-e'}$ where e, e' are rational numbers with denominator N, prime to l. This follows from the specialization of q-expansions to Tate curves, and the table of Lemma 6.2, Chapter 2, §6. Hence the orders of u_b, u'_b at $\mathfrak{p}$ are divisible by l. This proves (ii), and concludes the proof of the lemma.

We may now indicate the main line of argument in proving Theorem 3.1. Let X, Y be algebraic numbers such that

$$X^l = u_b \quad \text{and} \quad Y^l = u'_b.$$

Then (X, Y) is a point on the Fermat curve

$$X^l + Y^l = 1.$$

By Lemma 3.2, the coordinates X, Y lie in a finite extension K' of K, depending only on K and l. If one knew that the Fermat curve had only a finite number of points in K', then one would immediately conclude that there is only a finite number of N such that A has a point of order N, rational over K. Since we don't know such a property of the Fermat curve, we have to make a detour as follows.

The function

$$\varphi(X, Y) = (XY)^l$$

is a rational function on the Fermat curve. We have a corresponding (logarithmic) height function relative to K',

$$\begin{aligned} h_{\varphi,K'}(X, Y) &= \sum_{v'} \max(0, \log|\varphi(X, Y)|_{v'}) \\ &= \sum_{v'} \max(0, \log|u_b u'_b|_{v'}). \end{aligned}$$

The sum over v' is taken over all normalized absolute values of K', repeated with their local multiplicities. Cf. [L 9], Chapter 4. By the elementary relations of absolute values in a finite extension, we get

$$h_{\varphi,K'}(X, Y) = [K' : K] h_K(u_b u'_b),$$

where, for any element $\alpha \in K$, we have by definition

$$h_K(\alpha) = \sum_v \max(0, \log |\alpha|_v),$$

and the sum is taken over all normalized absolute values of K, repeated with their multiplicities.

We shall then estimate the heights of these points, which are given by the heights (relative to K) of the numbers $u_b u'_b$. On the one hand, we shall give estimates which show that for all choices of admissible r, the corresponding points b have "essentially" the same heights, which differ by a relatively slow growing function of N (essentially $N \log N$). These estimates can be summarized as follows.

Main Lemma. *Let K be a number field. There exists a constant C, depending only on K, having the following property. If A is an elliptic curve defined over K, with a point t of order N in A_K, then there is a positive number $\gamma = \gamma(A, t)$, such that for any t-admissible point b, we can write a decomposition*

$$h(u_b u'_b) = h_1(b) + h_2(b)$$

where h_1, h_2 satisfy inequalities:

$$\frac{\gamma}{4N} \leqq h_1(b) \leqq \gamma \quad \textit{and} \quad 0 \leqq h_2(b) \leqq CN \log N.$$

On the other hand, we quote the following theorem of Mumford [Mu],

Let V be a projective 1-dimensional variety of genus > 1, defined over a number field K. Let $h = h_K$ be the logarithmic height function on V. Order the K-points by increasing height, $P_0, P_1, \ldots$. There exists an integer m and a constant $C_1 > 1$ such that

$$h(P_{i+m}) > C_1 h(P_i)$$

for all positive integers i.

We may then conclude the proof of Theorem 3.1 as follows. Observe that there are at least

$$M = \frac{N-5}{2}$$

admissible values of r, whence at least that many admissible points b, which are obviously distinct. Let

$$b(1), \ldots, b(M)$$

be the family of t-admissible points b ordered by ascending height. By Mumford's theorem, there exists m depending only on K, l such that for N sufficiently large,

$$(*) \qquad h(b(M)) \geqq C_2^{M/m} h(b(m)) \geqq C_3^M h(b(m)).$$

Replacing the left hand side of $(*)$ by $h_1(b(M)) + CN \log N$ shows that we may assume without loss of generality that $h_1(b(M))$ grows exponentially with N, and in particular that

$$h_1(b(M)) \geqq 1.$$

On the other hand, the right hand side of $(*)$ is trivially $\geqq C_4^N h_1(b(0))$. Using the Main Lemma, we find

$$h_1(b(M)) + CN \log N \geqq C_5^N \frac{1}{4N} h_1(b(M)).$$

We divide both sides by $h_1(b(M))$ and multiply by $4N$. This yields a contradiction for N sufficiently large, thus proving Theorem 3.1.

The rest of this section is devoted to the proof of the Main Lemma.

Lemma 3.3. *Let $\mathfrak{p}$ be a prime ideal of K where j_A is not $\mathfrak{p}$-integral. Then either*

$$|u_b u_b'|_{\mathfrak{p}} = 1$$

for all choices of t-admissible points b, or

$$|u_b u_b'|_{\mathfrak{p}} \neq 1$$

for all choices of t-admissible points b. In the second case, we have the inequality

$$\frac{1}{N} \left|\log |j_A|_{\mathfrak{p}}\right| \leqq \left|\log |u_b u_b'|_{\mathfrak{p}}\right| \leqq \left|\log |j_A|_{\mathfrak{p}}\right|.$$

Proof. This follows from the two cases in Theorem 1.1, by specializing the generic units to the Tate curve. The algebraic numbers u_b and u'_b have q_A-expansions which correspond to the q-expansions of the generic units, and so the assertion of the lemma is clear.

We now turn to the archimedean absolute values. As in Chapter 2, §6, if A is isomorphic to the complex torus $\mathbf{C}/[\tau, 1]$, we also write $j_A = j(\tau)$. We represent t and b by

$$t = t_1\tau + t_2 \quad \text{and} \quad b = b_1\tau + b_2 \equiv rt \bmod [\tau, 1],$$

where

$$0 \leqq t_1, b_1 < 1,$$

and $t_1, t_2, b_1, b_2 \in (1/N)\mathbf{Z}$. We recall the q-expansion for the Klein forms from Chapter 2:

$$\mathfrak{k}(a) = -2\pi i q_\tau^{(1/2)a_1(a_1-1)} e^{2\pi i a_2(a_1-1)/2}(1-q_a)\prod_{n=1}^{\infty}(1-q_\tau^n q_a)(1-q_\tau^n/q_a)(1-q_\tau^n)^{-2}.$$

Since for $m \in \mathbf{Z}\tau + \mathbf{Z}$ and $z \in \dfrac{1}{N}(\mathbf{Z}\tau + \mathbf{Z})$, $z \notin \mathbf{Z}\tau + \mathbf{Z}$ we have

$$\mathfrak{k}(-z) = -\mathfrak{k}(z) \quad \text{and} \quad \mathfrak{k}(z+m) = \varepsilon\mathfrak{k}(z)$$

where ε is a root of unity, to compute $\log|\mathfrak{k}(z)|$ we may assume without loss of generality that z is normalized so that

$$0 \leqq z_1 \leqq \tfrac{1}{2} \quad \text{and} \quad 0 \leqq z_2 \leqq 1.$$

Lemma 3.4. *For z normalized as above and τ in the fundamental domain for* $\mathrm{SL}_2(\mathbf{Z})$, *we have*:

(i) $\log|\mathfrak{k}(z)| = 2\pi\,\tfrac{1}{2}z_1(z_1-1)\,\mathrm{Im}\,\tau + \log|1-q_z| + \psi(z)$

where $\psi(z)$ is a function such that $|\psi(z)| \leqq C_6$ for some absolute constant C_6.

(ii) *If $z_1 \neq 0$ then $1/2N \leqq |z_1(z_1-1)| \leqq 1$.*

(iii) $|\log|1-q_z|| \leqq C_7 \log N$ *for some absolute constant C_7.*

Proof. The term

$$\log\left|\prod (1-q^n q_z)(1-q^n/q_z)\right|$$

is bounded in absolute value because this function is continuous on the region of points (z, τ) with τ in the fundamental domain, $z \in [0, \tfrac{1}{2}] \times [0, 1]$, and tends to 0 uniformly as $\tau \to i\infty$. This makes the first assertion clear.

The second assertion is trivial. As to the third, the only way that $|\log|1 - q_z||$ can be large is that $|1 - q_z|$ is close to zero. If $\langle z_1 \rangle \neq 0$ then $|q_z| < 1$, and

$$|1 - q_z| \geqq 1 - |q_z|.$$

Then $|q_z|$ is maximized for $\tau = e^{2\pi i/6}$ and $z_1 = 1/N$, in which case $|q_z| = e^{-\pi\sqrt{3}/N}$. Then

$$|1 - q_z| \geqq 1 - e^{-\pi\sqrt{3}/N},$$

and the desired inequality follows at once. If $z_1 = 0$, then $|1 - q_z|$ is minimized for $z_2 = \pm 1/N$, in which case

$$|1 - q_z| = 2\sin(\pi/N) \geqq \pi/N.$$

This proves the lemma.

The lemma is applied when $z = a, b, a + b\ a - b$ for an admissible pair a, b of the form

$$a = t \text{ or } 2t, \qquad b = rt.$$

Let us write

$$-\log|u_b u_b'|_v = v_1(b) + v_2(b) + v_3(b),$$

where $v_1(b)$ is the contribution from the rational power of q in the q-product, $v_2(b)$ is the contribution from the terms of the form $(1 - q_z)$, and $v_3(b)$ is the contribution from the product

$$\prod_{n=1}^{\infty} (1 - q^n q_z)(1 - q^n/q_z).$$

Note that in the expression of $u_b u_b'$ in terms of the Klein forms, the product of type $\prod (1 - q^n)^{-2}$ cancels out. From Lemma 3.4 we obtain:

Lemma 3.5.

(i) *If* $\langle t_1 \rangle = 0$, *then* $v_1(b) = 0$. *If* $\langle t_1 \rangle \neq 0$, *then*

$$\frac{2\pi}{N} \operatorname{Im} \tau \leqq |v_1(b)| = 2\pi(\operatorname{Im} \tau)|\operatorname{ord}_q(uu')| \leqq 2\pi \operatorname{Im} \tau.$$

(ii) $|v_2(b)| \leqq C_8 \log N$
(iii) $|v_3(b)| \leqq C_9$.

Let $\alpha \in K$, $\alpha \neq 0$. By the product formula, we have

$$\sum_v \max(0, \log |\alpha|_v) = \sum_v - \min(0, \log |\alpha|_v).$$

Hence

$$2h_K(\alpha) = \sum_v \left|\log |\alpha|_v\right|.$$

We apply this to the case when $\alpha = u_b u_b'$ for an admissible point b. We then write

$$h(b) \quad \text{instead of} \quad h_K(u_b u_b')$$

for simplicity.

Let S_0 be the family of nonarchimedean absolute values v of K such that $|j_A|_v > 1$.

Let S_∞ be the family of archimedean absolute values of K such that, in the corresponding complex embedding, we have $t_1 \neq 0$ and

$$2N(C_8 \log N + C_9) \leqq 2\pi \operatorname{Im} \tau.$$

Let

$$S = S_0 \cup S_\infty.$$

Then S depends on A, N and t. Let us write

$$h(b) = h_1(b) + h_2(b),$$

where

$$h_1(b) = \frac{1}{2} \sum_{v \in S} \left|\log |u_b u_b'|_v\right|$$

$$h_2(b) = \frac{1}{2} \sum_{v \notin S} \left|\log |u_b u_b'|_v\right|.$$

We shall estimate $h_2(b)$ and $h_1(b)$ successively to prove the Main Lemma.

If $v \notin S$, then by Lemma 3.2(i) we may suppose that v is archimedean. In that case,

$$\text{either} \quad t_1 = 0 \quad \text{or} \quad 2\pi \operatorname{Im} \tau < 2N(C_9 + C_8 \log N).$$

Hence for $v \notin S$, we get from Lemma 3.4,

$$\left|\log |u_b u_b'|_v\right| \leqq C_{10} N \log N,$$

and therefore

$$0 \leqq h_2(b) \leqq [K:\mathbf{Q}] C_{10} N \log N = C_{11} N \log N.$$

To get an estimate for $v \in S$, define

$$\boxed{\gamma = \frac{1}{2} \sum_{v \in S_0} \left|\log |j_A|_v\right| + \frac{1}{2} \sum_{v \in S_\infty} 4\pi \operatorname{Im} \tau.}$$

Then from Lemma 3.3 and Lemma 3.5 we get for $v \in S$:

$$\left|\log |u_b u_b'|_v\right| \leqq 4\pi \operatorname{Im} \tau$$

and

$$\left|\log |u_b u_b'|_v\right| \geqq |v_1(b)| - |v_2(b)| - |v_3(b)| \geqq \frac{\pi}{N} \operatorname{Im} \tau.$$

Therefore

$$\frac{\gamma}{4N} \leqq h_1(b) \leqq \gamma,$$

which proves the Main Lemma, and also concludes the proof of Theorem 3.1.

Remark. The function $\gamma(A, t)$ is very natural. Indeed, suppose an elliptic curve A over K has integral j-invariant at some prime $\mathfrak{p}$ of K. Then A has essentially good reduction at $\mathfrak{p}$. Suppose that A actually has good reduction, and has a point t of order N prime to $\mathfrak{p}$ rational over K. Then t specializes mod $\mathfrak{p}$ to a point of order N over the residue class field, which therefore has to have enough elements to accommodate such a point. This means that roughly, the presence of a point of order N forces the primes "up to N" to divide the denominator of the j-invariant j_A. This is precisely the contribution to the finite part of the sum expressing γ.

Corollary. *Let K be a number field and l a prime number $\geqq 5$. There exists a constant $C(K, l)$ such that if A is an elliptic curve defined over K with a rational l^2-torsion point, then $|A_{\text{tor}}(K)| \leqq C(K, l)$.*

Proof. Immediate from the theorem by taking $A/(lP)$, where P is a rational point of order l^2.

APPLICATION TO COMPLEX MULTIPLICATION

Chapters 9 through 13 will be concerned with the specialization of the generic modular units to the case of complex multiplication. The more recent literature started with the Siegel Tata Institute Lecture Notes [Si], and continued with the papers of Ramachandra [Ra] and Robert [Ro].

We assume that the reader is acquainted with the complex multiplication of elliptic curves, for which we refer to Shimura's book [Sh 1] Chapter 5, or [L 5], especially the last part.

Notation

K is an imaginary quadratic field. In formulas for the number of roots of unity, discriminant, class number, etc., when we omit the subscript it is understood that the corresponding letter denotes the object associated with K. Thus:

$w_K = w =$ number of roots of unity in K.
$d_K = d =$ absolute value of the discriminant of K.
$h_K = h =$ class number of K.
$\mathfrak{o}_K = \mathfrak{o} =$ ring of algebraic integers in K.
$\mathrm{Cl}(1) =$ absolute ideal class group of K.
$K(1) =$ Hilbert class field of K.
$\mathrm{Cl}(\mathfrak{f}) =$ ray class group of conductor $\mathfrak{f}$.
$K(\mathfrak{f}) = K_{\mathfrak{f}} =$ ray class field of conductor $\mathfrak{f}$.
$w(\mathfrak{f}) =$ number of roots of unity in K which are $\equiv 1 \bmod \mathfrak{f}$.
$N(\mathfrak{f}) =$ smallest positive integer in the ideal $\mathfrak{f}$.
$C_0 =$ unit class in $\mathrm{Cl}(\mathfrak{f})$.

If $C \in \mathrm{Cl}(\mathfrak{f})$ and $\mathfrak{a}$ is an ideal prime to $\mathfrak{f}$, we denote by $C_{\mathfrak{a}}$ or $C(\mathfrak{a})$ the class containing $\mathfrak{a}$, and we let

$$C\mathfrak{a} = CC_{\mathfrak{a}}.$$

$\mathbf{N}\mathfrak{a} =$ absolute norm of $\mathfrak{a}$ for any ideal $\mathfrak{a}$.
$\mathbf{N}_{F'/F} =$ relative norm from an extension F' over F.
$\mathbf{N}_{\mathfrak{f}',\mathfrak{f}} =$ relative norm from $K(\mathfrak{f}')$ to $K(\mathfrak{f})$ if $\mathfrak{f} \mid \mathfrak{f}'$.
$H =$ abelian extension of K, of conductor $\mathfrak{f}$.
$w(H) = w_H =$ number of roots of unity in H.

$\mathrm{Cl}(H/K) =$ factor group of $\mathrm{Cl}(\mathfrak{f})$ by the subgroup corresponding to $\mathrm{Gal}(K(\mathfrak{f})/H)$ under the reciprocity law mapping. Then we have an isomorphism

$$\mathrm{Cl}(H/K) \approx \mathrm{Gal}(H/K).$$

CHAPTER 9

Unramified Units

In this chapter, we deal with the units in unramified extensions of an imaginary quadratic field. We define a group of units by means of values of the delta function, and determine the index in the group of all units. We follow Robert, and also give a refinement of the group of units leading to the precise best possible index due to Kersey.

The proofs of this chapter need no special elaborate machinery. This is the reason we have started with it. However, after the reader has read the theory of ramified units, and understood the role of distribution relations (relating the units at one level with the units at higher levels) then it will be clear that we could have obtained the results of this chapter as corollaries of the theory in the ramified case, using the Klein forms directly. But this more extensive theory requires considerably more knowledge about the modular forms in the generic case (for instance the quadratic relations and the transformation law of the Klein forms). For ease of reference and for those who are interested in more partial results, it should therefore be useful to have available the exposition that we have chosen here.

§1. The Invariants $\delta(c, c')$

Let K be an imaginary quadratic field, and $\mathrm{Cl}(1)$ the absolute ideal class group of K. Let $c, c' \in \mathrm{Cl}(1)$, and let $\mathfrak{a}, \mathfrak{a}'$ be ideals in c^{-1}, c'^{-1} respectively. Let $\alpha, \alpha' \in \mathfrak{o} = \mathfrak{o}_K$ be such that

$$\mathfrak{a}^h = (\alpha) \quad \text{and} \quad \mathfrak{a}'^h = (\alpha').$$

We define the **Siegel unit**

$$\delta(c, c') = \frac{\alpha^{12}\Delta(\mathfrak{a})^h}{\alpha'^{12}\Delta(\mathfrak{a}')^h}$$

and

$$g_{(1)}(c) = \delta(c) = \delta(c, c_0) = \frac{\alpha^{12}\Delta(\mathfrak{a})^h}{\Delta(\mathfrak{o})^h}$$

where c_0 is the unit class. Then $\delta(c, c')$ depends only on the classes c, c'. Note that $\delta(c_0) = 1$.

Lemma 1.1. *$\delta(c, c')$ is a unit.*

Proof. The ideal $\mathfrak{a}$ is linearly equivalent to a prime ideal $\mathfrak{p}$ above a prime number p which splits completely. The assertion is then immediate from the factorization of $p^{12}\Delta(\mathfrak{p})/\Delta(\mathfrak{o}) \sim \mathfrak{p}'^{12}$ which is standard (e.g. Theorem 5 of Chapter 12, §2, [L 5]).

Lemma 1.2. *For any ideals $\mathfrak{a}$, $\mathfrak{b}$ the quotient $\Delta(\mathfrak{a})/\Delta(\mathfrak{b})$ is a w-power in the Hilbert class field $K(1)$.*

Proof. A proof using the Shimura reciprocity law directly seems to be messy. Hence this lemma will be proved using higher level formulas as a corollary of Theorem 1.4, Chapter 11, following Robert's idea.

The transformation laws with respect to the Frobenius-Artin automorphism will be used as being standard, cf. for instance [L 5], Chapter 11, §2 and Chapter 12, §1. It is then immediate that we have the cocycle relations

$$\delta(\bar{b})\delta(c)^{\sigma(b)} = \delta(bc) \quad \text{and} \quad \delta(c, c')^{\sigma(b)} = \delta(c\bar{b}, c'\bar{b}).$$

In particular, $\delta(c)^{\sigma(b)}$ is a quotient of two units of the same type, so the group generated by the Siegel units is invariant by the Galois group over K.

Let H be an abelian unramified extension of K. We define $\mathrm{Cl}(H/K)$ to be the factor group of $\mathrm{Cl}(1)$ isomorphic to $\mathrm{Gal}(H/K)$ under the reciprocity law mapping. We define $\mathrm{Cl}(H(1)/H)$ to be the subgroup of $\mathrm{Cl}(1)$ isomorphic to $\mathrm{Gal}(K(1)/H)$. Thus

$$\mathrm{Cl}(H/K) = \mathrm{Cl}(1)/\mathrm{Cl}(K(1)/H).$$

For $c \in \mathrm{Cl}(H/K)$ we define

$$\delta_H(c) = \prod_{c'} \delta(cc', c'),$$

where c' ranges over the elements of $\mathrm{Cl}(K(1)/H)$.

Lemma 1.3. *For any class $C \in \mathrm{Cl}(1)$ above c, we have*

$$\mathrm{N}_{K(1)/H}\delta(C) = \delta_H(c).$$

Proof. Clear.

§2. The Index of the Siegel Group

Let Δ_H be the group of units generated by the units $\delta_H(c)$, with $c \in \mathrm{Cl}(H/K)$. We let $w = w_K$, $h = h_K$, and

$$v = [H:K] = r_2(H).$$

Theorem 2.1. *The units $\delta_H(c)$ for $c \neq c_0$ are independent, so the group Δ_H has finite index in E_H, given by*

$$(E_H : \Delta_H) = 12^{v-1}h^{v-2}w^v h_H.$$

Proof. By the Kronecker limit formula, if χ is a non-trivial character of $\mathrm{Cl}(H/K)$ we have

$$L(1, \chi) = \frac{-2\pi}{6w\sqrt{d}h} \sum_{c \in \mathrm{Cl}(H/K)} \bar{\chi}(c) \log|\delta_H(c)|.$$

Furthermore, if ρ denotes the residue of the zeta function, then

$$\rho_H = \rho_K \prod_{\chi \neq 1} L(1, \chi)$$

so

(1) $$\frac{(2\pi)^v h_H R_H}{w_H\sqrt{d_H}} = \frac{2\pi h}{w\sqrt{d}} \prod_{\chi \neq 1} \frac{-2\pi}{6w\sqrt{d}h} \sum_c \bar{\chi}(c) \log|\delta_H(c)|.$$

Since H is unramified over K, it follows from the multiplicativity of the different in towers that the $\sqrt{d_H}$ on the left hand side cancels the power of $\sqrt{d}$ on the right hand side.

By the Dedekind determinant formula, we have:

$$\begin{aligned}\prod_{\chi \neq 1} \sum_{c} \bar{\chi}(c) \log \delta_H(c) &= \det_{a,b \neq c_0} \log|\delta_H(ab^{-1})/\delta_H(a)| \\ &= \det_{a,b \neq c_0} \log|\delta_H(b^{-1})^{\sigma(a)}| \\ &= \frac{1}{2^{\nu-1}} \operatorname{Reg} \delta_H(c) \quad \text{for } c \neq c_0. \end{aligned} \tag{2}$$

On the other hand,

$$(E_H : \Delta_H) = w_H 2^{\nu-1}[\operatorname{Reg} \delta_H(c), c \neq c_0]/R_H. \tag{3}$$

Putting (1), (2), (3) together concludes the proof of the theorem.

§3. The Robert Group

Let H be an abelian unramified extension of K. We let

$$\Delta_H(c_0) = \{\prod \delta_H(c)^{n(c)} \text{ such that } \prod c^{n(c)} = c_0\}$$

where the products are taken over all elements c in $\mathrm{Cl}(H/K)$.

Lemma 3.1. *Let C_0 be the unit class of* Cl(1). *Then*

$$\Delta_H(c_0) = \mathbf{N}_{K(1)/H} \Delta_{K(1)}(C_0).$$

Proof. The inclusion $\supset$ is immediate. Conversely, let $u \in \Delta_H(c_0)$. For each c pick a class C_c in Cl(1) lying above c. Then

$$u = \mathbf{N}_{K(1)/H} \prod_{c} \delta_{K(1)}(C_c)^{n(c)}.$$

By hypothesis, we know that

$$\prod_{c \neq c_0} C_c^{n(c)} = C' \in \mathrm{Cl}(K(1)/H).$$

We could have picked $n(c_0)$ to have any value since $\delta_H(c_0) = 1$. Pick $n(c_0) = -1$ and $C_{c_0} = C'$. This proves the desired inclusion.

Lemma 3.2. *Every element of $\Delta_H(c_0)$ is a wh-power in H.*

Proof. Consider first the case $H = K(1)$. Let $\mathfrak{a}_0, \ldots, \mathfrak{a}_{h-1}$ with $\mathfrak{a}_0 = (1)$ be ideals of K representing the distinct ideal classes. Let $\mathcal{N} = \{n_i\}$ be a family of integers such that

$$\prod \mathfrak{a}_i^{n_i} = (\alpha)$$

is principal. Define

$$u(\mathcal{N}) = \alpha^{12} \prod_{i=1}^{h} \left(\frac{\Delta(\mathfrak{a}_i)}{\Delta(\mathfrak{o})}\right)^{n_i}.$$

Then $u(\mathcal{N})$ is a unit. Indeed, it suffices to show that u^h is a unit, and this follows at once from Lemma 1.1.

Let $R_{K(1)}^w$ be the group generated by all units $u(\mathcal{N})$ for all families $\mathcal{N}$ as above. Since each factor $\Delta(\mathfrak{a}_i)/\Delta(\mathfrak{o})$ is a w-power in $K(1)$ by Lemma 1.2, we let $R_{K(1)}$ be the w-th root of this group in $K(1)$. The next lemma is then clear.

Lemma 3.3. $R_{K(1)}^{wh} = \Delta_{K(1)}(C_0)$.

This proves Lemma 3.2 when $H = K(1)$. The general case follows from Lemma 3.1.

We define

$$\Delta_H(c_0)^{(1/wh)}$$

to be the maximal subgroup of E_H whose wh-power is equal to $\Delta_H(c_0)$.

It is then clear that $\Delta_H(c_0)^{(1/wh)}$ is a group of the same rank as $\Delta_H(c_0)$. Furthermore, Δ_H is generated by the independent units

$$\delta_H(c_i), \qquad i = 1, \ldots, v - 1.$$

We have a homomorphism

$$\Delta_H \to \mathrm{Cl}(H/K)$$

given by

$$\prod \delta_H(c_i)^{n_i} \mapsto \prod c_i^{n_i},$$

which is surjective. Consequently $\Delta_H(c_0)$ is the kernel, and

$$(\Delta_H : \Delta_H(c_0)) = v = [H : K].$$

Theorem 3.4. $(E_H : \boldsymbol{\mu}_H \Delta_H(c_0)^{(1/wh)}) = 12^{v-1} \frac{w}{w_H} \frac{h_H}{[K(1) : H]}$.

Proof. We have the following inclusions of subgroups, whose indices have already been computed and are indicated.

$$E_H \xrightarrow{12^{v-1}h^{v-2}w^v h_H} \Delta_H \xrightarrow{v} \Delta_H(c_0)$$

Since $\Delta_H(c_0)$ is a free group on $v - 1$ generators, it follows that the root group $\Delta_H(c_0)^{(1/wh)}$ is a direct product of a free group and a torsion group. Therefore

$$(\boldsymbol{\mu}_H \Delta_H(c_0)^{(1/wh)} : \Delta_H(c_0)) = (wh)^{v-1} w_H.$$

The theorem follows at once.

For a refinement of the above index, cf. Theorem 5.1.

§4. Lemmas on Roots of Unity

This section gives some lemmas on roots of unity which will be used constantly in the sequel.

Let m be a positive integer, and let n be a multiple of m. If $\mathfrak{a}$ is an ideal in K prime to n, then the Artin automorphism $\sigma(\mathfrak{a})$ on $K(\boldsymbol{\mu}_m)$ operates as

$$\zeta^{\sigma(\mathfrak{a})} = \zeta^{\mathbf{N}\mathfrak{a}}, \quad \text{for } \zeta \in \boldsymbol{\mu}_m.$$

From this and class field theory, we obtain the following two results concerning an abelian extension H of K.

Lemma 4.1. *Let $C \in \mathrm{Cl}(H/K)$. For all ideals $\mathfrak{a} \in C$ prime to w_H, the value*

$$\mathbf{N}\mathfrak{a} \bmod w_H$$

is the same.

Lemma 4.2. *The* **Z***-ideal* $\mathbf{Z}w_H$ *is the ideal generated by all values* $\mathbf{N}\mathfrak{a} - 1$, *where $\mathfrak{a}$ ranges over all ideals in the unit class C_0 of* $\mathrm{Cl}(H/K)$, *prime to any given integer n, and prime to w_H.*

As in Robert, one deduces from Lemma 4.1:

Lemma 4.3.

(i) *Let* $\mathfrak{f}$ *be the conductor of* H *over* K, *and* N *the smallest positive integer in* $\mathfrak{f}$. *Then* w_H *divides* $12N$.
(ii) *If* H *is the Hilbert class field of* K, *then* w_H *divides* 12. *If* $w_H \neq 2$ *then* d_K *is divisible by* 2 *or* 3.

Proof. For simplicity, suppose first that H is the Hilbert class field. If p is a prime number $\neq 2, 3$ and $\boldsymbol{\mu}_p \subset H$, then $\phi(p) = p - 1 > 2$, and the p-th roots of unity would generate a totally ramified extension of K, which is impossible. Now suppose $p = 2$ or 3. Essentially for the same reason as above (ramification), we see that $\boldsymbol{\mu}_8$ or $\boldsymbol{\mu}_9$ cannot be contained in H, so this leaves only $\boldsymbol{\mu}_4$ or $\boldsymbol{\mu}_3$, which are possible and do occur in the obvious cases, when d_K is divisible by 2 or 3.

Now let us give the proof in general for (i). It suffices to prove the lemma when $H = K(\mathfrak{f})$, or even when $H = K_{(N)}$, that is when $\mathfrak{f} = (N)$. Let p^a be the largest power of p such that $\boldsymbol{\mu}_{p^a} \subset K_{(N)}$. For any integer $\alpha \in \mathbf{Z}$ such that

$$\alpha \equiv 1 \bmod N,$$

we have $\mathbf{N}\alpha = \alpha^2$ and so $\alpha^2 \equiv 1 \bmod p^a$ by the action of the Artin automorphism. If $p \neq 2, 3$ this implies immediately that $p^a | N$, and proves the assertion in this case.

Suppose $p = 3$. If $3 \nmid N$ then $\boldsymbol{\mu}_9$ cannot be contained in $K_{(N)}$ and so we must have $a = 0$ or 1. If $3 | N$, then from the 3-adic expansion of α we see that $p^a | N$. In either case we see that $p^a | 3N$.

Suppose $p = 2$. A similar analysis shows that $p^a | 4N$, as desired.

We shall also need:

Lemma 4.4. *Suppose* $\mathfrak{f}$ *is prime to* $w_{K(1)}$, *and let* $\mathfrak{a}$, $\mathfrak{b}$ *lie in the same class in* $\mathrm{Cl}(\mathfrak{f})$. *Thus* $\mathfrak{a}$, $\mathfrak{b}$ *are prime to* $\mathfrak{f}$, *but we do not assume* $\mathfrak{a}$, $\mathfrak{b}$ *prime to* $w_{K(1)}$. *Then*

$$\mathbf{N}\mathfrak{a} \equiv \mathbf{N}\mathfrak{b} \bmod w_{K(\mathfrak{f})}/w_{K(1)}.$$

Proof. Since $\mathfrak{f}$ is prime to $w_{K(1)}$, then $w_{K(\mathfrak{f})}/w_{K(1)}$ is also prime to $w_{K(1)}$. Let ζ be a $w_{K(\mathfrak{f})}/w_{K(1)}$-th root of unity. Then

$$1 = \zeta^{\sigma(\mathfrak{a}) - \sigma(\mathfrak{b})} = \zeta^{\mathbf{N}\mathfrak{a} - \mathbf{N}\mathfrak{b}},$$

so $\mathbf{N}\mathfrak{a} \equiv \mathbf{N}\mathfrak{b} \bmod w_{K(\mathfrak{f})}/w_{K(1)}$.

§5. A Refined Index

We shall now refine the index of Theorem 3.4. This section is due to Kersey. We let H be an abelian unramified extension of K. We have to define a slight variation of the group $\Delta_H(c_0)$ so that we can take appropriate roots in H, thereby eliminating extraneous factors in the index formula.

First we note that if K has no non-trivial unramified extensions, then $H = K$ and $w = w_H$, while $12^{\nu-1} = 1$, so the previous index formula is satisfactory.

Suppose that $K \neq \mathbf{Q}(\sqrt{-1}), \mathbf{Q}(\sqrt{-3})$, so $w = 2$. An ideal $\mathfrak{a}$ will be said to be **H-admissible** if it is prime to w_H, and if it satisfies the following conditions:

$$\begin{cases} \text{If } 4 \nmid w_H, & \text{then } \mathbf{N}\mathfrak{a} \equiv 1 \bmod 4. \\ \text{If } 4 \mid w_H, & \text{then } \mathbf{N}\mathfrak{a} \equiv 1 \text{ or } 3 \bmod 8. \end{cases}$$

Such an ideal exists in any given class c, for the following reasons. Suppose $4 \nmid w_H$. Then $i \notin H$, and we have a diagram of fields:

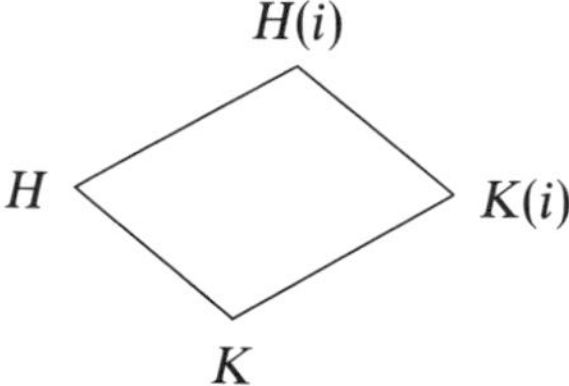

with $H \cap K(i) = K$. Given any automorphism σ of H over K, we can lift it to $H(i)$ so that it is the identity on i. An ideal $\mathfrak{a}$ in K such that $\sigma = \sigma(\mathfrak{a})$ on $H(i)$ will satisfy the desired condition.

If on the other hand $4 \mid w(H)$, let ζ be a primitive 8-th root of unity. Then $\zeta \notin H$, and $H \cap K(\zeta) = K(i)$. Any automorphism $\sigma \in \mathrm{Gal}(H/K)$ can be lifted to $H(\zeta)$ in such a way that $\sigma\zeta = \zeta$ or ζ^3. Then again an ideal $\mathfrak{a}$ such that $\sigma = \sigma(\mathfrak{a})$ satisfies the desired condition.

If $\mathfrak{a}_c$ is an H-admissible ideal in the class c, then the value

$$\mathbf{N}\mathfrak{a}_c \pmod{ww_H}$$

depends only on c, not on the choice of $\mathfrak{a}_c$. Indeed, $w = 2$ and the conditions of H-admissibility are precisely those needed to determine the value of $\mathbf{N}\mathfrak{a}_c$ mod this extra power of 2.

We **define**:

$\Delta_H(c_0, ww_H) =$ group of all elements

$$\prod \delta_H(c)^{n(c)}$$

satisfying the conditions

$$\prod c^{n(c)} = c_0 \quad \text{and} \quad \sum n(c)(\mathbf{N}\mathfrak{a}_c - 1) \equiv 0 \bmod ww_H,$$

where $\mathfrak{a}_c$ is H-admissible. It is clear that this group is stable under the action of $\mathrm{Gal}(K(1)/K)$.

If $K = \mathbf{Q}(\sqrt{-1})$ or $\mathbf{Q}(\sqrt{-3})$, then we **define**

$$\Delta_H(c_0, ww_H) = 1.$$

It will be convenient to give a variation of the definition of $\Delta(c_0, ww_H)$ in the non-exceptional cases as follows.

Let $\mathbf{I}$ be the free abelian group on the (non-zero) ideals of K. An element of $\mathbf{I}$ can be written $\mathbf{a} = \sum n(\mathfrak{a})\mathfrak{a}$. As usual, we define the **degree** $\deg \mathbf{a} = \sum n(\mathfrak{a})$. We let:

$\mathbf{I}_0$ = subgroup of elements $\mathbf{a} \in \mathbf{I}$ satisfying

(1) $$\deg \mathbf{a} = 0.$$

$\mathbf{I}[\mathfrak{o}]$ = subgroup of $\mathbf{I}$ consisting of those $\mathbf{a}$ such that

(2) $$\prod \mathfrak{a}^{n(\mathfrak{a})} = \mathfrak{o}.$$

$\mathbf{I}_0[\mathfrak{o}]$ = subgroup of $\mathbf{I}[\mathfrak{o}]$ consisting of the elements of degree 0.

$\mathbf{I}_0(\mathfrak{o}, ww_H)$ = subgroup of $\mathbf{I}_0[\mathfrak{o}]$ consisting of those elements satisfying the condition

(3) $$\sum_{\mathfrak{a}} n(\mathfrak{a})(\mathbf{N}\mathfrak{a}' - 1) \equiv 0 \bmod ww_H,$$

where $\mathfrak{a}'$ is an H-admissible ideal in the same class as $\mathfrak{a}$ in $\mathrm{Cl}(H/K)$.

We have a surjective homomorphism

$$\mathbf{I}_0[\mathfrak{o}] \to \Delta_H(c_0)$$

given by

$$\mathbf{a} \mapsto \prod \delta_H(c(\mathfrak{a}))^{n(\mathfrak{a})}.$$

Under this homomorphism the group $\mathbf{I}_0(\mathfrak{o}, ww_H)$ maps onto $\Delta(c_0, ww_H)$. Therefore we may also define $\Delta(c_0, ww_H)$ as the group of all elements

$$\prod \delta_H(c(\mathfrak{a}))^{n(\mathfrak{a})}$$

where $\mathbf{a} = \sum n(\mathfrak{a})\mathfrak{a}$ satisfies conditions (1), (2), (3) defining $\mathbf{I}_0(\mathfrak{o}, ww_H)$.

The next theorem gives us the index of a group of units devoid of all undesirable factors.

Theorem 5.1. *Let H be an unramified abelian extension of K. Then every element of $\Delta_H(c_0, ww_H)$ is a $12wh$-power in H, and we have the index*

$$(E_H : \Delta_H(c_0, ww_H)^{1/12wh}) = \frac{h_H}{[K(1):H]},$$

where $\Delta_H(c_0, ww_H)^{1/12wh}$ is the group of all elements in H whose $12wh$-power lies in $\Delta_H(c_0, ww_H)$.

Remark. We have $\boldsymbol{\mu}_H \subset \Delta_H(c_0, ww_H)^{1/12wh}$ because $w_H | 12$ by Lemma 4.3. The theorem is obvious by definitions if $K = \mathbf{Q}(\sqrt{-1})$ or $\mathbf{Q}(\sqrt{-3})$, so we assume that this is not the case.

The proof occupies the rest of this section.

First, from the alternative definition of $\Delta(c_0, ww_H)$ we have an isomorphism

$$\mathbf{I}_0[\mathfrak{o}]/\mathbf{I}_0(\mathfrak{o}, ww_H) \approx \Delta_H(c_0)/\Delta_H(c_0, ww_H).$$

Lemma 5.2. $(\mathbf{I}_0[\mathfrak{o}] : \mathbf{I}_0(\mathfrak{o}, ww_H)) = w_H/w = (\Delta_H(c_0) : \Delta_H(c_0, ww_H))$.

Proof. We may define a homomorphism

$$\mathbf{I}_0[\mathfrak{o}] \to \mathbf{Z}(ww_H)$$

by

$$\sum n(\mathfrak{a})\mathfrak{a} \mapsto \sum n(\mathfrak{a})(\mathbf{N}\mathfrak{a}' - 1) \bmod ww_H,$$

and it obviously has kernel $\mathbf{I}_0(\mathfrak{o}, ww_H)$. We show first that the image is contained in $w^2\mathbf{Z}/w_H\mathbf{Z}$.

If $4 \nmid w_H$, then the condition of H-admissibility guarantees that the image is contained in $w^2\mathbf{Z}/ww_H\mathbf{Z}$.

If $4 | w_H$, then $i \in H$, and

$$i = i^\sigma = i^s, \quad \text{where } \sigma = \sigma\left(\prod c^{n(c)}\right) \quad \text{and} \quad s = \prod \mathbf{N}\mathfrak{a}_c^{n(c)},$$

so

$$0 \equiv \prod \mathbf{N}\mathfrak{a}_c^{n(c)} - 1 \equiv \sum n(c)(\mathbf{N}\mathfrak{a}_c - 1) \bmod 4.$$

Thus in either case, the image is contained in $w^2\mathbf{Z}/ww_H\mathbf{Z}$.

Now let p be a prime of $\mathbf{Q}$ which splits completely in K, say $(p) = \mathfrak{p}\bar{\mathfrak{p}}$, and such that:

if $4 \nmid w_H$, then $p \equiv 5 \bmod 12$;
if $4 | w_H$, then $p \equiv 11 \bmod 24$.

(These conditions can be satisfied, provided some prime other than 2, 3 ramifies in K. We have eliminated $\mathbf{Q}(\sqrt{-1})$ and $\mathbf{Q}(\sqrt{-3})$ from consideration. If $K = \mathbf{Q}(\sqrt{-2})$, we take $p = 17$. If $K = \mathbf{Q}(\sqrt{-6})$, then $p = 5$ will do.)

Let

$$L(\mathfrak{p}, \mathfrak{p}) = 2(\mathfrak{p}) - (\mathfrak{p}^2) - (\mathfrak{o}) \in \mathbf{I}_0[\mathfrak{o}].$$

Its image is

$$(\mathbf{N}\mathfrak{p}^2 - 1) - 2(\mathbf{N}\mathfrak{p} - 1) = (\mathbf{N}\mathfrak{p} - 1)^2 \bmod ww_H.$$

By our own choice of $\mathfrak{p}$, we see that $(\mathbf{N}\mathfrak{p} - 1)^2/w^2$ is prime to w_H/w, and so $(\mathbf{N}\mathfrak{p} - 1)^2$ generates $w^2\mathbf{Z}/ww_H\mathbf{Z}$. This completes the proof of Lemma 5.2.

The next lemma describes generators for $\mathbf{I}_0(\mathfrak{o}, ww_H)$. We shall use the following notation. For any pair of ideals $\mathfrak{a}_1, \mathfrak{a}_2$ we let $L(\mathfrak{a}_1, \mathfrak{a}_2)$ be the formal linear combination

$$L(\mathfrak{a}_1, \mathfrak{a}_2) = (\mathfrak{a}_1) + (\mathfrak{a}_2) - (\mathfrak{a}_1\mathfrak{a}_2) - (\mathfrak{o}).$$

Lemma 5.3. *The group* $\mathbf{I}_0(\mathfrak{o}, ww_H)$ *is generated by the elements*

$$n((\mathfrak{a}_1) + (\mathfrak{a}_2) - (\mathfrak{a}_1\mathfrak{a}_2) - (\mathfrak{o})) = nL(\mathfrak{a}_1, \mathfrak{a}_2)$$

where n is an integer such that

$$n(\mathbf{N}\mathfrak{a}_1 - 1)(\mathbf{N}\mathfrak{a}_2 - 1) \equiv 0 \bmod ww_H.$$

Proof. Let $\mathbf{I}'_0$ be the group generated by such elements. Since

$$(\mathbf{N}\mathfrak{a}_1 - 1)(\mathbf{N}\mathfrak{a}_2 - 1) = \mathbf{N}\mathfrak{a}_1\mathbf{N}\mathfrak{a}_2 - 1 - (\mathbf{N}\mathfrak{a}_1 - 1) - (\mathbf{N}\mathfrak{a}_2 - 1),$$

we have $\mathbf{I}'_0 \subset \mathbf{I}_0(\mathfrak{o}, ww_H)$. Choose $\mathfrak{p}$ as in the proof of Lemma 5.2, so that $L(\mathfrak{p}, \mathfrak{p})$ generates $\mathbf{I}_0[\mathfrak{o}]/\mathbf{I}_0(\mathfrak{o}, ww_H)$. We first show that $\mathbf{I}'_0$ and $L(\mathfrak{p}, \mathfrak{p})$ together generate $\mathbf{I}_0[\mathfrak{o}]$. It suffices to see that any element of the form $L(\mathfrak{a}_1, \mathfrak{a}_2)$ can be written as the sum of an element of $\mathbf{I}'_0$ and an integral multiple of $L(\mathfrak{p}, \mathfrak{p})$.

Let

$$d = \text{g.c.d.}\left(\frac{w_H}{w}, \frac{\mathbf{N}\mathfrak{a}_1 - 1}{w} \cdot \frac{\mathbf{N}\mathfrak{a}_2 - 1}{w}\right).$$

Since $d|(w_H/w)$, it takes one of the four values 1, 2, 3, 6. We consider each one of these cases.

$d = w_H/w$. In this case, $ww_H|(\mathbf{N}\mathfrak{a}_1 - 1)(\mathbf{N}\mathfrak{a}_2 - 1)$, so $L(\mathfrak{a}_1, \mathfrak{a}_2)$ itself lies in $\mathbf{I}'_0$.

$d = 3 \neq w_H/w$. Then $w_H/w = 6$. Say $3|\mathbf{N}\mathfrak{a}_1 - 1$. Since $d \neq 6$, it follows that $4 \nmid \mathbf{N}\mathfrak{a}_1 - 1$ and $4 \nmid \mathbf{N}\mathfrak{a}_2 - 1$. But then

$$4|(\mathbf{N}\mathfrak{a}_1\mathfrak{a}_2 - 1), \qquad 4|(\mathbf{N}\mathfrak{a}_2\mathfrak{p} - 1) \quad \text{and} \quad 4|(\mathbf{N}\mathfrak{p}^2 - 1).$$

We have the identity

$$\begin{aligned} L(\mathfrak{a}_1, \mathfrak{a}_2) - 3L(\mathfrak{p}, \mathfrak{p}) = &-2L(\mathfrak{a}_1, \mathfrak{a}_2) - 3L(\mathfrak{a}_1\mathfrak{a}_2, \mathfrak{p}) + 3L(\mathfrak{a}_1, \mathfrak{a}_2\mathfrak{p}) \\ &- 3L(\mathfrak{a}_2\mathfrak{p}, \mathfrak{p}) + 3L(\mathfrak{a}_2, \mathfrak{p}^2) \end{aligned}$$

and we see that each term on the right lies in $\mathbf{I}'_0$.

$d = 2 \neq w_H/w$. Again $w_H/w = 6$. The argument is as in the preceding case, using the identity

$$\begin{aligned} L(\mathfrak{a}_1, \mathfrak{a}_2) + 2L(\mathfrak{p}, \mathfrak{p}) = &\ 3L(\mathfrak{a}_1, \mathfrak{a}_2) + 2L(\mathfrak{a}_1\mathfrak{a}_2, \mathfrak{p}) - 2L(\mathfrak{a}_1, \mathfrak{a}_2\mathfrak{p}) \\ &+ 2L(\mathfrak{a}_2\mathfrak{p}, \mathfrak{p}) - 2L(\mathfrak{a}_2, \mathfrak{p}^2). \end{aligned}$$

$d = 1$. Here we use the identity

$$L(\mathfrak{a}_1, \mathfrak{a}_2) - L(\mathfrak{p}, \mathfrak{p}) = -L(\mathfrak{a}_1\mathfrak{a}_2, \mathfrak{p}) + L(\mathfrak{a}_1, \mathfrak{a}_2\mathfrak{p}) - L(\mathfrak{a}_2\mathfrak{p}, \mathfrak{p}) + L(\mathfrak{a}_2, \mathfrak{p}^2).$$

Thus $\mathbf{I}'_0$ and $L(\mathfrak{p}, \mathfrak{p})$ generate $\mathbf{I}_0[\mathfrak{o}]$, as claimed.

Since for any positive integer k we have

$$kL(\mathfrak{p}, \mathfrak{p}) \in \mathbf{I}_0(\mathfrak{o}, ww_H) \Rightarrow (w_H/w)|k \Rightarrow kL(\mathfrak{p}, \mathfrak{p}) \in \mathbf{I}'_0,$$

it follows that $\mathbf{I}'_0 = \mathbf{I}_0(\mathfrak{o}, ww_H)$, thus proving the lemma.

Lemma 5.4. *The group $\Delta_H(c_0, ww_H)$ is generated by all elements of the form*

$$(\delta_H(c_1c_2)/\delta_H(c_1)\delta_H(c_2))^n,$$

where $n(\mathbf{N}\mathfrak{a}_{c_1} - 1)(\mathbf{N}\mathfrak{a}_{c_2} - 1) \equiv 0 \bmod ww_H$.

Proof. Immediate from Lemma 5.3.

Lemma 5.5. *Let $H \subset K(1)$ be an abelian unramified extension of K. Then*

$$\mathbf{N}_{K(1)/H}\Delta_{K(1)}(c_0, ww_{K(1)}) = \Delta_H(c_0, ww_H).$$

Proof. We leave it to the reader to verify that the inclusion $\subset$ is satisfied. Conversely, it will suffice to prove that the generators of Lemma 5.4 are norms from $\Delta_{K(1)}(c_0, ww_{K(1)})$. Let

$$\left(\frac{\delta_H(cc')}{\delta_H(c)\delta_H(c')}\right)^n \quad \text{lie in} \quad \Delta_H(c_0, ww_{K(1)}).$$

We may pick $\mathfrak{a}_c$ and $\mathfrak{a}_{c'}$ to be H-admissible. Also, if $3 \mid w_{K(1)}$ but $3 \nmid w_H$, then we may choose $\mathfrak{a}_c$, say, so that $\mathbf{N}\mathfrak{a}_c \equiv 1 \bmod 3$, by an argument similar to that used in proving admissibility. If we do this, we then find that

$$ww_H \mid n(\mathbf{N}\mathfrak{a}_c - 1)(\mathbf{N}\mathfrak{a}_{c'} - 1) \Rightarrow ww_{K(1)} \mid n(\mathbf{N}\mathfrak{a}_c - 1)(\mathbf{N}\mathfrak{a}_{c'} - 1).$$

Thus if $C = C(\mathfrak{a}_c)$ and $C' = C(\mathfrak{a}_{c'})$ are the classes of $\mathfrak{a}_c$, $\mathfrak{a}_{c'}$ in $\mathrm{Cl}(1)$, then

$$\left(\frac{\delta(CC')}{\delta(C)\delta(C')}\right)^n \quad \text{lies in} \quad \Delta_{K(1)}(C_0, ww_{K(1)}),$$

and its norm to H is $(\delta_H(cc')/\delta_H(c)\delta_H(c'))^n$. This completes the proof.

Lemma 5.6. *Every element of $\Delta_H(c_0, ww_H)$ is a $12wh$-power in H.*

Proof. This lemma, like Lemma 1.2, will be proved in Chapter 11, Corollary of Theorem 4.4, when $H = K(1)$. Another proof will be given in Theorem 3.1 of Chapter 12. It follows for arbitrary H by Lemma 5.5.

We now return to the proof of Theorem 5.1 proper. By Lemma 5.2 and Theorem 3.4 we get

$$\begin{aligned}(E_H : \Delta_H(c_0, ww_H)) &= (E_H : \Delta_H(c_0))(\Delta_H(c_0) : \Delta_H(c_0, ww_H)) \\ &= \frac{12^{v-1}h^{v-1}w^{v-1}w(H)}{[K(1):H]} h_H.\end{aligned}$$

On the other hand

$$(\Delta_H(c_0, ww_H)^{1/12wh} : \Delta_H(c_0, ww_H)) = (12wh)^{v-1}w_H.$$

The index formula of Theorem 5.1 follows at once.

In Chapter 12, Theorem 3.1, we shall give a more natural definition of the group

$$\Delta_H(c_0, ww_H)^{1/12wh}.$$

CHAPTER 10

More Units in the Modular Function Field

In this chapter we give some brief preliminaries in the modular function field, for application to the complex multiplication case afterward.

First we derive another transformation law for the Klein forms. We then apply this to get an expression of certain elliptic functions in terms of Klein forms. The prototype of these expressions (in terms of the Siegel functions) stems from Robert [Ro 1].

These analytic expressions are then used in the next chapter in the context of complex multiplication. The special expressions for Klein forms have values which constitute generators for certain interesting groups of algebraic numbers, and especially units.

§1. Transformation of the Klein Forms

In this section, we follow [K-L 13].

By Kronecker's first limit formula or otherwise, we know that

$$E(\tau, s) = \sum \frac{y^s}{|m\tau + n|^{2s}} = \frac{\pi}{s-1} + O(1), \quad \text{for } s \text{ near } 1.$$

Let $L = [\omega_1, \omega_2]$, and let $\tau = \omega_1/\omega_2 \in \mathfrak{H}$. Let

$$E(L, s) = \sum_{\omega \neq 0} \frac{1}{|\omega|^{2s}}.$$

Then in a neighborhood of $s = 1$ we have

$$E(L, s) \sim \frac{\pi}{\mathbf{M}L} \frac{1}{s - 1},$$

where

$$\mathbf{M}L = \frac{1}{2i} (\omega_1 \bar{\omega}_2 - \bar{\omega}_1 \omega_2)$$

is the area (measure) of the fundamental domain. The residue of $E(L, s)$ at $s = 1$ is therefore $\pi/\mathbf{M}L$.

Remark. If L is a fractional ideal in the imaginary quadratic field K, and $\mathfrak{o} = \mathfrak{o}_K$, then $\mathbf{M}(L) = \mathbf{N}L \cdot \mathbf{M}(\mathfrak{o})$, where $\mathbf{N}$ is the absolute norm. If $L = [\tau, 1]$ then $\mathbf{M}L = \operatorname{Im} \tau$.

Theorem 1.1.

(i) *The function* $\sum \frac{1}{\omega^2 |\omega|^{2s}}$ *is holomorphic at* $s = 0$ *(it is even an entire function).*

(ii) *The function defined for* $z \notin L$ *by*

$$G(z, L, s) = \sum_{\omega \neq 0} \frac{z + \omega}{|z + \omega|^{2s}} \quad \textit{for } \operatorname{Re} s > \tfrac{3}{2}$$

has an analytic continuation to $\operatorname{Re} s > \frac{1}{2}$, *and is holomorphic at* $s = 1$.

Proof. These are essentially well known, and can be proved by standard techniques using Poisson's summation formula. Cf. Siegel [Si 2], Theorem 3, p. 69.

In particular, we can define $s_2(L)$ as the value of the function in (i), at $s = 0$.

Following Birch-Swinnerton Dyer [B-SwD] we define

$$\begin{aligned}\zeta_s(z, L) &= \frac{\bar{z}}{|z|^{2s}} + \sum_{\omega \neq 0} \left\{ \frac{\overline{z + \omega}}{|z + \omega|^{2s}} - \frac{\bar{\omega}}{|\omega|^{2s}} \left(1 - \frac{sz}{\omega} + \frac{\bar{z}}{\bar{\omega}} (1 - s) \right) \right\} \\ &= \frac{\bar{z}}{|z|^{2s}} + \sum_{\omega \neq 0} \left\{ \frac{z + \omega}{|z + \omega|^{2s}} - \frac{\bar{\omega}}{|\omega|^{2s}} + \frac{sz}{|\omega|^{2s-2} \omega^2} + \frac{\bar{z}}{|\omega|^{2s}} (s - 1) \right\}\end{aligned}$$

The series converges absolutely for Re $s > \frac{1}{2}$ by the usual argument, and we have

$$\lim_{s \to 1} \zeta_s(z, L) = \zeta(z, L)$$

taking the limit for s real > 1. For $s > \frac{3}{2}$ we can rearrange the terms in $\{\ \}$. Combining ω with $-\omega$ shows

$$\sum_{\omega \neq 0} \frac{\bar{\omega}}{|\omega|^{2s}} = 0.$$

Since $E(L, s)$ has only a simple pole at $s = 1$, it follows that

$$\sum \frac{\bar{z}}{|\omega|^{2s}}(s-1) \sim \bar{z}\frac{\pi}{\mathbf{M}L} \quad \text{for } s \to 1.$$

Therefore the function

$$\sum_{\omega \neq 0} \left\{ \frac{\overline{z+\omega}}{|z+\omega|^{2s}} + \frac{sz}{|\omega|^{2s-2}\omega^2} \right\}$$

is holomorphic at $s = 1$. From the definition of $s_2(L)$ we then find:

$$\boxed{\zeta(z, L) = \frac{1}{\bar{z}} + G(z, L, 1) + s_2(L)z + \frac{\pi}{\mathbf{M}L}\bar{z}.}$$

Theorem 1.2. $\eta(z, L) = s_2(L)z + \dfrac{\pi}{\mathbf{M}L}\bar{z}$.

Proof. The expression

$$\frac{1}{z} + G(z, L, 1) = \frac{\bar{z}}{|z|^{2s}} + G(z, L, s) \quad \text{at } s = 1$$

is periodic in L. Hence

$$\begin{aligned} \zeta(z+\omega, L) - \zeta(z, L) &= s_2(L)\omega + \frac{\pi}{\mathbf{M}L}\bar{\omega}. \\ &= \eta(\omega, L). \end{aligned}$$

The theorem follows by **R**-linearity. It may be viewed as a generalization of the Legendre relation, which is seen to result from the above by putting

$z = \omega_1$ and $z = \omega_2$. The relation is known in the case of complex multiplication cf. Damerell, Acta Arith. XVII (1970) pp. 294 and 299, but we could find no reference for it in general.

We can now write the **Klein form** as the expression

$$\boxed{\mathfrak{k}(z, L) = e^{-s_2(L)z^2/2 - (\pi/\mathbf{M}L)z\bar{z}/2}\sigma(z, L).}$$

It satisfies the transformation law (probably known to Klein, compare with [Ro 1], formula (3)):

Theorem 1.3.

$$\boxed{\mathfrak{k}(z + \omega, L) = \mathfrak{k}(z, L)e^{-(\pi i/\mathbf{M}L)\,\mathrm{Im}(z\bar{\omega})}\psi(\omega, L),}$$

where

$$\begin{aligned} \psi(\omega, L) &= 1 && \text{if } \omega/2 \in L \\ \psi(\omega, L) &= -1 && \text{if } \omega/2 \notin L. \end{aligned}$$

Proof. Immediate from Theorem 1.2 and the transformation law for the sigma function.

§2. Klein Forms and Weierstrass Functions

Robert's construction of units in the complex multiplication case can be applied directly to the Klein forms as follows. Let $L' \supset L$ be lattices. Define

$$\mathfrak{k}(z, L'/L) = \mathfrak{k}(z, L)^{(L':L)}/\mathfrak{k}(z, L').$$

Lemma 2.1.

(i) *If $(L' : L)$ is odd, then the function $\mathfrak{k}(z, L'/L)$ is elliptic and periodic with respect to L. It is homogeneous of degree $(L' : L) - 1$ in (z, L, L').*

(ii) *If $(L' : L)$ is even, then $\mathfrak{k}^2(z, L'/L)$ is elliptic and periodic with respect to L.*

(iii) *Let L', L'' be two lattices containing L such that $(L' : L)$ and $(L'' : L)$ are both even. Assume that L'/L and L''/L (as subgroups of $\mathbf{C}/L$) have exactly the same points of order 2. Then the product*

$$\mathfrak{k}(z, L'/L)\mathfrak{k}(z, L''/L)$$

is elliptic and periodic with respect to L.

Proof. The exponential term occurring in the definition of $\mathfrak{k}(z, L'/L)$ is

$$-\eta(z, L)\frac{z}{2}(L' : L) + \eta(z, L')\frac{z}{2} = -\frac{z^2}{2}[s_2(L)(L' : L) - s_2(L')].$$

This shows that $\mathfrak{k}(z, L'/L)$ is meromorphic. (The anti-holomorphic exponent drops out.) Next, for the periodicity, note that

$$(L' : L)\frac{1}{\mathbf{M}L} = \frac{1}{\mathbf{M}L'}.$$

In the transformation law, it follows that the term arising from the exponential factor for $\mathfrak{k}(z, L'/L)$ is equal to 1. As to the term corresponding to ψ, it will also be equal to 1 in each of the three cases. Indeed, when $(L' : L)$ is odd, then $\psi(\omega, L) = \psi(\omega, L')$; in the other two cases, either the square makes the minus sign disappear, or the products of the ψ-functions will give 1 in light of the assumption on L' and L''. Of course, (ii) is a special case of (iii). This proves the Lemma.

We have the property

$$\mathfrak{k}(-z, L'/L) = (-1)^{(L' : L) - 1}\mathfrak{k}(z, L'/L),$$

so $\mathfrak{k}(z, L'/L)$ is an even function of z if $(L' : L)$ is odd. Otherwise, $\mathfrak{k}^2(z, L'/L)$ is an even function of z. This implies that they should be expressible in terms of the Weierstrass function, which is the case explicitly as follows.

Theorem 2.2.

(i) *If* $(L' : L)$ *is odd, then we have the expression*

$$\mathfrak{k}(z, L'/L) = \prod \frac{1}{\wp(z, L) - \wp(a, L)},$$

where the product is taken over $a \in (L'/L)/\pm 1$ *and* $a \neq 0$.

(ii) *If* $(L' : L)$ *is even, then* $\mathfrak{k}^2(z, L'/L)$ *is equal to the same expression, where the product is now taken over all* $a \in L'/L$, $a \neq 0$.

Proof. Suppose first that $(L' : L)$ is odd. Both the right and left hand side have zeros at the elements of L of order $(L' : L) - 1$, and poles at $L'/L - \{0\}$, of order 1, and no other zeros or poles. The power series in z at the origin starts with the same term, as one sees from the power series expansion for the Weierstrass sigma and $\wp$ functions. Hence the theorem is clear. The even case is proved the same way.

For refined applications involving the prime 2, it is necessary to give a third part to Theorem 2.2. We state it separately because of its more elaborate formulation.

Theorem 2.3. *Let L', L'' be two lattices containing L, such that $(L' : L)$ and $(L'' : L)$ are both even. Assume that L'/L and L''/L (as subgroups of $\mathbf{C}/L$) have exactly the same points of order 2. Then*

$$\mathfrak{k}(z, L'/L)\mathfrak{k}(z, L''/L) =$$

$$\prod_{a'} \frac{1}{\wp(z, L) - \wp(a', L)} \prod_{a''} \frac{1}{\wp(z, L) - \wp(a'', L)} \prod_{a} \frac{1}{\wp(z, L) - \wp(a, L)}$$

where the products are taken for:

$$a' \in (L'/L)/\pm 1, \quad 2a' \neq 0; \qquad a'' \in (L''/L)/\pm 1, \quad 2a'' \neq 0;$$
$$a \in L'/L, \quad 2a = 0, \quad a \neq 0.$$

Proof. The proof is entirely analogous to that of Theorem 2.2. We leave the verification of the steps to the reader. Of course, Theorem 2.2(ii) is a special case of the present result.

§3. More Expressions for Modular Units

This section will not be used in the sequel, but is included to emphasize once more the relation between generic units (that is, units in the function field), and the units in the complex multiplication case.

Let $\alpha \in \mathrm{Mat}_2^+(\mathbf{Z})$ be an integral matrix with odd determinant, and let $\mathbf{N}\alpha = \det \alpha$. Let $W = \begin{pmatrix} \omega_1 \\ \omega_2 \end{pmatrix}$ and let

$$\varphi_\alpha(z, W) = \varphi(z, \alpha, W) = \mathfrak{k}(z, W)^{\mathbf{N}\alpha}/\mathfrak{k}\left(z, \frac{\alpha W}{\mathbf{N}\alpha}\right).$$

The above value depends only on the left coset of α with respect to $\mathrm{SL}_2(\mathbf{Z})$, so we may assume that

$$\alpha = \begin{pmatrix} a & b \\ 0 & d \end{pmatrix}$$

is triangular. Let

$$\mathscr{A} = \{\alpha_j, n_j\}$$

be a finite family, where α_j are matrices as above, and n_j are integers such that

$$\sum n_j(\mathbf{N}\alpha_j - 1) = 0.$$

we let

$$\varphi(z, \mathscr{A}, W) = \prod_j \varphi(z, \alpha_j, W)^{n_j}.$$

Then φ has weight 0 in (z, W), and is an even elliptic function. For $z = \mathbf{a} \cdot W$ (where $\mathbf{a} = (a_1, a_2) \in \mathbf{Q}^{(2)}$, $\mathbf{a} \notin \mathbf{Z}^{(2)}$ its q-expansion is easily determined from that of the Klein forms. We let

$$\varphi_{\mathbf{a}}(\alpha, W) = \varphi(\mathbf{a} \cdot W, \alpha, W),$$

and similarly for $\varphi_{\mathbf{a}}(\mathscr{A}, W)$, replacing α by α_j and taking the product raising each term to the power n_j.

Theorem 3.1. *Assume that the* $\mathbf{N}\alpha_j$ *are prime to* $2N$ *for all* j, *where* N *is the denominator of* $\mathbf{a}$ *and* $\mathbf{b}$. *Then the function*

$$\varphi_{\mathbf{a}}(\mathscr{A}, W)/\varphi_{\mathbf{b}}(\mathscr{A}, W)$$

is a unit over $\mathbf{Z}$.

Proof. We return to the considerations of §1, §2 of Chapter 2, and the q-expansions. Let $f = \sum a_n q^{n/N}$ be a power series in $q^{1/N}$, having possibly a finite number of polar terms. If g is another such power series, we write

$$f \sim g$$

to mean that the coefficients of the lowest terms in f and g have a quotient which is a unit in a cyclotomic field. In the applications, f and g are modular forms, whose Fourier coefficients are algebraic integers in a cyclotomic field. If $f \sim g$ then f/g is an invertible power series over these integers. Furthermore, if f, g have zeros and poles only at the cusps, and their conjugates $f \circ \gamma$ and $g \circ \gamma$ for $\gamma \in \mathrm{SL}_2(\mathbf{Z})$ have the same properties as above, then f/g is a unit over $\mathbf{Z}$.

We recall the q-expansion for the Klein forms is given by

$$\begin{aligned}\Delta(\tau)^{1/12}\mathfrak{k}(a_1\tau + a_2; [\tau, 1]) \\ = -q_\tau^{(1/2)\mathbf{B}_2(a_1)} e^{2\pi i a_2(a_1 - 1)/2}(1 - q_z) \prod_{n=1}^{\infty} (1 - q_\tau^n q_z)(1 - q_\tau^n/q_z).\end{aligned}$$

As usual, we put $z = a_1\tau + a_2$, and a_1, a_2 are taken with

$$0 \leqq a_1, a_2 < 1.$$

Also, $\mathbf{B}_2(X) = X^2 - X + \frac{1}{6}$. We then find, putting $q_1 = e^{2\pi i}$:

$$\mathfrak{k}(a_1\tau + a_2; [\tau, 1]) \sim (1 - q_\tau^{a_1} q_1^{a_2})$$

$$\mathfrak{k}\left(a_1\tau + a_2; \frac{1}{\mathbf{N}\alpha}[a\tau + b, d]\right) = \frac{1}{a}\,\mathfrak{k}(da_1\tau' + aa_2 - ba_1, [\tau', 1])$$

$$\sim \frac{1}{a}(1 - q_{\tau'}^{\langle da_1\rangle} q_1^{aa_2 - ba_1}),$$

where $\tau' = (a\tau + b)/d$. We have assumed that $\mathbf{N}\alpha$ is prime to the denominator of a_1, a_2. Then

$$\varphi_\alpha(a_1\tau + a_2; [\tau, 1]) \sim a\,\frac{(1 - q_\tau^{a_1} q_1^{a_2})^{\mathbf{N}\alpha}}{1 - q_{\tau'}^{\langle da_1\rangle} q_1^{aa_2 - ba_1}},$$

where $\langle t\rangle$ denotes the smallest real number $\geqq 0$ in the residue class of t mod $\mathbf{Z}$. Therefore, we find

$$\varphi_\alpha(a_1\tau + a_2; [\tau, 1]) \sim \begin{cases} a & \text{if } a_1 \neq 0 \\ a\,\dfrac{(1 - q_1^{\langle a_2\rangle})^{\mathbf{N}\alpha}}{(1 - q_1^{\langle aa_2\rangle})} & \text{if } a_1 = 0. \end{cases}$$

Let $\mathscr{A} = \{\alpha_j, n_j\}$ be a family as before. Then

$$\prod \varphi_{\alpha_j}^{n_j} \sim \prod a_j^{n_j}$$

and is of weight 0. We assume throughout that $\mathbf{N}\alpha_j$ is prime to the denominators of a_1, a_2 for all j.

Under conjugation by $\gamma \in \mathrm{SL}_2(\mathbf{Z})$, we have

$$\varphi_{\mathbf{a}}\left(\alpha, \gamma\begin{pmatrix}\tau\\1\end{pmatrix}\right) = \varphi_{\mathbf{a}\gamma}\left(\alpha', \begin{pmatrix}\tau\\1\end{pmatrix}\right),$$

where

$$\alpha' = \begin{pmatrix} a' & b' \\ 0 & d' \end{pmatrix}$$

is some matrix in the left coset of $\alpha\gamma^{-1}$ with respect to $SL_2(\mathbf{Z})$; so we obtain a similar equivalence for such conjugates. Therefore:

If $\mathbf{N}\alpha$ *is odd and prime to the denominator of* $\mathbf{a} = (a_1, a_2)$ *and* $\mathbf{b} = (b_1, b_2)$, *then*

$$\prod \varphi_{\mathbf{a}}(\alpha_j, \tau)^{n_j} / \varphi_{\mathbf{b}}(\alpha_j, \tau)^{n_j} \sim 1.$$

A similar relation holds for the conjugates, so the theorem is proved.

Let us put

$$\delta(\tau) = \Delta^{1/12}(\tau) = q^{1/12} \prod (1 - q^n)^2.$$

Then for $\alpha = \begin{pmatrix} a & b \\ 0 & d \end{pmatrix}$ we find

$$\delta\left(\frac{\alpha}{\mathbf{N}\alpha}\begin{pmatrix} \tau \\ 1 \end{pmatrix}\right) \sim a.$$

Let

$$\delta_\alpha(\tau) = \delta(\tau)^{\mathbf{N}\alpha} \Big/ \delta\left(\frac{\alpha\tau}{\mathbf{N}\alpha}\right) = \delta(\tau, \alpha)$$

so that

$$\delta_\alpha(\tau) \sim 1/a.$$

Then

$$\delta(\tau, \mathscr{A}) = \prod \delta(\tau, \alpha_j)^{n_j} \sim 1 / \prod a_j^{n_j}.$$

Theorem 3.2. *Assume that the* $\mathbf{N}\alpha_j$ *are prime to* $2N$ *for all* j, *where* N *is the denominator of* $\mathbf{a}$. *Then the product* $\delta(\tau, \mathscr{A})\varphi_{\mathbf{a}}(\tau, \mathscr{A})$ *is a unit over* $\mathbf{Z}$.

CHAPTER 11

Siegel-Robert Units in Arbitrary Class Fields

In this chapter, we shall define units in arbitrary class fields by means of values of modular functions, following again Siegel, Ramachandra, and Robert. We follow the latter especially, with the modifications which arise from having a more systematic theory of the modular forms used to construct units. It has been apparent from the beginning that the 12th power was used in the definition of the Siegel functions in order to catch the delta function and not one of its roots. In the present chapter, we keep this 12th power (and the additional N-th power) which give rise to relatively simple explicit formulas. In the next chapter, we shall develop a more general systematic approach, resulting in a refinement of the group of modular units in class fields of K by using the Klein forms.

In each case, we use Shimura's reciprocity law to determine the action of the Galois group on these units more easily than in the above cited literature.

We let $\mathfrak{f}$ be an arbitrary ideal of K, usually $\neq (1)$, and we use systematically the notation listed at the beginning of the chapters on complex multiplication. We let in addition:

$$r(\mathfrak{f}) = w(\mathfrak{f})N(\mathfrak{f}) \quad \text{(because this product occurs frequently).}$$

§1. Siegel-Ramachandra Invariants as Distributions

Let $\mathfrak{f} \neq (1)$ be an ideal of the imaginary quadratic field K. We let $\mathrm{Cl}(\mathfrak{f})$ denote the generalized ideal class group of conductor $\mathfrak{f}$. Let C be a ray class in $\mathrm{Cl}(\mathfrak{f})$. Let f be a form of weight 0 on $\Gamma_1(N)$, where N is the least positive integer in $\mathfrak{f}$, that is $N = N(\mathfrak{f})$. We define

$$\boxed{f(C) = f(1, \mathfrak{f}\mathfrak{c}^{-1})}$$

where $\mathfrak{c}$ is any ideal in C (and in particular, $\mathfrak{c}$ is prime to $\mathfrak{f}$). This value is independent of the choice of $\mathfrak{c}$. Indeed, if $\mathfrak{c}_1$ is another such ideal, there exists $\alpha \in K^*$ such that $\mathfrak{c}_1 = \alpha\mathfrak{c}$ and $\alpha \equiv 1 \bmod^* \mathfrak{f}$. Since $\mathfrak{c}_1 \subset \mathfrak{o}$ (by definition of an ideal), it follows that $\alpha \in \mathfrak{c}^{-1}$. It is then immediate that

$$\alpha - 1 \in \mathfrak{f}\mathfrak{c}^{-1},$$

whence $f(1, \mathfrak{f}\mathfrak{c}^{-1}) = f(\alpha, \mathfrak{f}\mathfrak{c}^{-1})$, thus proving our assertion, since $f(\lambda t, \lambda L) = f(t, L)$ because f is of weight 0.

Theorem 1.1. *Let $\{f_a\}$ be a Fricke family of weight 0 and level N, where N is the smallest positive integer in $\mathfrak{f}$. Then*

$$f(C) \in K(\mathfrak{f}), \quad \textit{and} \quad f(C)^{\sigma(C')} = f(CC').$$

Proof. We shall use Shimura's reciprocity law, cf. [Sh], and [L 5], Chapter 11, §1, Theorem 1. We assume that the reader is acquainted with the notation of this reference.

Let $\mathfrak{c} \in C$ and let $\mathfrak{a} \in C'$ be such that $\mathbf{N}\mathfrak{a}$, $\mathbf{N}\mathfrak{c}$ are relatively prime to $\mathbf{N}\mathfrak{f}$. Then by definition,

$$f(C) = f(1, \mathfrak{f}\mathfrak{c}^{-1}) \quad \text{and} \quad f(CC') = f(1, \mathfrak{f}\mathfrak{c}^{-1}\mathfrak{a}^{-1}).$$

Let $\mathfrak{f}\mathfrak{c}^{-1} = [z_1, z_2]$ with $z = z_1/z_2$ in the upper half plane as usual. Let $\alpha \in \mathrm{Mat}_2^+(\mathbf{Z})$ be an integral matrix with positive determinant such that

$$\alpha^{-1}\begin{pmatrix} z_1 \\ z_2 \end{pmatrix} \quad \text{is a basis of } \mathfrak{f}\mathfrak{c}^{-1}\mathfrak{a}^{-1}.$$

Then $\det \alpha = \mathbf{N}\mathfrak{a}$. Let s be an idele of K such that:

$$\begin{aligned} s_p &= 1 \quad && \text{if } p \mid \mathbf{N}\mathfrak{f} \\ s_p\mathfrak{o}_p &= \mathfrak{a}_p \quad && \text{if } p \nmid \mathbf{N}\mathfrak{f}. \end{aligned}$$

Then we also have $s_p^{-1}\mathfrak{o}_p = \mathfrak{a}_p^{-1}$. Furthermore $(s, K) = \sigma(C')$ on the ray class field $K(\mathfrak{f})$ because for all $\mathfrak{p} \nmid \mathbf{N}\mathfrak{f}$, $\mathrm{ord}_\mathfrak{p}\, s_\mathfrak{p} = \mathrm{ord}_\mathfrak{p}\, \mathfrak{a}$. Write

$$1 = a_1 z_1 + a_2 z_2 \quad \text{with } a_1, a_2 \in \frac{1}{N}\mathbf{Z}.$$

Then $a = (a_1, a_2)$ is primitive of order N mod $\mathbf{Z}^2$. By the Shimura Reciprocity Law, we get

$$f(C)^{(s,K)} = f_a^\sigma(z) \quad \text{where } \sigma = \sigma(q_z(s^{-1})).$$

Since for all primes p,

$$q_{z,p}(s_p^{-1})\begin{pmatrix} z_1 \\ z_2 \end{pmatrix} \quad \text{and} \quad \alpha^{-1}\begin{pmatrix} z_1 \\ z_2 \end{pmatrix}$$

are bases of $(\mathfrak{f}\mathfrak{c}^{-1}\mathfrak{a}^{-1})_p$, it follows that there exists $u_p \in \mathrm{GL}_2(\mathbf{Z}_p)$ such that

$$q_{z,p}(s_p^{-1}) = u_p\alpha^{-1}.$$

We let $u = (u_p) \in \prod \mathrm{GL}_2(\mathbf{Z}_p)$, so that

$$q_z(s^{-1}) = u\alpha.$$

Then

$$\begin{aligned} f_a^\sigma(z) &= f_a^{\sigma(u\alpha^{-1})}(z) \\ &= f_{au}(\alpha^{-1}(z)) \\ &= f_{a\alpha}(\alpha^{-1}(z)) \end{aligned}$$

[because for $p \mid N\mathfrak{f}$, we have $1 = u_p\alpha^{-1}$, so $u_p = \alpha$ and $au = a\alpha$]

$$\begin{aligned} &= f\left(a\alpha\alpha^{-1}\begin{pmatrix} z_1 \\ z_2 \end{pmatrix}, \alpha^{-1}\begin{pmatrix} z_1 \\ z_2 \end{pmatrix}\right) \\ &= f(1, \mathfrak{f}\mathfrak{c}^{-1}\mathfrak{a}^{-1}) \\ &= f(CC'). \end{aligned}$$

This shows that $f(C)^{(s,K)} = f(CC')$, and hence that $f(C)^{(s,K)}$ depends only on $\sigma(C')$. Hence $f(C)$ lies in $K(\mathfrak{f})$, and we know that $(s, K) = \sigma(C')$ on $K(\mathfrak{f})$. This concludes the proof of the theorem.

We shall apply Theorem 1.1 to the case of the Siegel functions (distribution) g. We let:

$\{g_a\}$ = family of Siegel functions, where a has exact denominator N.

Then $\{g_a^{12N}\}$ is a Fricke family of level N. If $C \in \mathrm{Cl}(\mathfrak{f})$, we may then define the **Siegel-Ramachandra invariant**

$$g_\mathfrak{f}(C) = g^{12N}(C),$$

and Theorem 1.1 applies, so that

$$\boxed{g_{\mathfrak{f}}(C)^{\sigma(C')} = g_{\mathfrak{f}}(CC').}$$

Cf. the addendum to this section.

Theorem 1.2. *For any two classes C, $C' \in \mathrm{Cl}(\mathfrak{f})$, the algebraic number $g_{\mathfrak{f}}(C)/g_{\mathfrak{f}}(C')$ is a unit. If N is composite, then $g_{\mathfrak{f}}(C)$ is a unit.*

Proof. If N is composite, then the functions g_a, where a has period exactly N, are units over $\mathbf{Z}$ in the modular function field, and $g_{\mathfrak{f}}(C)$ is obtained by evaluating such a function at an appropriate number, as described above. Hence $g_{\mathfrak{f}}(C)$ is a unit. Suppose N is arbitrary. Let $\mathfrak{c} \in C$ be a prime ideal relatively prime to N, and of degree 1 over $\mathbf{Q}$. By definition, we have

$$g_{\mathfrak{f}}(C_0) = g^{12N}(1, \mathfrak{f}) \quad \text{and} \quad g_{\mathfrak{f}}(C) = g^{12N}(1, \mathfrak{f}\mathfrak{c}^{-1}).$$

By the distribution relation, we get:

$$g_{\mathfrak{f}}(C) = \prod_{a \in \mathfrak{c}^{-1} \bmod \mathfrak{o}} g^{12N}(1 + a, \mathfrak{f}),$$

and therefore

$$\frac{g_{\mathfrak{f}}(C)}{g_{\mathfrak{f}}(C_0)} = \prod_{a \neq 0} g^{12N}(1 + a, \mathfrak{f}).$$

But $1 + a$ has composite level mod $\mathfrak{f}$, so each term of the product on the right is a unit, thereby proving the theorem.

We shall now study the behavior of these invariants from the point of view of distributions.

The groups $\mathrm{Cl}(\mathfrak{f})$ form a projective system under divisibility of the conductor $\mathfrak{f}$. We shall see that the functions

$$C \mapsto g_{\mathfrak{f}}(C)^{1/w(\mathfrak{f})N(\mathfrak{f})}$$

whose values are taken in the multiplicative group modulo roots of unity satisfy the distribution relation relative to two levels $\mathfrak{f} | \mathfrak{f}'$, provided that $\mathfrak{f}$ and $\mathfrak{f}'$ have the same prime factors. When $\mathfrak{f}$, $\mathfrak{f}'$ have different prime factors, then an additional perturbation factor enters in the relation.

In the projective system $\{\mathrm{Cl}(\mathfrak{f})\}$, if a class C' in $\mathrm{Cl}(\mathfrak{f}')$ lies above C, then we shall denote this relation by

$$C' | C.$$

Theorem 1.3. *Let* $\mathfrak{f} \mid \mathfrak{f}'$ *and assume that* $\mathfrak{f}$, $\mathfrak{f}'$ *have the same prime factors. Then*

$$g_{\mathfrak{f}}(C)^{N(\mathfrak{f}')/N(\mathfrak{f})} = \prod_{C'|C} g_{\mathfrak{f}'}(C')^{w(\mathfrak{f})/w(\mathfrak{f}')}.$$

Proof. Without loss of generality we may assume that $\mathfrak{f}' = \mathfrak{f}\mathfrak{p}$, where $\mathfrak{p}$ is a prime dividing $\mathfrak{f}$. Let

$$u_1, \ldots, u_{\mathbf{N}\mathfrak{p}},$$

be a complete system of residue classes of $\mathfrak{o}$ mod $\mathfrak{f}\mathfrak{p}$ such that

$$u_i \equiv 1 \bmod \mathfrak{f} \quad \text{all } i.$$

Having assumed that $\mathfrak{p}$ divides $\mathfrak{f}$, it follows that u_i is prime to $\mathfrak{p}$. Let $\mathfrak{c}$ be an ideal of C'. Then $u_i\mathfrak{c}$ for $i = 1, \ldots, \mathbf{N}\mathfrak{p}$ represent all the classes lying above C. Then applying Theorem 4.1 of Chapter 2, we have:

$$\begin{aligned}\prod_{C'|C} g_{\mathfrak{f}\mathfrak{p}}(C')^{w(\mathfrak{f})/w(\mathfrak{f}\mathfrak{p})} &= \prod_{i=1}^{\mathbf{N}\mathfrak{p}} g^{12N(\mathfrak{f}\mathfrak{p})}(u_i, \mathfrak{f}\mathfrak{p}\mathfrak{c}^{-1}) \\ &= g^{12N(\mathfrak{f})}(1, \mathfrak{f}\mathfrak{c}^{-1})^{N(\mathfrak{f}\mathfrak{p})/N(\mathfrak{f})}\end{aligned}$$

which proves the theorem.

Theorem 1.4. *Assume that* $\mathfrak{p} \nmid \mathfrak{f}$, *and* $\mathfrak{f}' = \mathfrak{f}\mathfrak{p}$.

(i) *If* $\mathfrak{f} \neq (1)$, *then*

$$\prod_{C'|C} g_{\mathfrak{f}\mathfrak{p}}(C')^{w(\mathfrak{f})/w(\mathfrak{f}')} = \left(\frac{g_{\mathfrak{f}}(C)}{g_{\mathfrak{f}}(C\mathfrak{p}^{-1})}\right)^{N(\mathfrak{f}')/N(\mathfrak{f})}$$

(ii) *If* $\mathfrak{f} = (1)$, *then for any ideal* $\mathfrak{c} \in C$,

$$\prod_{C'|C} g_{\mathfrak{p}}(C')^{w/w(\mathfrak{p})} = \frac{\Delta^p(\mathfrak{c}^{-1})}{\Delta^p(\mathfrak{p}\mathfrak{c}^{-1})}.$$

Proof. We select u_i as before, but now suppose that

$$u_1 \equiv 0 \bmod \mathfrak{p}.$$

Then $u_2, \ldots, u_{\mathbf{N}\mathfrak{p}}$ are prime to $\mathfrak{p}$. Thus

$$g_{\mathfrak{f}}(C\mathfrak{p}^{-1}) = g^{12N(\mathfrak{f})}(u_1, \mathfrak{f}\mathfrak{p}\mathfrak{c}^{-1})$$

and

$$\prod_{C'|C} g_{\mathfrak{f}\mathfrak{p}}(C')^{w(\mathfrak{f})/w(\mathfrak{f}\mathfrak{p})} = \prod_{i=2}^{\mathbf{N}\mathfrak{p}} g^{12N(\mathfrak{f}\mathfrak{p})}(u_i, \mathfrak{f}\mathfrak{p}\mathfrak{c}^{-1}).$$

Since $u_2, \ldots, u_{\mathbf{N}\mathfrak{p}}$ form a complete system of non-zero residue classes of $\mathfrak{f}\mathfrak{c}^{-1} \bmod \mathfrak{f}\mathfrak{p}\mathfrak{c}^{-1}$, we obtain from the distribution relations of Theorem 4.1, Chapter 2,

$$\prod_{i=1}^{\mathbf{N}\mathfrak{p}} g^{12N(\mathfrak{f}\mathfrak{p})}(u_i, \mathfrak{f}\mathfrak{p}/\mathfrak{c}) = g^{12N(\mathfrak{f}\mathfrak{p})}(t, \mathfrak{f}/\mathfrak{c})$$

$$= g_{\mathfrak{f}}(C)^{N(\mathfrak{f}\mathfrak{p})/N(\mathfrak{f})} \quad \text{if } \mathfrak{f} \neq (1)$$

$$\prod_{i=2}^{\mathbf{N}\mathfrak{p}} g^{12p}(u_i, \mathfrak{p}\mathfrak{c}^{-1}) = \frac{\Delta^p(\mathfrak{c}^{-1})}{\Delta^p(\mathfrak{p}\mathfrak{c}^{-1})} \quad \text{if } \mathfrak{f} = (1).$$

Note also that $N(\mathfrak{f}\mathfrak{p})/N(\mathfrak{f}) = p$. The theorem follows at once.

Corollary 1. *For any ideal $\mathfrak{a}$ the quotient $\Delta(\mathfrak{a})/\Delta(\mathfrak{o})$ is a w-power in the Hilbert class field $K(1)$.*

Proof. Replacing $\mathfrak{a}$ by any ideal $\lambda\mathfrak{a}$ with $\lambda \in K^*$ changes the desired quotient by a 12-th power, and w divides 12. Hence by the existence theorem for primes in arithmetic progressions, we may assume that $\mathfrak{a} = \mathfrak{p}$ is prime, and that $w(\mathfrak{p}) = 1$. We use Formula (ii) with $\mathfrak{f}' = \mathfrak{p}$, $\mathfrak{c} = \mathfrak{o}$. Let

$$\alpha = \Delta(\mathfrak{o})/\Delta(\mathfrak{p}).$$

Then

$$\alpha^p = \prod_{C'} g_{\mathfrak{p}}(C')^w.$$

But $g_{\mathfrak{p}}(C') = g^{12p}(1, \mathfrak{p}/\mathfrak{c}')$, which is the p-th power of $g^{12}(1, \mathfrak{p}/\mathfrak{c}')$. From the generic theory, we know that the Siegel functions g_a^{12} have level N^2 whenever $a \in (1/N)\mathbf{Z}^2/\mathbf{Z}^2$. Hence $g^{12}(1, \mathfrak{p}/\mathfrak{c}')$ lies in the ray class field of conductor p^2 by complex multiplication. Hence there exists an element $\beta_p \in K(p^2)$ such that

$$\alpha^p = \beta_p^{wp}.$$

But the p-th roots of unity are contained in $K(p^2)$. Hence α itself is a w-th power in $K(p^2)$. Thus the extension $K(\alpha^{1/w})$ is contained in the ray class field $K(p^2)$ for two distinct primes p, whence it is contained in the Hilbert class field. This concludes the proof.

Corollary 2. *Let $c, c' \in \mathrm{Cl}(1)$, and let $\delta(c)$ be the number defined in Chapter 9, §1. Let $\mathfrak{p}$ be a prime in c. Then*

$$\left(\frac{\prod_{C|cc'} g_{\mathfrak{p}}(C)}{\prod_{C|c} g_{\mathfrak{p}}(C)}\right)^{hw/w(\mathfrak{p})} = \left(\frac{\delta(cc')}{\delta(c)\delta(c')}\right)^{p},$$

where the products on the left are taken over classes C in $\mathrm{Cl}(\mathfrak{p})$, as indicated.

Proof. The left hand side is equal to

$$\left(\frac{\Delta^h(\mathfrak{c}\mathfrak{c}')}{\Delta^h(\mathfrak{c}\mathfrak{c}'\mathfrak{p})}\bigg/\frac{\Delta^h(\mathfrak{c})}{\Delta^h(\mathfrak{c}\mathfrak{p})}\right)^{N(\mathfrak{p})}$$

by Theorem 1.4(ii), where $\mathfrak{c}$ is an ideal in c^{-1}. Note that for a prime $\mathfrak{p}$, $N(\mathfrak{p}) = p$. Then $\mathfrak{c}\mathfrak{p}$ is principal, and after using the homogeneity of Δ, the desired formula on the right drops out.

Addendum. In order to facilitate comparison with the literature, we note that the values $g_{\mathfrak{f}}(C)$ as we defined them are twisted from those in the paper of Ramachandra, but coincide with those of Robert. The next theorem makes the link, in the context of Fricke families.

Theorem 1.5. *Let $\{f_a\}$ be a Fricke family, of weight 0, and f the associated modular function. Let $\mathfrak{c}$ be an ideal in a ray class C mod $\mathfrak{f}$. Let $\mathfrak{d}$ be the different of $K/\mathbf{Q}$, and let*

$$\mathfrak{c}\mathfrak{d}^{-1}\mathfrak{f}^{-1} = \mathfrak{a} = [z_1, z_2], \qquad z_1/z_1 \in \mathfrak{H}.$$

Then

$$f((\mathrm{tr}\ z_2)z_1 - (\mathrm{tr}\ z_1)z_2, \mathfrak{a}) = f(\overline{C}),$$

where $\overline{C}$ is the complex conjugate of C.

Proof. We need a lemma.

Lemma. *Let $\mathfrak{a} = [z_1, z_2]$ be a fractional ideal with $\mathrm{Im}(z_1/z_2) > 0$. Let $D = D(\mathfrak{o}_K)$ be the discriminant. Then*

$$(\mathrm{tr}\ z_2)z_1 - (\mathrm{tr}\ z_1)z_2 = \sqrt{D}\mathbf{N}\mathfrak{a}.$$

Proof. If we replace (z_1, z_2) by $(\lambda z_1, \lambda z_2)$ with $\lambda \in K^*$, then both sides change by $\lambda\bar{\lambda} = \mathbf{N}\lambda$. Hence it suffices to prove the lemma for $\mathfrak{a} = [z, 1]$, i.e. $z_2 = 1$, and $z = x + y\sqrt{D}$, $y > 0$. The left hand side of our formula is equal to

$$2z - 2x = 2y\sqrt{D}.$$

Hence we have only to show that $2y = \mathbf{N}\mathfrak{a}$. But

$$D(\mathfrak{a}) = \mathbf{N}\mathfrak{a}^2 D(\mathfrak{o}) = \mathbf{N}\mathfrak{a}^2 \cdot D,$$

and

$$D(\mathfrak{a}) = \begin{vmatrix} 1 & z \\ 1 & \bar{z} \end{vmatrix}^2 = (2y)^2 D.$$

Since $y > 0$, our lemma is proved.

For the theorem, observe that $\mathfrak{d} = \mathfrak{o}\sqrt{D}$. Then we just substitute the expression of the lemma in the left hand side of the formula to be proved, and we find the value

$$f(-1, \bar{\mathfrak{f}}\bar{\mathfrak{c}}^{-1}) = f(1, \bar{\mathfrak{f}}\bar{\mathfrak{c}}^{-1}) = f(\bar{C}),$$

as was to be shown.

Note that $\bar{C}$ is a ray class for the conductor $\bar{\mathfrak{f}}$. In the above mentioned paper, there is a corresponding twist in the formula giving the expression for the Frobenius automorphism acting on $g_{\mathfrak{f}}(C)$, but no such twist appears in Theorem 1.1.

Finally, we explain a general algebraic framework for the Siegel-Ramachandra invariants.

Let M be a commutative monoid, and X a set on which M operates. Let A be a commutative monoid written additively. A function

$$f : X \to A$$

will be called an M-**distribution** if for all $m \in M$ and $x \in X$ we have

$$\sum_{my = x} f(y) = f(x).$$

We apply this notion in the complex multiplication case as follows. Let $\mathfrak{I}'$ be the group of fractional ideals of K. Write a fractional ideal I in the form

$$I = \mathfrak{c}\mathfrak{f}^{-1} \quad \text{with } \mathfrak{c}, \mathfrak{f} \text{ integral}, \quad (\mathfrak{c}, \mathfrak{f}) = 1.$$

Define an equivalence relation $I \sim I'$ if and only if

$$\mathfrak{f} = \mathfrak{f}' \quad \text{and} \quad \mathfrak{c}'\mathfrak{c}^{-1} \in P_1(\mathfrak{f}),$$

where $P_1(\mathfrak{f})$ is the group of principal ideals (α) such that $\alpha \equiv 1 \bmod^* \mathfrak{f}$. We let

$X = \mathfrak{I} = \mathfrak{I}'$ modulo the above equivalence relation.

Note that if $I \sim I'$ then $\mathfrak{a}I \sim \mathfrak{a}I'$ for all ideals $\mathfrak{a}$. Observe that $P_1(\mathfrak{o})$ by definition is the group of principal fractional ideals, so the equivalence classes of integral ideals are the ordinary ideal classes of K. A distribution is said to be punctured if it satisfies the distribution relations except for x equal to such an ideal class (i.e. x represented by an integral ideal).

Theorem 1.6. *Let $\mathfrak{I}$ be as above, and for $I \in \mathfrak{I}$ not integral define*

$$g(I) = g^{12/w(\mathfrak{f})}(1, \mathfrak{f}\mathfrak{c}^{-1})$$

with values in the multiplicative group modulo roots of unity. Let M be the monoid of all ideals, operating on $\mathfrak{I}$ by multiplication. Then

$$I \mapsto g(I)$$

is a punctured M-distribution.

This is verified directly from Theorem 1.4. We shall see later that this distribution has essentially maximal rank. Contrary to the situation of Chapter 1 on $\mathbf{Q}^k/\mathbf{Z}^k$, the universal distribution in the present case may have torsion. Cf. Kersey's thesis [Ke 2]. In the case of prime power conductor, one can in fact show that a variant of this distribution is universal. See Theorem 3.1 of Chapter 13.

If one wants to omit the "punctured" condition from the statement of the theorem, then one has either to refine the notion of distribution at the base level, or one has to factor something further in the group of values, for instance $K(1)^*$. This is in analogy with the generic case, of the modular units in the function field, where one factors out the constants.

§2. Stickelberger Elements

Let $\mathfrak{f} \neq (1)$ Let χ be a non trivial character of $\mathrm{Cl}(\mathfrak{f})$, not necessarily primitive. We also write $\chi_{\mathfrak{f}}$ if we emphasize the dependence on $\mathfrak{f}$. Define

$$S_{\mathfrak{f}}(\chi, g_{\mathfrak{f}}) = \sum_{C} \chi(C) \log |g_{\mathfrak{f}}(C)|,$$

where the sum is taken over $C \in \mathrm{Cl}(\mathfrak{f})$. We define

$$T_{\mathfrak{f}}(\chi, g_{\mathfrak{f}}) = \frac{1}{r(\mathfrak{f})} S_{\mathfrak{f}}(\chi, g_{\mathfrak{f}}),$$

and recall that $r(\mathfrak{f}) = w(\mathfrak{f})N(\mathfrak{f})$.

Let $\mathfrak{f} = (1)$. Then we define $S_{(1)}$ in the same way, putting

$$g_{(1)}(c) = \delta(c) \qquad \text{for } c \in \mathrm{Cl}(1),$$

and we define

$$T_{(1)}(\chi, g_{(1)}) = \frac{1}{wh} S_{(1)}(\chi, g_{(1)}).$$

If $f(\chi)$ or f_χ is the conductor of χ, we define the sum (without a subscript)

$$S(\chi, g) = S_{\mathfrak{f}(\chi)}(\chi_0, g_{\mathfrak{f}(\chi)}).$$

where χ_0 is the character corresponding to χ on $\mathrm{Cl}(\mathfrak{f}_\chi)$.

Theorem 2.1. *Let $\mathfrak{f} | \mathfrak{f}'$, let χ be a non-trivial character on* $\mathrm{Cl}(\mathfrak{f})$, *and let χ' be the corresponding character on* $\mathrm{Cl}(\mathfrak{f}')$. *Then*

$$T_{\mathfrak{f}'}(\chi', g_{\mathfrak{f}'}) = \prod_{\substack{\mathfrak{p} | \mathfrak{f}' \\ \mathfrak{p} \nmid \mathfrak{f}}} (1 - \chi(\mathfrak{p})) T_{\mathfrak{f}}(\chi, g_{\mathfrak{f}}).$$

In particular, if $\mathfrak{f}$ and $\mathfrak{f}'$ have the same prime factors, then

$$T_{\mathfrak{f}'}(\chi', g_{\mathfrak{f}'}) = T_{\mathfrak{f}}(\chi, g_{\mathfrak{f}}).$$

Proof. It suffices to prove the theorem when $\mathfrak{f}' = \mathfrak{f}\mathfrak{p}$ for some prime $\mathfrak{p}$. We have:

$$S_{\mathfrak{f}\mathfrak{p}}(\chi_{\mathfrak{f}\mathfrak{p}}, g_{\mathfrak{f}\mathfrak{p}}) = \sum_{C \in \mathrm{Cl}(\mathfrak{f})} \chi(C) \sum_{C' | C} \log |g_{\mathfrak{f}\mathfrak{p}}(C')|.$$

Case 1. $\mathfrak{p} | \mathfrak{f}$.

In this case, the desired relation follows immediately from Theorem 1.3.

Case 2. $\mathfrak{p} \nmid \mathfrak{f}$.

We leave the case $\mathfrak{f} \neq (1)$ to the reader, and carry out the case $\mathfrak{f} = (1)$. Let $\mathfrak{b}$ be an ideal in C^{-1}. Then:

$$S_{\mathfrak{p}}(\chi, g_{\mathfrak{p}}) = \frac{w(\mathfrak{p})}{w} \sum_{C \in \mathrm{Cl}(1)} \chi(C) \frac{w}{w(\mathfrak{p})} \sum_{C' | C} \log |g_{\mathfrak{p}}(C')|$$

$$= \frac{w(\mathfrak{p})}{w} p \sum_{C \in \mathrm{Cl}(1)} \chi(C) \log \frac{|\Delta(\mathfrak{b})|}{|\Delta(\mathfrak{b}\mathfrak{p})|} \quad \text{(by Theorem 1.4)}$$

$$= \frac{w(\mathfrak{p})}{w} p \sum_{C} \chi(C) \log |g_{(1)}(C) \mathrm{N}\mathfrak{p}^{6} / g_{(1)}(C\mathfrak{p}^{-1})|,$$

$$= \frac{w(\mathfrak{p})}{w} p(1 - \chi(\mathfrak{p})) \frac{1}{h} S_{(1)}(\chi, g_{(1)})$$

by making the transformation of variables $C \mapsto C\mathfrak{p}$. This proves the theorem.

The sums $S_{\mathfrak{f}}(\chi, g_{\mathfrak{f}})$ arise in the Kronecker limit formula and its application to L-series. Because of the way we defined the values $g_{\mathfrak{f}}(C)$, and the normalization by means of Klein forms, we need statements which are slightly differently normalized from those in the literature, so we repeat these statements here.

KRONECKER LIMIT FORMULA. *Let u, v be real numbers which are not both integers. Define*

$$E_{u,v}(\tau, s) = \sum_{(m,n) \neq (0,0)} e^{2\pi i(mu + nv)} \frac{y^{s}}{|m\tau + n|^{2s}}.$$

Then

$$E_{u,v}(\tau, 1) = -2\pi \log |g_{-v,u}(\tau)|$$

where g_{a_1,a_2} is the Siegel function.

The proof is the same as that given in Siegel [Sie], see also [L 5], Chapter 20, §5. This is then applied to the L-series as in the following theorem.

Theorem 2.2. *Let $\mathfrak{d}$ be the different of $K/\mathbf{Q}$, and let γ be an element of K such that $\gamma\mathfrak{d}\mathfrak{f}$ is an ideal prime to $\mathfrak{f}$. Let the Gauss sum for a non-trivial character χ of $\mathrm{Cl}(\mathfrak{f})$ be*

$$\tau(\chi) = -\sum_{x \in \mathfrak{o}(\mathfrak{f})} \chi(x\gamma\mathfrak{d}\mathfrak{f}) e^{2\pi i \operatorname{Tr}(x\gamma)}.$$

Then:

LF 1. $$L(1,\chi_0) = \frac{2\pi}{6r(\mathfrak{f}_\chi)\tau(\bar{\chi}_0)\sqrt{d_K}} S_{\mathfrak{f}(\chi)}(\bar{\chi}_0, g_{\mathfrak{f}(\chi)})$$

This formula is originally due to Meyer [Me], Cf. also Siegel [Si] and Ramachandra [Ra], and [L 5], Chapter 21, Theorem 2. Again, the normalizations are slightly different because of the twist already mentioned in the addendum of §1. In fact, the theorem should have been stated and proved as we are now giving it, since the normalization of the Siegel functions that we are using, and the definition of the invariants seem most natural. However, the present version is also easily deduced formally from Theorem 2 of [L 5], Chapter 21, by using the addendum of §1, and Proposition 1.3 of Chapter 2.

If we multiply the equation **LF 1** with the Euler factor as in Theorem 2.1, then we find the relation

LF 2. $$\prod_{\mathfrak{p}|\mathfrak{f}} (1 - \bar{\chi}_0(\mathfrak{p}))L(1,\chi_0) = \frac{2\pi}{6r(\mathfrak{f}_\chi)\tau(\bar{\chi}_0)} S_{\mathfrak{f}}(\bar{\chi}, g_{\mathfrak{f}}).$$

The non-vanishing of the value of the L-function at 1 is given by the usual factorization of zeta functions, which yields:

LF 3. $$\frac{h_H R_H}{w_H} = \frac{h}{w} \prod_{\chi \neq 1} \frac{\mathbf{N}\mathfrak{f}(\chi)^{1/2}}{2r(\mathfrak{f}_\chi)\tau(\bar{\chi}_0)} S_{\mathfrak{f}(\chi)}(\bar{\chi}_0, g_{\mathfrak{f}(\chi)}).$$

Formula **LF 2** then shows exactly when the sum $S_{\mathfrak{f}}$ vanishes.

We define for $C \in \mathrm{Cl}(\mathfrak{f})$:

$$\boxed{\varphi_{\mathfrak{f}}(C) = g_{\mathfrak{f}}(C)/g_{\mathfrak{f}}(C_0).}$$

Then $\varphi_{\mathfrak{f}}(C)$ is a unit, and its absolute values at infinity satisfy the product formula.

Let H be an abelian extension of K of conductor $\mathfrak{f} \neq (1)$.

We define for $C \in \mathrm{Cl}(H/K)$:

$$\varphi_H(C) = \mathbf{N}_{K(\mathfrak{f})/H}\varphi_{\mathfrak{f}}(C')$$

where C' is any class in $\mathrm{Cl}(\mathfrak{f})$ lying above C. Similarly,

$$g_H(C) = \mathbf{N}_{K(\mathfrak{f})/H} g_{\mathfrak{f}}(C').$$

We let

Φ_H = group generated by the units $\varphi_H(C)$ for $C \in \mathrm{Cl}(H/K)$.

We abbreviate $\Phi_{K_{\mathfrak{f}}} = \Phi_{\mathfrak{f}}$. Let

$$G_H = \mathrm{Gal}(H/K) \quad \text{and} \quad G_{\mathfrak{f}} = \mathrm{Gal}(K(\mathfrak{f})/K).$$

We write σ_C instead of $\sigma(C)$ for typographical reasons. We let

$$\rho : E_H \to \mathbf{C}[G_H]$$

be the usual "regulator" map,

$$u \mapsto \rho(u) = \sum_C \log |u^{\sigma(C)}| \sigma_C^{-1}.$$

where E_H as usual is the group of units in H. Then the image of E_H is a discrete subgroup contained in the hyperplane on which the trivial character vanishes. The kernel consists of the roots of unity.

The regulator map should really be indexed ρ_H. We then have a commutative diagram:

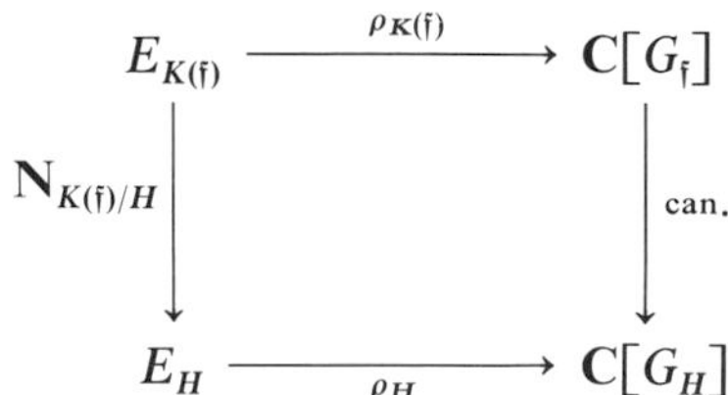

The vertical arrow on the right is the canonical homomorphism arising from the natural map $\mathrm{Cl}(\mathfrak{f}) \to \mathrm{Cl}(H/K)$, or in terms of Galois groups, the restriction map from $\mathrm{Gal}(K(\mathfrak{f})/K)$ to $\mathrm{Gal}(H/K)$.

The rank of Φ_H is the same as the $\mathbf{C}$-dimension of the vector space $\mathbf{C}\rho(\Phi_H)$ by the Dirichlet unit theorem. As in [Ro 1], §4.6, we shall determine this rank by determining the eigenspace decomposition with respect to non-trivial characters of G_H.

Theorem 2.3. *Let H be an abelian extension of K with conductor $\mathfrak{f} \neq (1)$. Then:*

(i) *We have the eigenspace decomposition*

$$\mathbf{C}\rho(\Phi_H) = \bigoplus_{\chi \neq 1} \mathbf{C}S_{\mathfrak{f}}(\bar{\chi}, g_{\mathfrak{f}})e_{\chi}$$

where the sum is over non-trivial characters of G_H, identified with $\mathrm{Cl}(H/K)$, *and e_χ is the usual idempotent associated with χ.*

(ii) *Let m be the number of characters χ such that*

$$\prod_{\mathfrak{p}|\mathfrak{f}} (1 - \chi_0(\mathfrak{p})) = 0.$$

Then

$$\text{rank } \Phi_H = v - m, \quad \textit{where } v = [H:K].$$

Proof. The first assertion is immediate from the commutative diagram giving the compatibility of the image of the regulator map for H and $K(\mathfrak{f})$. The second follows from formula **LF 2** and the fact that $L(1, \chi_0) \neq 0$.

§3. Ideal Factorization of the Siegel Numbers

We determine the ideal factorization of the algebraic numbers $g_{\mathfrak{f}}(C)$.

Let $x \sim y$ denote the fact that (xy^{-1}) is the unit ideal.

Theorem 3.1. *For any ideal $\mathfrak{c}$, and ideal $\mathfrak{a}$, we have*

$$\frac{\Delta(\mathfrak{c}\mathfrak{a})}{\Delta(\mathfrak{a})} \sim \mathfrak{c}^{-12}.$$

Proof. Let $\mathfrak{b}$ be an ideal prime to $\mathbf{N}\mathfrak{c}$ such that $\mathfrak{b}\mathfrak{c}$ is principal, say $\mathfrak{b}\mathfrak{c} = \lambda\mathfrak{o}$. Then

$$\mathbf{N}\mathfrak{b}^{12} \frac{\Delta(\mathfrak{b}\mathfrak{c}\mathfrak{a})}{\Delta(\mathfrak{c}\mathfrak{a})} \mathbf{N}\mathfrak{c}^{12} \frac{\Delta(\mathfrak{c}\mathfrak{a})}{\Delta(\mathfrak{a})} = \mathbf{N}\mathfrak{b}^{12}\mathbf{N}\mathfrak{c}^{12}\lambda^{-12}.$$

By [L 5], Theorem 4 of Chapter 12, §2 we know that the first factor on the left divides $\mathbf{N}\mathfrak{b}^{12}$, which is prime to $\mathbf{N}\mathfrak{c}$, and the second factor on the left divides $\mathbf{N}\mathfrak{c}^{12}$. On the other hand, $\mathfrak{c}$ appears with multiplicity 1 in the factorization of λ. This proves the theorem.

Theorem 3.2. *Let ϕ be the Euler function for ideals, that is*

$$\phi(\mathfrak{p}^n) = (\mathbf{N}\mathfrak{p} - 1)\mathbf{N}\mathfrak{p}^{n-1}$$

for prime powers. For $C \in \mathrm{Cl}(\mathfrak{p}^n)$, we have

$$g_{\mathfrak{p}^n}(C) \sim \mathfrak{p}^{12N(\mathfrak{p}^n)/\phi(\mathfrak{p}^n)}.$$

Proof. For any classes C, C' we know by Theorem 1.2 that

$$g_{\mathfrak{p}^n}(C) \sim g_{\mathfrak{p}^n}(C').$$

First consider the case $n = 1$. Given a class $c \in \mathrm{Cl}(1)$, the number of classes $C \in \mathrm{Cl}(\mathfrak{p})$ lying above c is equal to

$$[K(\mathfrak{p}) : K(1)] = \frac{\mathbf{N}\mathfrak{p} - 1}{w/w(\mathfrak{p})} = v(\mathfrak{p}, \mathfrak{o}), \quad \text{say.}$$

Cf. [L 4], Theorem 1 of Chapter VI, §1 (the elementary formula for the orders of the generalized ideal class groups). By the distribution relation of Theorem 1.4(ii) we get

$$g_{\mathfrak{p}}(c)^{v(\mathfrak{p},\mathfrak{o})w/w(\mathfrak{p})} = \frac{\Delta^p(\mathfrak{c})}{\Delta^p(\mathfrak{p}\mathfrak{c})} \sim \mathfrak{p}^{12p}$$

by Theorem 3.1. This proves the desired formula in the present case.

Next, from level $\mathfrak{p}^{n+1}$ to level $\mathfrak{p}^n$, we note that the number of classes C' in $\mathrm{Cl}(\mathfrak{p}^{n+1})$ lying above a given class C in $\mathrm{Cl}(\mathfrak{p}^n)$ is equal to

$$[K(\mathfrak{p}^{n+1}) : K(\mathfrak{p}^n)] = \frac{\mathbf{N}\mathfrak{p}}{w(\mathfrak{p}^n)/w(\mathfrak{p}^{n+1})}.$$

Using the distribution relation of Theorem 1.3, we find that

$$g_{\mathfrak{p}^{n+1}}(C')^{\mathbf{N}\mathfrak{p}} \sim g_{\mathfrak{p}^n}(C)^{N(\mathfrak{p}^{n+1})/N(\mathfrak{p}^n)}$$

The theorem follows at once.

An equivalent formulation is given in Robert [Ro 1], *Remarque*, p. 18, 19.

Let $\mathfrak{f}$ be an ideal $\neq (1)$, and $\mathfrak{a}$ an ideal such that $\mathfrak{f} \nmid \mathfrak{a}$. For any ideal $\mathfrak{c}$ prime to $\mathfrak{f}$, define

$$\mathfrak{k}_{\mathfrak{a}}^{(\mathfrak{f})}(\mathfrak{c}) = \frac{\mathfrak{k}(1, \mathfrak{f}\mathfrak{c}^{-1})^{\mathbf{N}\mathfrak{a}}}{\mathfrak{k}(1, \mathfrak{f}\mathfrak{c}^{-1}\mathfrak{a}^{-1})}.$$

If C is the class of $\mathfrak{c}$ in $\mathrm{Cl}(\mathfrak{f})$, and $\mathbf{N}\mathfrak{a} \equiv 1 \bmod 12$, we let

$$v_{\mathfrak{a}}^{(\mathfrak{f})}(C) = v_{\mathfrak{a}}(C) = k_{\mathfrak{a}}^{(\mathfrak{f})}(\mathfrak{c})\,\Delta(\mathfrak{f}\mathfrak{c}^{-1})^{(\mathbf{N}\mathfrak{a}-1)/12}.$$

Thus

$$v_{\mathfrak{a}}(C_0) = \frac{\mathfrak{k}(1, \mathfrak{f})^{\mathbf{N}\mathfrak{a}}}{\mathfrak{k}(1, \mathfrak{f}\mathfrak{a}^{-1})} \Delta(\mathfrak{f})^{(\mathbf{N}\mathfrak{a}-1)/12}.$$

Indeed, by Chapter 10, §2 we see that the expression for $v_{\mathfrak{a}}^{(\mathfrak{f})}(\mathfrak{c})$ depends only on the class C, using an argument similar to the beginning of §1. It will be shown in the next section that $v_{\mathfrak{a}}(C)$ lies in $K(\mathfrak{f})$, and satisfies the usual formula

$$v_{\mathfrak{a}}(C)^{\sigma(C')} = v_{\mathfrak{a}}(CC').$$

This is irrelevant for the considerations in the rest of this section.

Theorem 3.3. *Let* $N = N(\mathfrak{f})$, *let* $(\mathfrak{a}, \mathfrak{f}) = 1$, *and* $\mathbf{N}\mathfrak{a} \equiv 1 \bmod 12$. *Then*

(i) $$v_{\mathfrak{a}}^{(\mathfrak{f})}(C)^{12N} = \frac{g_{\mathfrak{f}}(C)^{\mathbf{N}\mathfrak{a}}}{g_{\mathfrak{f}}(C\mathfrak{a}^{-1})} \frac{\Delta(\mathfrak{f}\mathfrak{c}^{-1})^N}{\Delta(\mathfrak{f}\mathfrak{c}^{-1}\mathfrak{a}^{-1})^N}.$$

Furthermore, if $\mathfrak{f}$ *is not a prime power, then*

(ii) $$v_{\mathfrak{a}}^{(\mathfrak{f})}(C) \sim \mathfrak{a}.$$

If $\mathfrak{f} = \mathfrak{p}^n$ *is a prime power, then*

(iii) $$v_{\mathfrak{a}}^{(\mathfrak{f})}(C) \sim \mathfrak{p}^{(\mathbf{N}\mathfrak{a}-1)/\phi(\mathfrak{f})}\mathfrak{a}.$$

Proof. The expression (i) is an immediate consequence of the definition $g^{12} = \mathfrak{k}^{12}\Delta$. The ideal factorizations of (ii) and (iii) are then corollaries of Theorems 3.1 and 3.2.

We shall now prove for modular units the analogue of the classical congruence formula for cyclotomic units

$$\frac{\zeta^a - 1}{\zeta - 1} \equiv a \bmod \mathfrak{P}.$$

No use will be made of this in the rest of the book.

Let F be a field complete under a discrete valuation, with ring of integers $\mathfrak{o}$, maximal ideal $\mathfrak{p}$, prime element π. Let A be an elliptic curve defined over F, with good reduction at $\mathfrak{p}$. We denote by a tilde the operation of reduction, so

$$\tilde{A}, \quad \tilde{\mathfrak{o}}, \quad \tilde{f}, \quad \text{etc.}$$

denote respectively the reduced curve, the origin on $\tilde{A}$, a reduced function $\tilde{f}$ for $f \in F(A)$, etc.

We may view A as a scheme over $\mathfrak{o}$. If A is defined by an equation in Weierstrass form, and (x, y) are the coordinates of a Weierstrass model (cf. [L 1], Appendix 1, §3), let $R = \mathfrak{o}[x, y]$. The point $\tilde{O}$ on $\tilde{A}$ corresponds to a maximal ideal $M \in \operatorname{spec} R$, and the local ring R_M is the local ring of $\mathfrak{p}$-finite functions which have no pole at $\tilde{O}$. We call R_M the **local ring of** $\tilde{O}$ in the function field $F(A)$. Let z be a local parameter in $F(A)$ at $\tilde{O}$. This is equivalent to saying that π, z generate the maximal ideal of R_M. The completion of this local ring can then be identified with the power series ring $\mathfrak{o}[[z]]$. With respect to the Weierstrass model, we may take $z = -x/y$.

The formal group of A in F will be denoted by $A_1(F)$, or A_1. It consists of those points P such that $z(P) \in \mathfrak{p}$. We let A_i be the subgroup of elements P such that

$$\operatorname{ord}_{\mathfrak{p}} z(P) \geqq i.$$

For simplicity of notation, we shall identify the formal group with $\mathfrak{p}$ itself, if the local parameter z has been fixed. The group $A_1(F)$ consists precisely of the points in $A(F)$ whose reduction is $\tilde{O}$.

Let ω be a differential of first kind, defined over F. We shall assume that the elements of $\operatorname{End}(A)$ are all defined over F. We assume fixed an embedding

$$\varkappa : \operatorname{End}(A) \to \mathfrak{o}$$

which is **normalized**, that is $\omega \circ \alpha = \varkappa(\alpha)\omega$.

Theorem 3.4. *Let $f \in F(A)$ be a function such that $\tilde{f} \neq 0, \infty$. Assume that $\operatorname{ord}_O f = r(f) \geqq 1$ and that any point in the divisor of f other than O does not lie in the formal group (in any finite extension of F). Let $Q \in A_i(F)$ but $Q \notin A_{i+1}(F)$. Then*

$$\operatorname{ord}_{\mathfrak{p}} f(Q) = i \cdot r(f).$$

Furthermore, if $\alpha \in \operatorname{End}(A)$ and $\varkappa(\alpha) \in \mathfrak{o}^*$ is a unit at $\mathfrak{p}$, then

$$\frac{f(\alpha Q)}{f(Q)} \equiv \varkappa(\alpha)^{r(f)} \bmod \mathfrak{p}.$$

Proof. The function field $F(A)$ is embedded in the quotient field of $\mathfrak{o}[[z]]$. The hypothesis on the poles of f implies that $f \in R_M$, and $R_M \subset \mathfrak{o}[[z]]$, so f has a power series expansion

$$f = a_r z^r + \text{higher terms}.$$

The hypothesis $\tilde{f} \neq 0$ implies that for some coefficient, we have $\tilde{a}_v \neq 0$, so

$$f(z) = a_r z^r + \cdots + a_m z^m + \text{higher terms},$$

such that a_m is a unit, and $a_n \equiv 0 \bmod \mathfrak{p}$ for $n < m$. If $m > r$ then f has a zero in the formal group (namely, a zero of the Weierstrass polynomial of the above power series). Hence $m = r$ and $a_r \in \mathfrak{o}^*$, that is a_r is a unit. The first statement is then clear since the term $a_r z^r$ dominates the power series if $\operatorname{ord}_{\mathfrak{p}} z > 0$.

As to the second statement, let us take $z = -x/y$. As in [L 1], Appendix 1, §3, the invariant differential (suitably normalized) has an expansion

$$\omega = H(z)\,dz,$$

where $H(z)$ is a power series with integral coefficients, and leading coefficient 1 (the constant term), $H(z) = 1 +$ higher terms. Then

$$H(z \circ \alpha) \equiv 1 \bmod z,$$
$$\varkappa(\alpha)\omega = \omega \circ \alpha = H(z \circ \alpha)\,d(z \circ \alpha) \equiv \varkappa(\alpha)\,dz \bmod z\,dz.$$

If $z \circ \alpha \equiv \alpha_1 z \bmod z^2$, then $d(z \circ \alpha) \equiv \alpha_1\,dz \bmod z\,dz$, and therefore $\varkappa(\alpha) = \alpha_1$. This proves the second statement.

We apply the local result to get a congruence property for the invariants $v_{\mathfrak{a}}(C)$.

Theorem 3.5. *Let* $\mathfrak{f} = \mathfrak{p}^n$, $(\mathfrak{a}, \mathfrak{f}) = 1$, $\mathbf{N}\mathfrak{a} \equiv 1 \bmod 12$. *Let α be a $\mathfrak{p}$-unit in K, and an algebraic integer. Then for any prime $\mathfrak{P} | \mathfrak{p}$ in K^{ab}, we have*

$$\frac{v_{\mathfrak{a}}(C\alpha)}{v_{\mathfrak{a}}(C)} \equiv 1 \bmod \mathfrak{P}.$$

Proof. Without loss of generality, we may assume that $C = C_0$. We then have

$$\begin{aligned}\frac{v_{\mathfrak{a}}(C_0\alpha)}{v_{\mathfrak{a}}(C_0)} &= \alpha^{1-\mathbf{N}\mathfrak{a}} \frac{\mathfrak{k}(\alpha, \mathfrak{f})^{\mathbf{N}\mathfrak{a}}}{\mathfrak{k}(\alpha, \mathfrak{f}\mathfrak{a}^{-1})} \frac{\mathfrak{k}(1, \mathfrak{f}\mathfrak{a}^{-1})}{\mathfrak{k}(1, \mathfrak{f})^{\mathbf{N}\mathfrak{a}}} \\ &= \alpha^{1-\mathbf{N}\mathfrak{a}} \prod_t \frac{x(\varphi(1)) - x(\varphi(t))}{x(\varphi(\alpha)) - x(\varphi(t))}.\end{aligned}$$

But the function $\prod (x - x(\varphi(t)))^{-1}$ satisfies the hypotheses of Theorem 3.4, and x has a zero of order 2 at infinity (the origin of the elliptic curve). Furthermore,

$$(\mathfrak{f}\mathfrak{a}^{-1} : \mathfrak{f}) = (L' : L) = \mathbf{N}\mathfrak{a},$$

and there are $(\mathbf{N}\mathfrak{a} - 1)/2$ elements in the product, taken over $t \in (L'/L)/\pm 1$, $t \neq 0$. By the above mentioned theorem, the product is congruent to $\alpha^{\mathbf{N}\mathfrak{a}-1}$, and the desired result follows.

Finally we apply this to $g_{\mathfrak{f}}$.

Theorem 3.6. *Let $\mathfrak{f} = \mathfrak{p}^n$, $\mathfrak{p} \nmid 6$ and let α be an algebraic integer in K, prime to $\mathfrak{f}$. Let $N = N(\mathfrak{f})$. Then for any prime $\mathfrak{P}|\mathfrak{p}$ in $K(\mathfrak{f})$, we have*

$$\frac{g_{\mathfrak{f}}(C_0\alpha)}{g_{\mathfrak{f}}(C_0)} \equiv \alpha^{12N} \bmod \mathfrak{P}.$$

Proof. Choose an ideal $\mathfrak{b}$ satisfying the following conditions:

(1) $\mathfrak{b} = (\beta)$ is principal, and $\mathbf{N}\mathfrak{b} \equiv 1 \bmod 12\mathfrak{f}$;
(2) $(\mathbf{N}\mathfrak{b} - 1)/12$ is prime to $\mathbf{N}\mathfrak{P}_1 - 1$, where $\mathfrak{P}_1$ is any prime of $K(1)$ lying above $\mathfrak{p}$.

The existence of $\mathfrak{b}$ follows from the lemma on roots of unity, Lemma 4.2 of Chapter 9. Then

$$v_{\mathfrak{b}}(C_0) = \frac{\mathfrak{k}(1, \mathfrak{f})^{\mathbf{N}\mathfrak{b}}}{\mathfrak{k}(1, \mathfrak{f}\beta^{-1})} \Delta(\mathfrak{f})^{(\mathbf{N}\mathfrak{b}-1)/12}$$

and therefore, using the fact that $\mathfrak{k}(z, L)$ is homogeneous of degree 1 in (z, L), we find:

$$\begin{aligned} v_{\mathfrak{b}}(C_0)^N &= \beta^N \mathfrak{k}^{12N}(1, \mathfrak{f})^{(\mathbf{N}\mathfrak{b}-1)/12} \Delta(\mathfrak{f})^{N(\mathbf{N}\mathfrak{b}-1)/12}. \\ &= \beta^N g_{\mathfrak{f}}(C_0)^{(\mathbf{N}\mathfrak{b}-1)/12}, \end{aligned}$$

because $g_{\mathfrak{f}}(C_0) = \mathfrak{k}^{12N}(1, \mathfrak{f})\Delta(\mathfrak{f})^N$. Similarly we obtain

$$v_{\mathfrak{b}}(C_0\alpha)^N = \beta^N g_{\mathfrak{f}}(C_0\alpha)^{(\mathbf{N}\mathfrak{b}-1)/12} \alpha^{-12N(\mathbf{N}\mathfrak{b}-1)/12}.$$

By Theorem 3.5, we obtain

$$\left(\frac{g_{\mathfrak{f}}(C_0\alpha)}{g_{\mathfrak{f}}(C_0)}\right)^{(\mathbf{N}\mathfrak{b}-1)/12} \equiv \alpha^{12N(\mathbf{N}\mathfrak{b}-1)/12} \bmod \mathfrak{P}.$$

Note that α lies in K. We chose $\mathfrak{b}$ such that $(\mathbf{N}\mathfrak{b} - 1)/12$ is prime to the order of the multiplicative group of the residue class field of $\mathfrak{P}_1$ in $K(1)$, and we know that $K(\mathfrak{p}^n)$ is totally ramified over $K(1)$, so there is no further residue class field extension in that extension. Hence we may extract the $(\mathbf{N}\mathfrak{b} - 1)/12$ root of the above relation to conclude the proof of the theorem.

Remark. The theorem extends in the same manner to Klein invariants as defined in Chapter 12.

§4. The Robert Group in the Ray Class Field

Let $\mathfrak{f} \neq (1)$. Let $C \in \mathrm{Cl}(\mathfrak{f})$, and let $\mathfrak{c}$ be an ideal in C. Let $\mathfrak{a}$ be an ideal not divisible by $\mathfrak{f}$. We define the **Robert invariant**

$$\boxed{u_{\mathfrak{a}}(\mathfrak{c}) = u_{\mathfrak{a}}^{(\mathfrak{f})}(\mathfrak{c}) = \frac{g^{12}(1, \mathfrak{f}\mathfrak{c}^{-1})^{\mathbf{N}\mathfrak{a}}}{g^{12}(1, \mathfrak{f}\mathfrak{a}^{-1}\mathfrak{c}^{-1})}.}$$

By Chapter 10, §2 we see that this depends only on the class C, using an argument similar to Chapter 11, §1 for the definition of the invariants $g_{\mathfrak{f}}(C)$. Note that $u_{\mathfrak{a}}(C)$ is not necessarily a unit, but we shall take products of such numbers with a linear condition to get units. By definition, we have

$$u_{\mathfrak{a}}(C_0) = \frac{g^{12}(1, \mathfrak{f})^{\mathbf{N}\mathfrak{a}}}{g^{12}(1, \mathfrak{f}\mathfrak{a}^{-1})}.$$

Theorem 4.1. *Let* $\mathfrak{a}$ *be an ideal,* $\mathfrak{f} \nmid \mathfrak{a}$. *Then* $u_{\mathfrak{a}}(C)$ *lies in* $K(\mathfrak{f})$, *If* $\mathfrak{b}$ *is an ideal prime to* $\mathfrak{f}$, *then*

$$\sigma(\mathfrak{b}) u_{\mathfrak{a}}(C) = u_{\mathfrak{a}}(C\mathfrak{b}).$$

Similarly, if $\mathbf{N}\mathfrak{a} \equiv 1 \bmod 12$, *then* $v_{\mathfrak{a}}(C) \in K(\mathfrak{f})$ *and*

$$v_{\mathfrak{a}}(C)^{\sigma(C')} = v_{\mathfrak{a}}(CC').$$

Proof. Let us begin with $u_{\mathfrak{a}}(C)$. We first determine how this element transforms under an appropriately chosen automorphism of the complex numbers over K. For this, we use Shimura's commutative diagram for complex multiplication, cf. [L 5], Chapter 10, Theorem 3, §2.

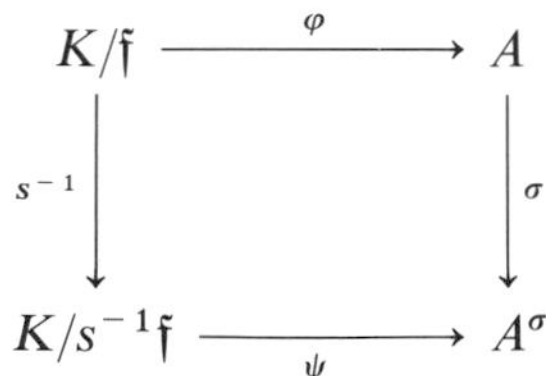

In this diagram, A is an elliptic curve in Weierstrass form, defined over $K(j(\mathfrak{o}))$, say by the equation

$$y^2 = 4x^3 - g_2 x - g_3.$$

Furthermore, s is an idele of K, and σ an automorphism of the complex numbers such that α restricted to K^{ab} is equal to (s, K). We select s as follows. If $\mathfrak{p}$ is a prime not dividing $\mathfrak{b}$, we let the $\mathfrak{p}$-component $s_\mathfrak{p} = 1$. If $\mathfrak{p} | \mathfrak{b}$ we let $\text{ord}_\mathfrak{p}\, s_\mathfrak{p} = \text{ord}_\mathfrak{p}\, \mathfrak{b}$.

The curve A^σ is defined by the equation

$$y_\sigma^2 = 4x_\sigma^3 - g_2^\sigma x_\sigma - g_3^\sigma,$$

where y_σ, x_σ are new variables.

The maps φ, ψ are analytic parametrizations of A and A^σ respectively. There exists $\lambda \in \mathbf{C}^*$ such that for all $t \in K/\mathfrak{f}$, $t \neq 0$, we have

$$x \circ \varphi(t) = \wp(\lambda t, \lambda\mathfrak{f}).$$

Furthermore,

$$\Delta(A) = \Delta(\lambda\mathfrak{f}) = g_2^3 - 27g_3^2.$$

We recall the formula for the Klein form, Theorem 2.2 of Chapter 10:

$$\mathfrak{k}^{12}(z, L'/L) = \prod_t \frac{1}{(\wp(z, L) - \wp(t, L))^6}$$

where the product is taken over $t \in L'/L$ and $t \neq 0$. We apply this to the lattices

$$L = \mathfrak{f} \quad \text{and} \quad L' = \mathfrak{f}\mathfrak{a}^{-1}.$$

Let

$$F(x) = \frac{\Delta(\mathfrak{f})}{\Delta(\mathfrak{f}\mathfrak{a}^{-1})} \prod_t \Delta(A)(x - x \circ \varphi(t))^{-6}.$$

Then

$$(*) \qquad u_\mathfrak{a}(C_0) = u_\mathfrak{a}(1, \mathfrak{f}) = F(x(\varphi(1))).$$

On the other hand, we may repeat the same thing for the curve A^σ. There exists an element $\lambda' \in \mathbf{C}^*$ such that

$$x_\sigma(\psi(t')) = \wp(\lambda' t', \lambda' s^{-1}\mathfrak{f}) \quad \text{and} \quad \Delta(A^\sigma) = \Delta(\lambda' s^{-1}\mathfrak{f}).$$

Note that $s^{-1}\mathfrak{f} = \mathfrak{f}\mathfrak{b}^{-1}$. We now use the lattices $\mathfrak{f}\mathfrak{b}^{-1}$ and $\mathfrak{f}\mathfrak{a}^{-1}\mathfrak{b}^{-1}$. We let

$$F_\sigma(x_\sigma) = \frac{\Delta(\mathfrak{f}\mathfrak{b}^{-1})}{\Delta(\mathfrak{f}\mathfrak{a}^{-1}\mathfrak{b}^{-1})} \prod_{t'} \Delta(A^\sigma)(x_\sigma - x_\sigma \circ \psi(t'))^{-6},$$

where the product is taken over $t' \in \mathfrak{f}\mathfrak{a}^{-1}\mathfrak{b}^{-1}/\mathfrak{f}\mathfrak{b}^{-1}$, $t' \neq 0$. Then

$$(**) \qquad u_\mathfrak{a}(C_0\mathfrak{b}) = u_\mathfrak{a}(1, \mathfrak{f}\mathfrak{b}^{-1}) = F_\sigma(x_\sigma(\psi(1))).$$

Now we shall apply the automorphism σ to $u_\mathfrak{a}(C_0)$ in $(*)$, and we shall see that it yields $(**)$. Since

$$\left(\frac{\Delta(\mathfrak{f})}{\Delta(\mathfrak{f}\mathfrak{a}^{-1})}\right)^\sigma = \frac{\Delta(\mathfrak{f}\mathfrak{b}^{-1})}{\Delta(\mathfrak{f}\mathfrak{a}^{-1}\mathfrak{b}^{-1})}$$

by classical complex multiplication (see e.g. [L 5], Corollary of Theorem 6 of Chapter 11, §2), we get

$$\begin{aligned} F^\sigma(x_\sigma) &= \left(\frac{\Delta(\mathfrak{f})}{\Delta(\mathfrak{f}\mathfrak{a}^{-1})}\right)^\sigma \prod_t \Delta(A^\sigma)(x_\sigma - x_\sigma(\varphi(t)^\sigma))^{-6} \\ &= \frac{\Delta(s^{-1}\mathfrak{f})}{\Delta(s^{-1}\mathfrak{f}\mathfrak{a}^{-1})} \prod_t \Delta(\lambda' s^{-1}\mathfrak{f})(x_\sigma - \wp(\lambda' s^{-1}t, \lambda' s^{-1}\mathfrak{f}))^{-6} \end{aligned}$$

because $\varphi(t)^\sigma = \psi(s^{-1}t)$ by Shimura's diagram.

We substitute $x(\varphi(1))$ for x. We obtain:

$$F(x(\varphi(1)))^\sigma = \frac{\Delta(s^{-1}\mathfrak{f})}{\Delta(s^{-1}\mathfrak{f}\mathfrak{a}^{-1})} \prod \Delta(A^\sigma)(x^\sigma(\psi(s^{-1}1)) - \wp(\lambda' s^{-1}t, \lambda' s^{-1}\mathfrak{f}))^{-6}.$$

But

$$x^\sigma(\psi(s^{-1}1)) = \wp(\lambda'\psi(s^{-1}1), \lambda' s^{-1}\mathfrak{f}).$$

The conditions on s imply that $s^{-1}1 \equiv 1 \bmod s^{-1}\mathfrak{f}$, so

$$x^\sigma(\psi(s^{-1}1)) = \wp(\lambda'\psi(1), \lambda' s^{-1}\mathfrak{f}).$$

The map $t \mapsto s^{-1}t$ transforms the family $\{t\}$ to $\{t'\}$. Thus we see that $F(x(\varphi(1)))^\sigma = F_\sigma(x_\sigma(\psi(1)))$, and we use $(**)$.

This proves that $\sigma(\mathfrak{b})u_{\mathfrak{a}}(C_0) = u_{\mathfrak{a}}(C_0\mathfrak{b})$, whence it follows that for any class C we also have $\sigma(\mathfrak{b})u_{\mathfrak{a}}(C) = u_{\mathfrak{a}}(C\mathfrak{b})$. It also follows that $u_{\mathfrak{a}}(C) \in K(\mathfrak{f})$ by class field theory. This proves the theorem for $u_{\mathfrak{a}}(C)$.

The proof for $v_{\mathfrak{a}}(C)$ is entirely similar. We work with

$$\mathfrak{k}(z, L'/L) = \prod_t \frac{1}{\wp(z, L) - \wp(t, L)}$$

where the product is taken over $t \in (L'/L)/\pm 1$, $t \neq 0$, and $L = \mathfrak{f}$, $L' = \mathfrak{f}\mathfrak{a}^{-1}$. We use the rational function $F(x)$ given by

$$F(x) = \Delta(A)^{(\mathbf{N}\mathfrak{a}-1)/12} \prod_t \frac{1}{x - x \circ \varphi(t)},$$

and otherwise, the proof proceeds step by step exactly as before.

The elements $u_{\mathfrak{a}}(C)$ will turn out to yield generators for a certain group of units. It is convenient to establish the following notation. Let $\mathbf{I}$ be the free abelian group on the ideals of K. We write an element of $\mathbf{I}$ as

$$\mathbf{a} = \sum n(\mathfrak{a})\mathfrak{a}.$$

As usual, we define the **degree**

$$d(\mathbf{a}) = \deg \mathbf{a} = \sum n(\mathfrak{a}),$$

and we extend the norm by linearity, that is we put

$$\mathbf{N}\mathbf{a} = \sum n(\mathfrak{a})\mathbf{N}\mathfrak{a}.$$

For any ideal $\mathfrak{b}$, we define

$$\mathfrak{b} \cdot \mathbf{a} = \mathbf{a} \cdot \mathfrak{b} = \sum n(\mathfrak{a})(\mathfrak{a}\mathfrak{b}).$$

Let $\mathfrak{f} \neq (1)$. We assume (at least) throughout that all elements $\mathbf{a}$ as above are such that if $\mathfrak{a}$ occurs in $\mathbf{a}$ (that is, $n(\mathfrak{a}) \neq 0$) then $\mathfrak{a}$ is not divisible by $\mathfrak{f}$. Let $\mathfrak{c}$ be an ideal prime to $\mathfrak{f}$. We then define

$$g_{\mathfrak{f}}^{12}(\mathbf{a}; \mathfrak{c}) = \prod_{\mathfrak{a}} g^{12}(1, \mathfrak{f}\mathfrak{a}^{-1}\mathfrak{c}^{-1})^{n(\mathfrak{a})}$$

$$u(\mathbf{a}; C) = \prod_{\mathfrak{a}} u_{\mathfrak{a}}(C)^{n(\mathfrak{a})}.$$

There is a simple relation between these elements, as follows.

Let $\mathbf{a}$ satisfy the conditions $\mathbf{Na} = \deg \mathbf{a} = 0$. Then

$$g_{\mathfrak{f}}^{12}(\mathbf{a}; \mathfrak{c}) = u(\mathbf{a}; \mathfrak{c}) = u(\mathbf{a}\mathfrak{c}; \mathfrak{o}). \tag{1}$$

Conversely, let $\mathbf{a} = \sum m(\mathfrak{a})\mathfrak{a}$ satisfy the condition

$$\sum m(\mathfrak{a})(\mathbf{N}\mathfrak{a} - 1) = 0, \quad \text{i.e. } \mathbf{Na} = \deg \mathbf{a}.$$

Then we get

$$u(\mathbf{a}; \mathfrak{c}) = g_{\mathfrak{f}}^{12}(\mathbf{a}', \mathfrak{c}) \tag{2}$$

where $\mathbf{a}' = -\mathbf{a} + \mathbf{Na} \cdot \mathfrak{o}$ and $\mathbf{a}'$ satisfies $\mathbf{Na}' = \deg \mathbf{a}' = 0$. Indeed, we can write

$$u(\mathbf{a}; \mathfrak{c}) = \prod_{\mathfrak{a}} g^{12}(1, \mathfrak{f}\mathfrak{a}^{-1}\mathfrak{c}^{-1})^{-m(\mathfrak{a})} \prod_{\mathfrak{a}} g^{12}(1, \mathfrak{f}\mathfrak{c}^{-1})^{m(\mathfrak{a})\mathbf{N}\mathfrak{a}}.$$

From (1), we get a corollary to Theorem 4.1.

Corollary. *Assume that* $\mathbf{Na} = \deg \mathbf{a} = 0$. *Then* $g_{\mathfrak{f}}(\mathbf{a}; \mathfrak{c})$ *depends only on the ray class of* $\mathfrak{c}$ mod $\mathfrak{f}$, *and satisfies the usual rule*

$$g_{\mathfrak{f}}(\mathbf{a}; C)^{\sigma(C')} = g_{\mathfrak{f}}(\mathbf{a}; CC').$$

An element $\mathbf{a}$ will be said to be of **primitive Robert type (with respect to** $\mathfrak{f}$) if $\mathbf{a}$ is prime to $6N(\mathfrak{f})$ (that is, every $\mathfrak{a}$ occurring in $\mathbf{a}$ is prime to $6N(\mathfrak{f})$), and if $\mathbf{Na} = \deg \mathbf{a} = 0$. Note that from $\deg \mathbf{a} = 0$, we conclude that $g_{\mathfrak{f}}^{12}(\mathbf{a}; \mathfrak{c})$ is a unit by Theorem 1.2.

We define the **primitive Robert group** to be the group of all elements

$$g_{\mathfrak{f}}^{12}(\mathbf{a}; \mathfrak{o})$$

with all elements $\mathbf{a} \in \mathbf{I}$ of primitive Robert type. By Theorem 1.2, it follows that $\mathscr{R}_{\mathfrak{f}}^*$ is a group of units.

Theorem 4.2. *The Robert group* $\mathscr{R}_{\mathfrak{f}}^*$ *consists of all products*

$$\prod u_{\mathfrak{a}}(C_0)^{m(\mathfrak{a})}$$

taken with all elements $\mathbf{a} = \sum m(\mathfrak{a})\mathfrak{a} \in \mathbf{I}$ *prime to* $6N(\mathfrak{f})$ *satisfying*

$$\sum m(\mathfrak{a})(\mathbf{N}\mathfrak{a} - 1) = 0.$$

It is a subgroup of the units in $K(\mathfrak{f})$, *and is stable under the action of the Galois group.*

Proof. All the assertions of the theorem follow immediately from (1) and (2) above. Note that we may restrict ourselves to using C_0 (the unit class) instead of arbitrary classes C, because $u(\mathbf{a}; \mathfrak{c}) = u(\mathbf{a}\mathfrak{c}; \mathfrak{o})$. We may select $\mathfrak{c}$ in any class C to be prime to $6N(\mathfrak{f})$ to insure that the additional condition of Robert type is satisfied by $\mathbf{a}\mathfrak{c}$ as well as $\mathbf{a}$.

We shall now investigate the N-th power of the Robert group, where $N = N(\mathfrak{f})$, and see that it gives essentially another group obtained from a different point of view. **Define**:

$\Phi_{\mathfrak{f}}(w_{K(\mathfrak{f})})$ = group of units

$$\prod_C g^{12N}(C)^{n(C)}$$

where the product is taken over all classes $C \in \mathrm{Cl}(\mathfrak{f})$, and the exponents $n(C)$ satisfy the conditions:

$$\sum n(C) = 0 \quad \text{and} \quad \sum n(C)\mathbf{N}\mathfrak{a}(C) \equiv 0 \bmod w_{K(\mathfrak{f})},$$

where $\mathfrak{a}(C)$ is any ideal in the class C, prime to $6N$, cf. Lemma 4.1 of Chapter 9. These units can also be written in the form

$$\prod_{i=1}^{h(\mathfrak{f})-1} \left(\frac{g^{12N}(C_i)}{g^{12N}(C_0)}\right)^{n_i}$$

where the integers n_i are integers satisfying the congruence

$$\sum n_i(\mathbf{N}\mathfrak{a}(C_i) - 1) \equiv 0 \bmod w_{K(\mathfrak{f})}.$$

Here $\mathfrak{a}(C_i)$ denotes any ideal in the class C_i, prime to $6N$, and Lemma 4.1 of Chapter 9 shows that it does not matter which one we take. The classes C_i range over the non-trivial elements of $\mathrm{Cl}(\mathfrak{f})$.

Theorem 4.3. *We have* $\mathscr{R}_{\mathfrak{f}}^{*N} = \Phi_{\mathfrak{f}}(w_{K(\mathfrak{f})})$.

Proof. Let $u \in \mathcal{R}_{\mathfrak{f}}^*$. Let $\mathfrak{a}_i \in C_i$ and $\sum n_i(\mathbf{N}\mathfrak{a}_i - 1) = 0$. Then

$$u^N = \prod_i \left(\frac{g^{12N}(C_i)}{g^{12N}(C_0)^{\mathbf{N}\mathfrak{a}_i}} \right)^{n_i}$$
$$= \prod_i \left(\frac{g^{12N}(C_i)}{g^{12N}(C_0)} \right)^{n_i} \left(\frac{g^{12N}(C_0)}{g^{12N}(C_0)^{\mathbf{N}\mathfrak{a}_i}} \right)^{n_i}.$$

The second product on the right hand side is equal to 1 by the definition of $\mathcal{R}_{\mathfrak{f}}^*$, and the first product on the right hand side clearly lies in $\Phi_{\mathfrak{f}}(w_{K(\mathfrak{f})})$, by Lemma 3.2.

Conversely, given a unit in $\Phi_{\mathfrak{f}}(w_{K(\mathfrak{f})})$, write it in the form

$$\prod_i \left(\frac{g^{12N}(C_i)}{g^{12N}(C_0)} \right)^{n_i} = \prod \left(\frac{g^{12N}(C_i)}{g^{12N}(C_0)^{\mathbf{N}\mathfrak{a}_i}} \right)^{n_i} \prod \left(\frac{g^{12N}(C_0)^{\mathbf{N}\mathfrak{a}_i}}{g^{12N}(C_0)} \right)^{n_i}$$

where the integers n_i satisfy the congruence

$$\sum n_i(\mathbf{N}\mathfrak{a}_i - 1) \equiv 0 \bmod w_{K(\mathfrak{f})}.$$

By Lemma 3.2, there exist ideals $\mathfrak{c}_1, \ldots, \mathfrak{c}_s \in C_0$ prime to $6N$, and integers $m_1, \ldots, m_s$ such that

$$\sum m_j(\mathbf{N}\mathfrak{c}_j - 1) = \sum n_i(\mathbf{N}\mathfrak{a}_i - 1).$$

Then the second product on the right yields

$$g^{12N}(C_0)^{\sum n_i(\mathbf{N}\mathfrak{a}_i - 1)} = g^{12N}(C_0)^{\sum m_j(\mathbf{N}\mathfrak{c}_j - 1)}$$
$$= \prod \left(\frac{g^{12N}(C_0)^{\mathbf{N}\mathfrak{c}_j}}{g^{12N}(C_0)} \right)^{m_j}.$$

Hence the given unit lies in $\mathcal{R}_{\mathfrak{f}}^{*N}$, thus proving the converse inclusion.

Theorem 4.3 gives natural N-th roots in $\mathcal{R}_{\mathfrak{f}}^*$ for the group of units in the right hand side. Actually, we want to refine this result.

Theorem 4.4. *If* $K(\mathfrak{f}) \neq K(1)$, *then every element of* $\Phi_{\mathfrak{f}}(w_{K(\mathfrak{f})})$ *is a* $12w(\mathfrak{f})N(\mathfrak{f})$ *power in* $K(\mathfrak{f})$.

A proof is more delicate, and will be given in the next section. Getting the $12w(\mathfrak{f})$ in the power was obtained by Stark [St 1] and Gillard-Robert [Gi-Ro], Proposition A-2.

However, in the next chapter we shall define in a natural way a group of units which is larger than the Robert group, by using values of Klein forms taken at all levels dividing $\mathfrak{f}$, not just at the primitive level. This will provide another method to give natural roots, and allows for a more natural definition of the group of modular units in class fields over K.

The next corollary, due to Kersey, concludes the proof of Lemma 5.6 of Chapter 9.

Corollary 4.5. *Every element of* $\Delta_H(c_0, ww_H)$ *is a* $12wh$*-power in* H.

Proof. As remarked in Lemma 5.5, following Lemma 5.4 of Chapter 9, it suffices to prove this for $H = K(1)$. It is enough to check it for the generators in Lemma 5.4 of Chapter 9, so let $c, c' \in \mathrm{Cl}(1)$, and let

$$n(\mathbf{N}\mathfrak{a}_c - 1)(\mathbf{N}\mathfrak{a}_{c'} - 1) \equiv 0 \bmod ww_H.$$

Choose $K(1)$-admissible prime ideals $\mathfrak{p}$, $\mathfrak{p}'$ in c, c' respectively. Assume also that $\mathfrak{p}$ has degree 1, $\mathbf{N}\mathfrak{p} = p$. In Corollary 2 of Theorem 1.4, we have $w(\mathfrak{p}) = 1$, so we have the formula

$$\gamma^{hw} = \left(\frac{\delta(cc')}{\delta(c)\delta(c')}\right)^p \quad \text{where} \quad \gamma = \prod_{C|cc'} g_{\mathfrak{p}}(C) \Big/ \prod_{C|c} g_{\mathfrak{p}}(C).$$

We claim that

$$\gamma^n \in \Phi_{\mathfrak{p}}(w_{K(\mathfrak{p})}).$$

In fact, $w_{K(\mathfrak{p})} = w_{K(1)}$, since $\mathfrak{p}$ has degree 1. If ζ is a primitive $w_{K(1)}$-root of unity, then writing $\zeta[x]$ instead of ζ^x for typographical reasons, we get

$$\begin{aligned}\zeta\left[n\left(\sum_{C|cc'} \mathbf{N}\mathfrak{a}_c - \sum_{C|c} \mathbf{N}\mathfrak{a}_c\right)\right] &= \mathbf{N}_{K(\mathfrak{p})/K(1)}(\zeta^{\sigma(cc') - \sigma(c)})^n \\ &= (\zeta^{\sigma(cc') - \sigma(c)})^{n[K(\mathfrak{p}):K(1)]} \\ &= \zeta^{\sigma(c)(\mathbf{N}\mathfrak{p}' - 1)n(\mathbf{N}\mathfrak{p} - 1)/w} \\ &= 1\end{aligned}$$

because $ww(K(1))\,|\,n(\mathbf{N}\mathfrak{p} - 1)(\mathbf{N}\mathfrak{p}' - 1)$. Thus γ^n is a $12p$-power in $K(\mathfrak{p})$ by Theorem 4.4, and so

$$\left(\frac{\delta(cc')}{\delta(c)\delta(c')}\right)^{np} = \gamma^{nwh}$$

is a $12hwp$-power in $K(\mathfrak{p})$. Since $K(\mathfrak{p})$ has no p-th roots of 1, it follows that

$$\left(\frac{\delta(cc')}{\delta(c)\delta(c')}\right)^n$$

is a $12wh$-power in $K(\mathfrak{p})$. Since this is true for infinitely many $\mathfrak{p}$, it is a $12wh$-power in $K(1)$, as desired.

§5. Taking Roots

A proof of Theorem 4.4 will result from a sequence of lemmas. We first take care of the $w(\mathfrak{f})$ exponent.

Lemma 5.1. *Assume that $K(\mathfrak{f}) \neq K(1)$. If $w(\mathfrak{f}) \neq 1$, then $w(\mathfrak{f}) = 2$, $\mathfrak{f} = (2)$, and 2 remains prime in K, or ramifies in K, but cannot split completely.*

Proof. Suppose $w(\mathfrak{f}) > 1$. We first show $w(\mathfrak{f}) = 2$. Suppose otherwise. Then $K = \mathbf{Q}(\sqrt{-3})$ or $\mathbf{Q}(\sqrt{-1})$. Say $K = \mathbf{Q}(\sqrt{-3})$. Let ζ be a primitive cube root of unity. If $\pm\zeta \equiv 1 \bmod \mathfrak{f}$, then $\sqrt{-3}$ divides $\mathfrak{f}$, and since $1 \pm \zeta$ is a prime element in $\mathbf{Q}(\sqrt{-3})$, it follows that $\mathfrak{f} = (\sqrt{-3})$, which implies $K(\mathfrak{f}) = K$, a contradiction. Similarly we exclude the case when $w(\mathfrak{f}) = 4$, $\mathfrak{f} = (1 + i)$ and $K = \mathbf{Q}(\sqrt{-1})$.

Thus we must have $w(\mathfrak{f}) = 2$, and the only possible roots of unity are ± 1. But then $\mathfrak{f} | 2$. Suppose $\mathfrak{f} \neq (2)$. Then

$$(2) = \mathfrak{p}\bar{\mathfrak{p}} \quad \text{or} \quad (2) = \mathfrak{p}^2$$

for some prime ideal $\mathfrak{p}$, and say $\mathfrak{f} = \mathfrak{p}$. Since by assumption $K(\mathfrak{f}) \neq K(1)$, we must have some ideal $\mathfrak{a}$ in K which is prime to $\mathfrak{p}$, principal, $\mathfrak{a} = (\alpha)$, and $\alpha \not\equiv 1 \bmod \mathfrak{p}$. But since the residue class field at $\mathfrak{p}$ is $\mathbf{Z}/2\mathbf{Z}$, this is impossible. Thus $\mathfrak{f} = (2)$. By elementary algebraic number theory, we have

$$|\mathrm{Cl}(\mathfrak{f})| = h\phi(\mathfrak{f})w(\mathfrak{f})/w.$$

(See [L 4], Theorem 1 of Chapter VI, §1.) If 2 splits completely, then $\phi(\mathfrak{f}) = 1$, K cannot be $\mathbf{Q}(\sqrt{-1})$ or $\mathbf{Q}(\sqrt{-3})$, and so $w = 2$, whence $|\mathrm{Cl}(\mathfrak{f})| = h = |\mathrm{Cl}(1)|$, contrary to assumption. This proves the first part of the theorem.

The possibility of getting the $w(\mathfrak{f})$-power in Theorem 4.4 is then based on the following lemma, as in Gillard-Robert [Gi-Ro]. Lemma A-4. *We assume throughout that $\mathfrak{a}$ is odd, that is* $(\mathfrak{a}, 2) = (1)$. We shall pass to level 4 and use the distribution relations. We must then index $u_{\mathfrak{a}}(C_0)$ by 2 or 4, so we write

$$u_{\mathfrak{a}}^{(2)}(C_0^{(2)}) \quad \text{or} \quad u_{\mathfrak{a}}^{(4)}(C_0^{(4)})$$

to indicate the reference to conductor $\mathfrak{f} = 2$ or $\mathfrak{f} = 4$. We abbreviate the norm by

$$\mathbf{N}_{K(4)/K(2)} = \mathbf{N}_{(4)/(2)}.$$

Lemma 5.2. *Suppose that $w(\mathfrak{f}) \neq 1$ and $K(\mathfrak{f}) \neq K(1)$. Then:*

(i) *Every element $u_{\mathfrak{a}}(C_0)$ is equal to $\pm\alpha^2$ for some $\alpha \in K(\mathfrak{f})$, and in fact*

$$\pm\mathbf{N}_{(4)/(2)}(u_{\mathfrak{a}}^{(4)}(C_0^{(4)}))^2 = u_{\mathfrak{a}}^{(2)}(C_0^{(2)}).$$

(ii) *Every element of $\mathcal{R}_{\mathfrak{f}}^*$ is equal to $\pm\alpha^2$ for some $\alpha \in K(\mathfrak{f})$.*

Proof. Note that part (ii) follows from part (i) by Theorem 4.1. We are therefore reduced to proving part (i), which we now do. Let

$$\mathfrak{o} = \mathfrak{o}_K = [z, 1]$$

be the ring of algebraic integers in K. By definition we have

$$u_{\mathfrak{a}}^{(2)}(C_0^{(2)}) = g^{12}(1, 2\mathfrak{o})^{\mathbf{N}\mathfrak{a}}/g^{12}(1, 2\mathfrak{a}^{-1}).$$

Let $\mathfrak{b} = (\beta)$, where β ranges over the four numbers

$$\beta = 1,\ 3,\ 1 + 2z,\ 3 + 2z.$$

We have the tower of fields

$$K(4) \supset K(2) \supset K(1) \supset K.$$

By the elementary formula already used for the number of generalized ideal classes $|\mathrm{Cl}(4)|$ and $|\mathrm{Cl}(2)|$, we find

$$[K(4) : K(2)] = 2.$$

The elements $\sigma(1)$, $\sigma(3)$ leave $K(2)$ fixed, while $\sigma(1 + 2z)$ and $\sigma(3 + 2z)$ give the non-trivial automorphism of $K(4)$ over $K(2)$, since $1 + 2z$ and $3 + 2z$ are $\not\equiv \pm 1 \bmod 4$.

By Theorem 4.1 and the definition of $u_{\mathfrak{a}}^{(4)}(C_0^{(4)})$, we get:

$$\mathbf{N}_{(4)/(2)}(u_{\mathfrak{a}}^{(4)}(C_0^{(4)}))^2 = \prod_{\mathfrak{b}} \sigma(\mathfrak{b}) u_{\mathfrak{a}}^{(4)}(C_0^{(4)})$$
$$= \prod_{\beta} \frac{g^{12}(\beta, 4\mathfrak{o})^{\mathbf{N}\mathfrak{a}}}{g^{12}(\beta, 4\mathfrak{a}^{-1})}$$

But the elements 2, $2z$, $2 + 2z$ represent the non-trivial classes of L' mod L, where $L = 4\mathfrak{o}$ and $L' = 2\mathfrak{o}$. By the distribution relations of Theorem 4.3 of Chapter 2, we get

$$\mathbf{N}_{(4)/(2)}(u_{\mathfrak{a}}^{(4)}(C_0^{(4)}))^4 = u_{\mathfrak{a}}^{(2)}(C_0^{(2)})^2.$$

Taking square roots yields

$$\pm \mathbf{N}_{(4)/(2)}(u_{\mathfrak{a}}^{(4)}(C_0^{(4)}))^2 = u_{\mathfrak{a}}^{(2)}(C_0^{(2)}),$$

thereby proving Lemma 5.2.

Lemma 5.3. *Let $\mathfrak{f} \neq (1)$ be prime to $\mathfrak{a}$, and $C \in \mathrm{Cl}(\mathfrak{f})$, $N = N(\mathfrak{f})$. Then*

$$u_{\mathfrak{a}}(C)^N \in K(\mathfrak{f})^{12N}.$$

Stark's proof of this lemma can be reformulated as in [Gi-Ro], Lemma A-3:

Lemma 5.4. *Let K be a number field, H an abelian extension of conductor $\mathfrak{f}$. Let n be an integer > 0, and α an algebraic number such that:*

(i) *α lies in an abelian extension of K;*
(ii) *$\alpha^n \in H$.*

Let $\mathfrak{b}$ be an ideal in K prime to $n\mathfrak{f}$. Then

$$(\alpha^n)^{\mathbf{N}\mathfrak{b}}/(\alpha^n)^{(\mathfrak{b}, H/K)}$$

is an n-th power in H.

Proof. Let $H' = H(\alpha)$. Let m be an integer > 0 divisible by n and by the conductor of H' over K. For any $\sigma \in \mathrm{Gal}(H'/H)$ we have

$$\alpha^\sigma = \varepsilon(\sigma)\alpha$$

for some n-th root of unity $\varepsilon(\sigma)$, because $\alpha^n \in H$. Let $\mathfrak{b}'$ be an ideal in K prime to m satisfying:

the restriction of $(\mathfrak{b}', H'/K)$ to H is $(\mathfrak{b}, H/K)$;

$\mathbf{N}\mathfrak{b}' \equiv \mathbf{N}\mathfrak{b} \bmod n$.

Then:

$$\begin{aligned}(\alpha^{(\mathfrak{b}',H'/K)})^\sigma = (\alpha^\sigma)^{(\mathfrak{b}',H'/K)} &= (\varepsilon(\sigma)\alpha)^{(\mathfrak{b}',H'/K)}\\ &= \varepsilon(\sigma)^{\mathbf{N}\mathfrak{b}'}\alpha^{(\mathfrak{b}',H'/K)}\\ &= \varepsilon(\sigma)^{\mathbf{N}\mathfrak{b}}\alpha^{(\mathfrak{b}',H'/K)}.\end{aligned}$$

Hence $\alpha^{\mathbf{N}\mathfrak{b}}/\alpha^{(\mathfrak{b}',H'/K)}$ is fixed by σ, and therefore belongs to H. Since the restriction of $(\mathfrak{b}', H'/K)$ to H is $(\mathfrak{b}, H/K)$, it follows that the n-th power of this element is equal to

$$(\alpha^n)^{\mathbf{N}\mathfrak{b}}/(\alpha^n)^{(\mathfrak{b},H/K)},$$

thus proving the lemma.

Lemma 5.3 follows by putting $n = 12N$, since

$$u_{\mathfrak{b}}(C)^N = g_{\mathfrak{f}}(C)^{\mathbf{N}\mathfrak{b}}/g_{\mathfrak{f}}(C)^{(\mathfrak{b},H/K)}$$

by Theorem 1.1, and

$$g_{\mathfrak{f}}(C) = g^n(1, \mathfrak{f}\mathfrak{c}^{-1})$$

is the n-th power of the value of a modular function lying in an abelian extension of K by complex multiplication.

The functions $g^{12}(z, L)$ for any lattice L are products of the Klein form and the delta function. If $(L' : L)$ is odd, we recall the formula of Theorem 2.2, Chapter 10:

$$\mathfrak{k}(z, L'/L) = \frac{\mathfrak{k}(z, L)^{(L' : L)}}{\mathfrak{k}(z, L')} = \prod_t \frac{1}{\wp(z, L) - \wp(t, L)} \tag{5.5}$$

where t ranges over $(L'/L)/\pm 1$, $t \neq 0$.

Lemma 5.6. *Let L be a lattice admitting complex multiplication by the elements of $\mathfrak{o}$ $(= \mathfrak{o}_K)$, such that $g_2(L)$ and $g_3(L)$ lie in $K(1)$ (the Hilbert class field). Let $\mathfrak{a}$ be an odd ideal, that is $(\mathfrak{a}, 2) = (1)$. Let $L' = \mathfrak{a}^{-1}L$, let $\mathfrak{f}$ be an ideal prime to $\mathbf{N}\mathfrak{a}$, $\mathfrak{f} \neq (1)$, and let z be in $\mathfrak{f}^{-1}L$, a generator of $\mathfrak{f}^{-1}L$* mod L

over $\mathfrak{o}$. *Then*

$$\mathfrak{k}(z, \mathfrak{a}^{-1}L/L) \in K(\mathfrak{f}).$$

Proof. The polynomial

$$\prod_t (X - \wp(t, L))$$

has coefficients in $K(1)$. By complex multiplication, Corollary of Theorem 7, Chapter 10, §3 in [L 5], we know that

$$\wp(z, L) \in K(\mathfrak{f}).$$

This proves the lemma.

We shall use the lemma after raising both sides to the 12-th power. The next two lemmas are due to Robert.

Lemma 5.7. *Let* $\mathfrak{a}$ *be an ideal prime to* w_K. *Let* L *be a lattice which admits complex multiplication by* $\mathfrak{o}$, *and such that* $g_2(L), g_3(L) \in K(1)$. *Then*

$$u = \Delta(L)^{\mathbf{N}\mathfrak{a}}/\Delta(\mathfrak{a}^{-1}L)$$

is a 12-th *power in* $K(1)$.

Proof. The cases $K = \mathbf{Q}(\sqrt{-1})$ and $\mathbf{Q}(\sqrt{-3})$ will be left as an exercise, so we suppose K unequal to either of these two fields. Let p be a prime $\geqq 5$, $(p, \mathfrak{a}) = 1$, such that p splits in K, $p = \mathfrak{p}\bar{\mathfrak{p}}$, and $p \equiv -1 \bmod 12$. Then $(p-1)/2$ is odd and prime to 3. By the elementary formula for the number of generalized ideal classes, we have

$$[K(\mathfrak{p}) : K(1)] = \frac{p-1}{2}.$$

In particular, the p-th roots of unity are not contained in $K(\mathfrak{p})$ since they are totally ramified over $\mathbf{Q}$ of degree $p - 1$. After multiplying (5.5) with u on both sides, and applying Lemma 5.6 with $\mathfrak{f} = \mathfrak{p}$, we find

$$\frac{g^{12}(z, L)^{\mathbf{N}\mathfrak{a}}}{g^{12}(z, \mathfrak{a}^{-1}L)} = u\alpha^{12}$$

for some $\alpha \in K(\mathfrak{p})$. Since the left hand side is homogeneous in (z, L) of degree 0, and since $z^{-1}L$ is an ideal prime to $\mathfrak{p}$ by the way we selected z, we see that the left hand side has the form $u_{\mathfrak{a}}(C)$ for some class $C \in \mathrm{Cl}(\mathfrak{p})$, that is

$$u_{\mathfrak{a}}(C) = u\alpha^{12}. \tag{5.8}$$

By Stark's lemma 5.3, we know that $u_{\mathfrak{a}}(C)^p$ is a $12p$-power in $K(\mathfrak{p})$, that is there exists $\beta \in K(\mathfrak{p})$ such that

$$u^p = \beta^{12p}.$$

Then u/β^{12} is a p-th root of unity in $K(\mathfrak{p})$, whence by the assumptions on p, we get

$$u = \beta^{12}.$$

The expression (5.8) shows that $u \in K(\mathfrak{p})$ for all choices of $\mathfrak{p}$, whence $u \in K(1)$. Since $[K(\mathfrak{p}) : K(1)]$ is prime to 12, and u is 12-th power in $K(\mathfrak{p})$, it follows that u is 12-th power in $K(1)$, thus proving the lemma.

Lemma 5.9. *Let $\mathfrak{f}$ be an ideal of K, $\mathfrak{f} \neq (1)$. For any ideal $\mathfrak{a}$ prime to w_K, and $C \in \mathrm{Cl}(\mathfrak{f})$, we have*

$$u_{\mathfrak{a}}(C) \in K(\mathfrak{f})^{12}.$$

Proof. This is a direct consequence of Lemmas 5.6 and 5.7, together with the definition of $u_{\mathfrak{a}}(C)$.

We are now ready to prove Theorem 4.4. We are done if $w(\mathfrak{f}) = 1$, by Theorem 4.3 and Lemma 5.9. Suppose $w(\mathfrak{f}) \neq 1$. Then we may assume that we are in the special case of Lemma 5.1. By Lemma 5.2(i) and Lemma 5.9 (applied to $\mathfrak{f} = (4)$) we see that there exists $\gamma \in K(4)$ such that

$$u_{\mathfrak{a}}(C_0) = \pm N_{(4)/(2)}(\gamma^{12})^2,$$

so $u_{\mathfrak{a}}(C_0) = \pm\alpha^{24}$ for some $\alpha \in K(2)$. Since $N = 2$ and

$$\mathscr{R}^{*N}_{(2)} = \Phi_{(2)}(w_{K(2)})$$

by Theorem 4.3, we conclude that every element of $\Phi_{(2)}(w_{K(2)})$ is a 48-th power in $K(2)$. But

$$48 = 12w(2)N(2),$$

thereby concluding the proof of Theorem 4.4.

§6. The Robert Group under the Norm Map

We shall investigate the Robert group under the norm map in two cases:

(a) From the ray class field of conductor $\mathfrak{f}'$ to the ray class field of conductor $\mathfrak{f}$ when $\mathfrak{f}|\mathfrak{f}'$ and $\mathfrak{f}$, $\mathfrak{f}'$ have the same prime factors.

(b) From the ray class field of conductor $\mathfrak{f}$ to an abelian extension H of conductor $\mathfrak{f}$.

We let $\mathbf{N}_{\mathfrak{f}',\mathfrak{f}}$ denote the norm from $K(\mathfrak{f}')$ to $K(\mathfrak{f})$.

Theorem 6.1. *Let $\mathfrak{f}|\mathfrak{f}'$, and suppose $\mathfrak{f}$, $\mathfrak{f}'$ have the same prime factors. Then*:

(i) $$\mathbf{N}_{\mathfrak{f}',\mathfrak{f}}(\Phi_{\mathfrak{f}'})^{w(\mathfrak{f})/w(\mathfrak{f}')} = \Phi_{\mathfrak{f}}^{N(\mathfrak{f}')/N(\mathfrak{f})}$$

(ii) $$\mathbf{N}_{\mathfrak{f}',\mathfrak{f}}(\Phi_{\mathfrak{f}'}(w_{K(\mathfrak{f}')}))^{w(\mathfrak{f})/w(\mathfrak{f}')} = \Phi_{\mathfrak{f}}(w_{K(\mathfrak{f})})^{N(\mathfrak{f}')/N(\mathfrak{f})}.$$

Proof. The first assertion is immediate from the definitions and Theorem 1.3, recalling that

$$\varphi_{\mathfrak{f}}(C) = g_{\mathfrak{f}}(C)/g_{\mathfrak{f}}(C_0).$$

In fact, formula (i) holds with Φ replaced by φ, and we shall use this below.

Let us prove (ii). Let

$$u = \prod_{C' \neq C_0'} \varphi_{\mathfrak{f}'}(C')^{n(C')}$$

be an element of $\Phi_{\mathfrak{f}'}(w_{K(\mathfrak{f}')})$, so that

$$(*) \qquad \sum_{C' \neq C_0'} n(C')(\mathbf{N}\mathfrak{a}(C') - 1) \equiv 0 \bmod w_{K(\mathfrak{f}')}.$$

Then

$$\mathbf{N}_{\mathfrak{f}',\mathfrak{f}}(u) = \prod_{C \neq C_0} \mathbf{N}_{\mathfrak{f}',\mathfrak{f}}\varphi_{\mathfrak{f}'}(C')^{n(C)}$$

where C' inside the product is any class lying in C, and

$$n(C) = \sum_{C'|C} n(C').$$

From $(*)$ we get

$$\sum_{C \neq C_0} n(C)(\mathbf{N}\mathfrak{a}(C) - 1) \equiv 0 \bmod w_{K(\mathfrak{f})},$$

and hence $\mathbf{N}_{\mathfrak{f}',\mathfrak{f}}(u)^{w(\mathfrak{f})/w(\mathfrak{f}')}$ lies in $\Phi_{\mathfrak{f}}(w_{K(\mathfrak{f})})^{N(\mathfrak{f}')/N(\mathfrak{f})}$ by the formula of (i).

Conversely, let u be an element of $\Phi_{\mathfrak{f}}(w_{K(\mathfrak{f})})$, and write

$$u = \prod_{i=1}^{v-1} \varphi_{\mathfrak{f}}(C_i)^{n_i} \quad \text{with} \quad \sum_{i=1}^{v-1} n_i(\mathbf{N}\mathfrak{a}(C_i) - 1) \equiv 0 \bmod w_{K(\mathfrak{f})}.$$

By Lemma 2 of §3 there exist ideals $\mathfrak{a}_1, \ldots, \mathfrak{a}_s \in C_0$ prime to $6N(\mathfrak{f}')$ and integers $m_1, \ldots, m_s$ such that

$$\sum_{j=1}^{s} m_j(\mathbf{N}\mathfrak{a}_j - 1) + \sum_{i=1}^{v-1} n_i(\mathbf{N}\mathfrak{a}(C_i) - 1) = 0.$$

Let

$$u' = \prod_{j=1}^{s} \varphi_{\mathfrak{f}'}(C'_{\mathfrak{a}_j})^{m_j} \prod_{i=1}^{v-1} \varphi_{\mathfrak{f}'}(C'_{\mathfrak{a}(C_i)})^{n_i}.$$

Then, putting $\varepsilon = w(\mathfrak{f})/w(\mathfrak{f}')$,

$$\mathbf{N}_{\mathfrak{f}',\mathfrak{f}}(u'^{\varepsilon}) = \prod_{i=1}^{v-1} \varphi_{\mathfrak{f}}(C_i)^{n_i N(\mathfrak{f}')/N(\mathfrak{f})} = u^{N(\mathfrak{f}')/N(\mathfrak{f})}.$$

This proves the converse inclusion, and concludes the proof of the theorem.

Let H be an abelian extension of K with conductor $\mathfrak{f} \neq (1)$. Then $H \subset K(\mathfrak{f})$. We had defined the group Φ_H. We now let:

$\Phi_H(w_H)$ = group of elements

$$\prod g_H(C)^{n(C)}$$

where the product is taken over $C \in \mathrm{Cl}(H/K)$, and the exponents $n(C)$ satisfy the conditions

$$\sum_C n(C) = 0 \quad \text{and} \quad \sum n(C)\mathbf{N}\mathfrak{a}(C) \equiv 0 \bmod w_H.$$

These elements can also be written in the form

$$\prod \varphi_H(C_i)^{n_i}$$

where $i = 1, \ldots, v - 1$, and the C_i range over the distinct non-trivial classes of $\mathrm{Cl}(H/K)$, satisfying the congruence condition

$$\sum n_i(\mathbf{N}\mathfrak{a}(C_i) - 1) \equiv 0 \bmod w_H.$$

As usual, $\mathfrak{a}(C_i)$ is any ideal in the class C_i.

Theorem 6.2. $\Phi_H(w_H) = N_{K(\mathfrak{f})/H}\Phi_{\mathfrak{f}}(w_{K(\mathfrak{f})})$.

Proof. The proof is entirely similar to that of Theorem 6.1. The first part holds because w_H divides $w_{K(\mathfrak{f})}$. In the converse, putting $\mathfrak{f}' = \mathfrak{f}$, we define u' as before, relative to an element u of $\Phi_H(w_H)$. Then

$$u = \mathbf{N}_{K(\mathfrak{f})/H}(u'),$$

thus showing the converse, that $u \in \mathbf{N}_{K(\mathfrak{f})/H}\Phi_{\mathfrak{f}}(w_{K(\mathfrak{f})})$.

Let us define

$$\boxed{\mathscr{R}_H^* = \mathbf{N}_{K(\mathfrak{f})/H}\mathscr{R}_{\mathfrak{f}}^*.}$$

From Theorems 4.3 and 4.4, we conclude

Corollary. $\mathscr{R}_H^{*N(\mathfrak{f})} = \Phi_H(w_H)$,

and in fact, every element of $\Phi_H(w_H)$ *is a* 12 $w(\mathfrak{f})N(\mathfrak{f})$*-power in* H.

CHAPTER 12

Klein Units in Arbitrary Class Fields

In this chapter we define in a natural way a large group of units which may be called the modular units in class fields of K. This construction is done with the Klein forms exclusively, but because of the distribution relations, the group which we obtain also contains the unramified units constructed previously with the delta function.

As usual, we prove that they have the right behavior with respect to the Artin automorphism. We also relate this group with the Robert group of the preceding chapter, by showing that it provides appropriate roots.

This chapter was written in collaboration with Don Kersey, and the last section is entirely due to him.

§1. The Klein Invariants

Let $\mathbf{I}$ be as before the free abelian group on the (non-zero) ideals of K. Suppose

$$\mathbf{a} = \sum_{\mathfrak{a}} n(\mathfrak{a})\mathfrak{a} \in \mathbf{I}.$$

Let $\mathfrak{f}$ be an ideal which does not divide any of the ideals $\mathfrak{a}$ occurring in $\mathbf{a}$ (that is, such that $n(\mathfrak{a}) \neq 0$). For any ideal $\mathfrak{c}$ prime to $\mathfrak{f}$, we define

$$\boxed{\mathfrak{k}_{\mathfrak{f}}(\mathbf{a}; \mathfrak{c}) = \prod_{\mathfrak{a}} \mathfrak{k}(1, \mathfrak{f}\mathfrak{a}^{-1}\mathfrak{c}^{-1})^{n(\mathfrak{a})}.}$$

If $\mathfrak{f} | \mathfrak{a}$, then $1 \in \mathfrak{f}\mathfrak{a}^{-1}\mathfrak{c}^{-1}$, so $\mathfrak{k}(1, \mathfrak{f}\mathfrak{a}^{-1}\mathfrak{c}^{-1}) = 0$. This is the reason for the assumption that no ideal in **a** is divisible by $\mathfrak{f}$.

In this section, we determine the conditions which must be satisfied by **a** in order that the numbers $\mathfrak{k}_{\mathfrak{f}}(\mathbf{a}; \mathfrak{c})$ be ray class invariants, that is

$$\mathfrak{k}_{\mathfrak{f}}(\mathbf{a}; \mathfrak{c}) = \mathfrak{k}_{\mathfrak{f}}(\mathbf{a}; \mathfrak{c}')$$

whenever $\mathfrak{c}$, $\mathfrak{c}'$ lie in the same class in $\mathrm{Cl}(\mathfrak{f})$. In §2, we show that when this is the case, the invariants lie in the ray class field $K(\mathfrak{f})$, and transform appropriately under the Galois group.

The prime 2 will cause some complications in the statements of some conditions. We say that $\mathfrak{a}$ is **odd** if $(\mathfrak{a}, (2)) = \mathfrak{o}$. We say that **a** is **odd** if every ideal $\mathfrak{a}$ occurring in **a** is odd. We abbreviate $(\mathfrak{a}, (2))$ by $(\mathfrak{a}, 2)$.

We define the **degree** as usual by

$$d(\mathbf{a}) = \deg \mathbf{a} = \sum n(\mathfrak{a}),$$

and we extend the absolute norm by linearity as before, namely

$$\mathbf{N}\mathbf{a} = \sum n(\mathfrak{a})\mathbf{N}\mathfrak{a}.$$

Let $\mathfrak{d}$ be an ideal dividing (2), and define the **partial degree**

$$d_{\mathfrak{d}}(\mathbf{a}) = \sum_{(\mathfrak{a},2)=\mathfrak{d}} n(\mathfrak{a}).$$

Thus

$$d(\mathbf{a}) = \sum_{\mathfrak{d}|2} d_{\mathfrak{d}}(\mathbf{a}).$$

If **a** is odd, then the only term in this sum occurs for $\mathfrak{d} = \mathfrak{o}$.

With these notational conventions, we define a subgroup $\mathbf{I}(\mathfrak{f})$ of $\mathbf{I}$ by:

$\mathbf{a} \in \mathbf{I}(\mathfrak{f})$ if and only if the following conditions are satisfied:

$\mathbf{I}_{\mathfrak{f}}$ **1.** $\mathfrak{f}$ does not divide any of the ideals $\mathfrak{a}$ occurring in **a**;
$\mathbf{I}_{\mathfrak{f}}$ **2.** $\mathbf{N}\mathbf{a} \equiv 0 \bmod N(\mathfrak{f})$;
$\mathbf{I}_{\mathfrak{f}}$ **3(i)** If $\mathbf{N}\mathbf{a}/N(\mathfrak{f})$ is even, then $d_{\mathfrak{d}}(\mathbf{a})$ is even for $\mathfrak{d} \neq (2)$.

[*Note*: Condition **I 3(i)** is usually applied when $d(\mathbf{a})$ is even, or $d(\mathbf{a}) = 0$. in which case it then follows that $d_{(2)}(\mathbf{a})$ is also even.]

If $\mathbf{N}\mathbf{a}/N(\mathfrak{f})$ is odd, then precisely one of the following conditions is to be satisfied. Note that 2 cannot divide $N(\mathfrak{f})/\mathfrak{f}^{-1}$, otherwise $N(\mathfrak{f})/2 \in \mathfrak{f}$.

$\mathbf{I}_{\mathfrak{f}}$ **3(ii)** 2 ramifies, $(2) = \mathfrak{p}^2$, and $\mathfrak{p} | N(\mathfrak{f})\mathfrak{f}^{-1}$. Then:

$$d_{\mathfrak{o}}(\mathbf{a}) \text{ is even}; \qquad d_{\mathfrak{p}}(\mathbf{a}) \text{ is odd.}$$

$\mathbf{I}_{\mathfrak{f}}$ **3(iii)** 2 splits, $(2) = \mathfrak{p}\bar{\mathfrak{p}}$, and $N(\mathfrak{f})\mathfrak{f}^{-1}$ is odd. Then:

$$d_{\mathfrak{o}}(\mathbf{a}) \text{ is even}; \qquad d_{\mathfrak{p}}(\mathbf{a}) \quad \text{and} \quad d_{\bar{\mathfrak{p}}}(\mathbf{a}) \text{ are odd.}$$

$\mathbf{I}_{\mathfrak{f}}$ **3(iv)** 2 splits, $(2) = \mathfrak{p}\bar{\mathfrak{p}}$, and, say, $\mathfrak{p} | N(\mathfrak{f})\mathfrak{f}^{-1}$. Then:

$$d_{\mathfrak{o}}(\mathbf{a}) \quad \text{and} \quad d_{\mathfrak{p}}(\mathbf{a}) \text{ are even}; \qquad d_{\bar{\mathfrak{p}}}(\mathbf{a}) \text{ is odd.}$$

If $\mathfrak{f}$ is fixed throughout a discussion, we write **I 1, I 2, I 3** and omit $\mathfrak{f}$ from the notation.

Remark 1. The most important condition is **I 2.** Condition **I 3** is imposed to deal with problems arising from the prime 2. Condition **I 2** may be viewed as the **quadratic relations** mod $N(\mathfrak{f})$ in the context of complex multiplication.

Remark 2. There is some flexibility in the definition of $\mathfrak{k}_{\mathfrak{f}}(\mathbf{a}; \mathfrak{c})$ where the $\mathfrak{c}$ plays an auxiliary role which will be useful for the formulation of the reciprocity law, but which is actually irrelevant in the definition of that number. Indeed, for $\mathbf{a} = \sum n(\mathfrak{a})\mathfrak{a} \in \mathbf{I}$ and $\mathfrak{b}$ an ideal, let

$$\mathfrak{b} \cdot \mathbf{a} = \mathbf{a} \cdot \mathfrak{b} = \sum n(\mathfrak{a}) \cdot \mathfrak{a}\mathfrak{b} \in \mathbf{I}.$$

If no confusion can occur, we often omit the dot in $\mathbf{a} \cdot \mathfrak{b}$. With the above notation, we have the obvious formula

$$\mathfrak{k}_{\mathfrak{f}}(\mathbf{a}; \mathfrak{c}) = \mathfrak{k}_{\mathfrak{f}}(\mathbf{a}\mathfrak{c}; \mathfrak{o}).$$

Furthermore if $\mathfrak{c}$ *is divisible by* 2, *then for any ideal* $\mathfrak{b}$ *prime to* $\mathfrak{f}$,

$$\mathbf{I}(\mathfrak{f}) \cdot \mathfrak{c}\mathfrak{b} \subset \mathbf{I}(\mathfrak{f}\mathfrak{c}).$$

Proof. Let $\mathbf{a} \in \mathbf{I}(\mathfrak{f})$. Then it is immediate that $\mathbf{a} \cdot \mathfrak{c}\mathfrak{b}$ satisfies $\mathbf{I}_{\mathfrak{f}\mathfrak{c}}$ **1**, that is no ideal occurring in $\mathbf{a} \cdot \mathfrak{c}\mathfrak{b}$ is divisible by $\mathfrak{f}\mathfrak{c}$. For the second condition, note that

$$\frac{\mathbf{N}\mathbf{a}\mathbf{N}\mathfrak{b}\mathbf{N}\mathfrak{c}}{N(\mathfrak{f}\mathfrak{c})} = \frac{\mathbf{N}\mathbf{a}}{N(\mathfrak{f})} \cdot \frac{\mathbf{N}\mathfrak{c}N(\mathfrak{f})}{N(\mathfrak{f}\mathfrak{c})}\mathbf{N}\mathfrak{b}$$

and

$$\frac{\mathbf{N}\mathfrak{c}\cdot N(\mathfrak{f})}{N(\mathfrak{f}\mathfrak{c})} = 2\frac{N(\mathfrak{c}/2)N(\mathfrak{f})}{N(\mathfrak{f}\mathfrak{c}/2)} \quad \text{is even.}$$

Furthermore, we are in $\mathbf{I}_{\mathfrak{f}\mathfrak{c}}$ **3(i)**, which is trivially satisfied since $d_{\mathfrak{d}}(\mathbf{a}\cdot\mathfrak{c}\mathfrak{b}) = 0$ for $\mathfrak{d} \neq (2)$. This concludes the proof.

Remark 3. The situation of **I 3(ii)** through **(iv)** can frequently be reduced to **I 3(i)** as follows.

Assume that $\mathbf{Na}/N(\mathfrak{f})$ *is odd. Then trivially*

$$\mathfrak{k}_{\mathfrak{f}}(\mathbf{a};\mathfrak{c}) = \mathfrak{k}_{2\mathfrak{f}}((2)\mathbf{a};\mathfrak{c}).$$

Furthermore $(2)\mathbf{a} \in \mathbf{I}(2\mathfrak{f})$, *and satisfies* $\mathbf{I}_{2\mathfrak{f}}$ **3(i)**.

Proof. We note that

$$\mathbf{N}((2)\mathbf{a})/N(2\mathfrak{f}) = 4\mathbf{Na}/2N(\mathfrak{f}) = 2\mathbf{Na}/N(\mathfrak{f})$$

is even, and for $\mathfrak{d}|2$, $\mathfrak{d} \neq (2)$, we have $d_{\mathfrak{d}}((2)\mathbf{a}) = 0$. This proves our assertion.

Let n be a positive integer. We define subgroups of $\mathbf{I}(\mathfrak{f})$ as follows:

$\mathbf{I}_n(\mathfrak{f})$ = subgroup of elements $\mathbf{a} \in \mathbf{I}(\mathfrak{f})$ such that $d(\mathbf{a}) \equiv 0 \bmod n$;
$\mathbf{I}_0(\mathfrak{f})$ = subgroup of elements $\mathbf{a} \in \mathbf{I}(\mathfrak{f})$ of degree 0, that is $d(\mathbf{a}) = 0$.

Of course, $\mathbf{I}_0(\mathfrak{f})$ is contained in $\mathbf{I}_n(\mathfrak{f})$ for all n.

The goal of this section is to prove:

Proposition 1.1

(i) *Let* $\mathbf{a} \in \mathbf{I}$. *Then* $\mathbf{a} \in \mathbf{I}(\mathfrak{f})$ *if and only if for all* $\gamma \in \mathfrak{o}$, $\gamma \equiv 1 \bmod \mathfrak{f}$, *and all* $\mathfrak{c}$ *prime to* $\mathfrak{f}$, *we have*

$$\mathfrak{k}_{\mathfrak{f}}(\mathbf{a};\mathfrak{c}\gamma) = \gamma^{-d(\mathbf{a})}\mathfrak{k}_{\mathfrak{f}}(\mathbf{a};\mathfrak{c}).$$

(ii) *In particular, the numbers* $\mathfrak{k}_{\mathfrak{f}}(\mathbf{a};\mathfrak{c})$ *are ray class invariants if and only if* $\mathbf{a} \in \mathbf{I}_0(\mathfrak{f})$.

It is clear that (ii) in the proposition is an immediate consequence of (i). Thus when $\mathbf{a} \in \mathbf{I}_0(\mathfrak{f})$ and $C \in \mathrm{Cl}(\mathfrak{f})$, we may define the **Klein invariants**

$$\mathfrak{k}_{\mathfrak{f}}(\mathbf{a}; C) = \mathfrak{k}_{\mathfrak{f}}(\mathbf{a}; \mathfrak{c})$$

for any $\mathfrak{c} \in C$.

The rest of the section is devoted to the proof of Proposition 1.1.

We are going to study the variation $\mathfrak{k}_{\mathfrak{f}}(\mathbf{a}; \mathfrak{c}\gamma)/\mathfrak{k}_{\mathfrak{f}}(\mathbf{a}; \mathfrak{c})$ when $\mathbf{a} \in \mathbf{I}$ and $\gamma \in \mathfrak{o}$, $\gamma \equiv 1 \bmod \mathfrak{f}$. We shall use the transformation law of the Klein forms given in Chapter 10, Theorem 1.3. If L is a lattice in $\mathbf{C}$, we let $\mathbf{M}(L)$ be the lattice norm, that is the area (measure) of a fundamental domain. If I is a fractional ideal in K, then the lattice norm is related to the absolute norm by the formula

$$\mathbf{M}(I) = \mathbf{N}I \cdot \mathbf{M}(\mathfrak{o}).$$

Furthermore, if $\mathfrak{o} = [z, 1]$, then $\mathbf{M}(\mathfrak{o}) = \mathrm{Im}\ z$.

Let $\mathbf{a} = \sum n(\mathfrak{a})\mathfrak{a}$. Then we have

$$\begin{aligned}
\frac{\mathfrak{k}_{\mathfrak{f}}(\mathbf{a}; \mathfrak{c}\gamma)}{\mathfrak{k}_{\mathfrak{f}}(\mathbf{a}; \mathfrak{c})} &= \prod_{a} \left(\frac{\mathfrak{k}(1, \mathfrak{f}\mathfrak{a}^{-1}\mathfrak{c}^{-1}\gamma^{-1})}{\mathfrak{k}(1, \mathfrak{f}\mathfrak{a}^{-1}\mathfrak{c}^{-1})} \right)^{n(\mathfrak{a})} \\
&= \gamma^{-d(\mathfrak{a})} \prod_{\mathfrak{a}} \left(\frac{\mathfrak{k}(\gamma, \mathfrak{f}\mathfrak{a}^{-1}\mathfrak{c}^{-1})}{\mathfrak{k}(1, \mathfrak{f}\mathfrak{a}^{-1}\mathfrak{c}^{-1})} \right)^{n(\mathfrak{a})} \\
&= \gamma^{-d(\mathfrak{a})} \prod_{\mathfrak{a}} \left(e\left[\frac{-\pi i}{\mathbf{M}(\mathfrak{f}\mathfrak{a}^{-1}\mathfrak{c}^{-1})} \mathrm{Im}(\bar{\gamma} - 1) \right] \psi(\gamma - 1, \mathfrak{f}\mathfrak{a}^{-1}\mathfrak{c}^{-1}) \right)^{n(\mathfrak{a})}
\end{aligned}$$

by Chapter 10, Theorem 1.3. Since $\gamma - 1 \in \mathfrak{f}$ and $N(\mathfrak{f}) \in \bar{\mathfrak{f}}$, we have

$$N(\mathfrak{f})(\bar{\gamma} - 1) \in \mathfrak{f}\bar{\mathfrak{f}} = (\mathbf{N}\mathfrak{f}).$$

Hence there exists $\alpha = a + bz \in \mathfrak{o}$ (a, b rational integers) such that

$$N(\mathfrak{f})(\bar{\gamma} - 1) = \alpha \mathbf{N}\mathfrak{f},$$

and $\alpha \in N(\mathfrak{f})\mathfrak{f}^{-1}$. Conversely, if α and γ are elements of K related by this equation, and $\alpha \in N(\mathfrak{f})\mathfrak{f}^{-1}$, then $\gamma - 1 \in \mathfrak{f}$.

$$\mathrm{Im}(\bar{\gamma} - 1) = \frac{\mathbf{N}\mathfrak{f}}{N(\mathfrak{f})} \mathrm{Im}\ \alpha = \frac{b\mathbf{N}\mathfrak{f}}{N(\mathfrak{f})} \mathrm{Im}\ z.$$

Thus the exponential factor in the above expression for

$$\mathfrak{k}_{\mathfrak{f}}(\mathbf{a}, \mathfrak{c}\gamma)/\mathfrak{k}_{\mathfrak{f}}(\mathbf{a}; \mathfrak{c})$$

simplifies to

$$e[-\pi i \mathbf{N}\mathfrak{c} \cdot \mathfrak{b} \sum n(\mathbf{a})\mathbf{N}\mathfrak{a}/N(\mathfrak{f})] = e[-\pi i \mathbf{N}\mathfrak{c} \cdot b\mathbf{N}\mathfrak{a}/N(\mathfrak{f})].$$

Recall that for any lattice L, and $\omega \in L$, we have:

$$\psi(\omega, L) = \begin{cases} 1 & \text{if } \omega/2 \in L \\ -1 & \text{if } \omega/2 \notin L. \end{cases}$$

In our situation, if $(\mathfrak{a}, 2) = \mathfrak{d}$, then

$$\psi(\gamma - 1, \mathfrak{f}\mathfrak{a}^{-1}\mathfrak{c}^{-1}) = \psi(\gamma - 1, \mathfrak{f}\mathfrak{d}^{-1}\mathfrak{c}^{-1}),$$

so the ψ-factor is

$$(-1)^{\Sigma d_{\mathfrak{d}}(\mathbf{a})},$$

where the sum extends over those $\mathfrak{d} \mid (2)$ such that

$$\frac{\gamma - 1}{2} \notin \mathfrak{f}\mathfrak{c}^{-1}\mathfrak{d}^{-1}.$$

Since $N(\mathfrak{f})(\bar{\gamma} - 1) = \alpha\mathbf{N}\mathfrak{f}$, we see that

$$\frac{\gamma - 1}{2} \notin \mathfrak{f}\mathfrak{c}^{-1}\mathfrak{d}^{-1} \quad \text{if and only if} \quad \alpha \notin N(\mathfrak{f})\bar{\mathfrak{f}}^{-1}\bar{\mathfrak{c}}^{-1}2\bar{\mathfrak{d}}^{-1}.$$

(Note that in that case $\mathfrak{d} \neq (2)$, because $\alpha \in N(\mathfrak{f})\bar{\mathfrak{f}}^{-1}$.) Thus we obtain:

Lemma 1.2. *Let $\gamma \equiv 1 \bmod \mathfrak{f}$. Let α be defined by the formula*

$$N(\mathfrak{f})(\bar{\gamma} - 1) = \alpha\mathbf{N}\mathfrak{f},$$

and $\alpha = a + bz$ with rational integers a, b. Let $\mathbf{a} \in \mathbf{I}$, $\mathfrak{c}$ prime to $\mathfrak{f}$. Then

$$\frac{\mathfrak{k}_{\mathfrak{f}}(\mathbf{a}; \mathfrak{c}\gamma)\gamma^{d(\mathbf{a})}}{\mathfrak{k}_{\mathfrak{f}}(\mathbf{a}; \mathfrak{c})} = e^{-\pi i \mathbf{N}\mathfrak{c} b \mathbf{N}\mathbf{a}/N(\mathfrak{f})}(-1)^{\Sigma d_{\mathfrak{d}}(\mathbf{a})},$$

where the sum extends over those $\mathfrak{d}$ for which $\alpha \notin N(\mathfrak{f})\bar{\mathfrak{f}}^{-1}\bar{\mathfrak{c}}^{-1}2\bar{\mathfrak{d}}^{-1}$.

To analyze further the right hand side of the formula in Lemma 1.2, we need a technical remark, expressed as the next lemma.

Lemma 1.3. *Let* $\mathfrak{b}$ *be an ideal in* K. *The following are equivalent*:

(1) $\mathbf{Z}/N(\mathfrak{b})\mathbf{Z} \approx \mathfrak{o}/\mathfrak{b}$;
(2) $\mathbf{N}\mathfrak{b} = N(\mathfrak{b})$;
(3) $\mathfrak{b} = N(\mathfrak{f})\mathfrak{f}^{-1}$ *for some ideal* $\mathfrak{f}$.

Proof. The equivalence of (1) and (2) is clear, since we have an injection

$$\mathbf{Z}/N(\mathfrak{b})\mathbf{Z} \to \mathfrak{o}/\mathfrak{b}$$

and

$$|\mathbf{Z}/N(\mathfrak{b})\mathbf{Z}| = N(\mathfrak{b}), \qquad |\mathfrak{o}/\mathfrak{b}| = \mathbf{N}\mathfrak{b}.$$

(2) $\Rightarrow$ (3). If $\mathbf{N}\mathfrak{b} = N(\mathfrak{b})$, then

$$\mathfrak{b} = \mathbf{N}\mathfrak{b} \cdot \bar{\mathfrak{b}}^{-1} = N(\mathfrak{b})\bar{\mathfrak{b}}^{-1} = N(\bar{\mathfrak{b}})\bar{\mathfrak{b}}^{-1},$$

so we can take $\mathfrak{f} = \bar{\mathfrak{b}}$.

(3) $\Rightarrow$ (2). We have

$$N(\mathfrak{f}) = N(\bar{\mathfrak{f}}) = N(\mathbf{N}\mathfrak{f} \cdot \mathfrak{f}^{-1}) = \frac{\mathbf{N}\mathfrak{f}}{N(\mathfrak{f})} N(N(\mathfrak{f}) \cdot \mathfrak{f}^{-1}),$$

so

$$N(N(\mathfrak{f})\mathfrak{f}^{-1}) = N(\mathfrak{f})^2/\mathbf{N}\mathfrak{f} = \mathbf{N}(N(\mathfrak{f})\mathfrak{f}^{-1}).$$

This proves the lemma.

We now come to the proof proper of Proposition 1.1(i). Suppose the right hand side of the formula in Lemma 1.2 is equal to 1 for all choices of $\alpha \in N(\mathfrak{f})\mathfrak{f}^{-1}$ and of $\mathfrak{c}$ prime to $\mathfrak{f}$. Then the exponential term must be ± 1, so

$$\mathbf{N}\mathfrak{c} \cdot b \cdot \mathbf{N}\mathbf{a}/N(\mathfrak{f})$$

must be integral for all choices of $\mathfrak{c}$ and α. In particular, let $\mathfrak{c} = (1)$, and let $a \in \mathbf{Z}$ be such that

$$a \equiv -z \bmod N(\mathfrak{f})\mathfrak{f}^{-1}.$$

This is possible, by Lemma 1.3. Then let

$$\alpha = a + z \in N(\mathfrak{f})\mathfrak{f}^{-1}.$$

With this choice of α, we have $b = 1$. Hence we see that

$$\mathbf{Na}/N(\mathfrak{f})$$

is integral, which is condition **I 2** in the definition of $\mathbf{I}(\mathfrak{f})$.

We must now have

$$1 = (-1)^{b\mathbf{N}\mathfrak{c}\cdot\mathbf{Na}/N(\mathfrak{f}) + \Sigma d_{\mathfrak{d}}(\mathbf{a})}$$

where the sum in the exponent is still taken for $\mathfrak{d}\,|\,(2)$, with $\alpha \notin N(\mathfrak{f})\mathfrak{f}^{-1}\bar{\mathfrak{c}}^{-1}2\bar{\mathfrak{d}}^{-1}$. Therefore the exponent must be even for any choice of α and $\mathfrak{c}$. We show that this implies condition **I 3** in the definition of $\mathbf{I}(\mathfrak{f})$.

If $\mathbf{Na}/N(\mathfrak{f})$ is even, then the sum

$$\sum d_{\mathfrak{d}}(\mathbf{a}) \quad \text{taken for} \quad \alpha \notin N(\mathfrak{f})\mathfrak{f}^{-1}\bar{\mathfrak{c}}^{-1}2\bar{\mathfrak{d}}^{-1}$$

is even for all α, and it follows easily by choosing appropriate α that all of the $d_{\mathfrak{d}}(\mathbf{a})$, $\mathfrak{d}\,|\,(2)$, $\mathfrak{d} \neq (2)$, must be even. Now suppose $\mathbf{Na}/N(\mathfrak{f})$ is odd. Then

$$b\mathbf{N}\mathfrak{c} + \sum d_{\mathfrak{d}}(\mathbf{a}) \quad (\text{sum taken for } \alpha \notin N(\mathfrak{f})\mathfrak{f}^{-1}\bar{\mathfrak{c}}^{-1}2\bar{\mathfrak{d}}^{-1})$$

is always even. We consider separately the cases where 2 stays prime, ramifies, or splits, and the cases $N(\mathfrak{f})\mathfrak{f}^{-1}$ prime to 2 or not prime to 2. In each case, we choose $\mathfrak{c} = \mathfrak{o}$ and appropriate α to determine the parity of the various $d_{\mathfrak{d}}(\mathbf{a})$. The cases can conveniently be arranged as in the displayed table on the next page. We write $d_{\mathfrak{o}}$ instead of $d_{\mathfrak{o}}(\mathbf{a})$, and similarly for $d_{\mathfrak{p}}$, etc. The cases in the table exhaust the possibilities, since (2) cannot divide $N(\mathfrak{f})\mathfrak{f}^{-1}$, otherwise $N(\mathfrak{f})/2 \in \mathfrak{f}$.

That we can choose such α as in the table follows from Lemma 1.3. In the case where 2 splits, we should note that since $z + \bar{z} = 1$,

$$a + z \in \mathfrak{p} \Leftrightarrow a + \bar{z} \notin \mathfrak{p} \Leftrightarrow a + z \notin \bar{\mathfrak{p}}.$$

In the first two cases, we get a contradiction, so these do not arise. The remaining three cases lead to **I 3(ii)**, **(iii)**, **(iv)** respectively.

Hence it is necessary that $\mathbf{a} \in \mathbf{I}(\mathfrak{f})$ in order that the $\mathfrak{k}_{\mathfrak{f}}(\mathbf{a};\mathfrak{c})$ satisfy the right formula. On the other hand, the same kind of tedious checking of cases shows that when $\mathbf{a} \in \mathbf{I}(\mathfrak{f})$, then

$$b\mathbf{N}\mathfrak{c}\mathbf{Na}/N(\mathfrak{f}) + \sum d_{\mathfrak{d}}(\mathbf{a})$$

(with the sum taken for $\alpha \notin N(\mathfrak{f})\mathfrak{f}^{-1}\bar{\mathfrak{c}}^{-1}2\bar{\mathfrak{d}}^{-1}$) is indeed always even, so the $\mathfrak{k}_{\mathfrak{f}}(\mathbf{a};\mathfrak{c})$ satisfy the formula. This completes the proof of Proposition 1.1.

If 2	and $N(\mathfrak{f})\mathfrak{f}^{-1}$ is	then choosing $\alpha =$	shows that	is even
stays prime	prime to 2	$a + z \in N(\mathfrak{f})\mathfrak{f}^{-1}$ $\notin (2)$	$1 + d_{\mathfrak{o}}$	
		$a \in N(\mathfrak{f})\mathfrak{f}^{-1}$ $\notin (2)$	$d_{\mathfrak{o}}$	
ramifies $(2) = \mathfrak{p}^2$	not divisible by $\mathfrak{p}$	$a + z \in N(\mathfrak{f})\mathfrak{f}^{-1}$ $\notin \mathfrak{p}$	$1 + d_{\mathfrak{o}} + d_{\mathfrak{p}}$	
		$a \in N(\mathfrak{f})\mathfrak{f}^{-1}$ $\notin \mathfrak{p}$	$d_{\mathfrak{o}} + d_{\mathfrak{p}}$	
	divisible by $\mathfrak{p}$ but not by $\mathfrak{p}^2$	$a + z \in N(\mathfrak{f})\mathfrak{f}^{-1}$ (then $a + z \notin \mathfrak{p}N(\mathfrak{f})\mathfrak{f}^{-1}$)	$1 + d_{\mathfrak{o}} + d_{\mathfrak{p}}$	
		$a \in \mathfrak{p}N(\mathfrak{f})\mathfrak{f}^{-1}$ $a \notin 2N(\mathfrak{f})\mathfrak{f}^{-1}$	$d_{\mathfrak{o}}$	
splits $(2) = \mathfrak{p}\bar{\mathfrak{p}}$	prime to (2)	$a \in N(\mathfrak{f})\mathfrak{f}^{-1}$, a odd	$d_{\mathfrak{o}} + d_{\mathfrak{p}} + d_{\bar{\mathfrak{p}}}$	
		$a + z \in N(\mathfrak{f})\mathfrak{f}^{-1}, \notin \mathfrak{p}$ (then $a + z \in \bar{\mathfrak{p}}$)	$1 + d_{\mathfrak{o}} + d_{\mathfrak{p}}$	
		$a + z \in N(\mathfrak{f})\mathfrak{f}^{-1}, \notin \bar{\mathfrak{p}}$ (then $a + z \in \mathfrak{p}$)	$1 + d_{\mathfrak{o}} + d_{\bar{\mathfrak{p}}}$	
	divisible by $\mathfrak{p}$ but not $\bar{\mathfrak{p}}$	$a \in N(\mathfrak{f})\mathfrak{f}^{-1}$ $a \notin \mathfrak{p}N(\mathfrak{f})\mathfrak{f}^{-1}$: then a is even, so $a \notin \bar{\mathfrak{p}}N(\mathfrak{f})\mathfrak{f}^{-1}$	$d_{\mathfrak{o}} + d_{\mathfrak{p}}$	
		$a + z \in N(\mathfrak{f})\mathfrak{f}^{-1}$ $\notin \mathfrak{p}N(\mathfrak{f})\mathfrak{f}^{-1}$ (since $a + z \in \mathfrak{p}, \notin \bar{\mathfrak{p}}$)	$1 + d_{\mathfrak{o}} + d_{\bar{\mathfrak{p}}} + d_{\mathfrak{p}}$	
		$a + z \in \mathfrak{p}N(\mathfrak{f})\mathfrak{f}^{-1}$	$1 + d_{\mathfrak{o}} + d_{\bar{\mathfrak{p}}}$	

§2. Behaviour under the Artin Automorphism

Throughout this section we let $\mathfrak{f}$ be an ideal $\neq (1)$ (with $(1) = \mathfrak{o}$).

If $\mathbf{a} \in \mathbf{I}_0(\mathfrak{f})$ we know from the last section that $\mathfrak{k}_{\mathfrak{f}}(\mathbf{a}; \mathfrak{c})$ depends only on the class C, so we may write $\mathfrak{k}_{\mathfrak{f}}(\mathbf{a}; C)$. We want to prove the main transformation law, that for $C, C' \in \mathrm{Cl}(\mathfrak{f})$, then $\mathfrak{k}_{\mathfrak{f}}(\mathbf{a}; C)$ lies in $K(\mathfrak{f})$ and satisfies

$$\mathfrak{k}_{\mathfrak{f}}(\mathbf{a}; C)^{\sigma(C')} = \mathfrak{k}_{\mathfrak{f}}(\mathbf{a}; CC').$$

However, it will be necessary to consider invariants of a still more general type to include the delta function, without assuming that $\mathbf{a}$ has degree 0.

Let

$$\mathbf{b} = \sum m(\mathfrak{b})\mathfrak{b}$$

be an element in the free abelian group on the non-zero *fractional ideals* of K. If $\mathfrak{c}$ is an ideal, we define as usual

$$\Delta(\mathbf{b}; \mathfrak{c}) = \prod \Delta(\mathfrak{b}^{-1}\mathfrak{c}^{-1})^{m(\mathfrak{b})},$$

and $d(\mathbf{b}) = \sum m(\mathfrak{b})$. Suppose that

$$\mathbf{a} \in \mathbf{I}_{12}(\mathfrak{f}) \quad (\text{so } d(\mathbf{a}) \equiv 0 \bmod 12) \quad \text{and} \quad d(\mathbf{b}) = d(\mathbf{a})/12.$$

Under these conditions, we define the **modular number**

$$\beta_{\mathfrak{f}}(\mathbf{a}, \mathbf{b}; \mathfrak{c}) = \mathfrak{k}_{\mathfrak{f}}(\mathbf{a}; \mathfrak{c})\,\Delta(\mathbf{b}; \mathfrak{c}).$$

For any $\lambda \neq 0$ we have $\beta_{\mathfrak{f}}(\lambda\mathbf{a}, \lambda\mathbf{b}; \mathfrak{c}) = \beta_{\mathfrak{f}}(\mathbf{a}, \mathbf{b}; \lambda\mathfrak{c}) = \beta_{\mathfrak{f}}(\mathbf{a}, \mathbf{b}; \mathfrak{c})$ because the total weight is zero. By Proposition 1.1, it follows that $\beta_{\mathfrak{f}}(\mathbf{a}, \mathbf{b}; \mathfrak{c})$ is a ray class invariant, and we may thus write it in the form

$$\beta_{\mathfrak{f}}(\mathbf{a}, \mathbf{b}; C).$$

The invariants $\mathfrak{k}_{\mathfrak{f}}(\mathbf{a}, C)$ are special cases, when $d(\mathbf{a}) = 0$, taking $\mathbf{b} = 0$.

Theorem 2.1 *Let* $\mathbf{a} \in \mathbf{I}_{12}(\mathfrak{f})$ *and* $d(\mathbf{a}) = 12d(\mathbf{b})$. *Let* $C, C' \in \mathrm{Cl}(\mathfrak{f})$. *Then*

$$\beta_{\mathfrak{f}}(\mathbf{a}, \mathbf{b}; C) \in K(\mathfrak{f}),$$

and

$$\beta_{\mathfrak{f}}(\mathbf{a}, \mathbf{b}; C)^{\sigma(C')} = \beta_{\mathfrak{f}}(\mathbf{a}, \mathbf{b}; CC').$$

The proof will need a sequence of lemmas. We shall prove inhomogeneous transformation properties for certain power products of values of Klein forms, similar to those of the last chapter, but more elaborate and using deeper properties of the Klein forms, especially the quadratic relations.

The power products arising from conditions **I 1 I 2 I 3** will be seen to decompose into more special types, and the lemmas will concern the transformation laws of these simpler expressions.

We fix an elliptic curve A defined by an equation

$$y^2 = 4x^3 - g_2x - g_3$$

with $g_2, g_3 \in K(1)$ such that A is parametrized by $\mathbf{C}/\mathfrak{f}$, so we have a complex analytic isomorphism

$$\varphi : \mathbf{C}/\mathfrak{f} \to A_{\mathbf{C}}.$$

As in Theorem 4.2 of Chapter 11, we use here the fact that there exists $\lambda \in \mathbf{C}^*$ such that for all $z \in \mathbf{C}/\mathfrak{f}$ we have

$$x \circ \varphi(z) = \wp(\lambda z, \lambda\mathfrak{f}) \quad \text{and} \quad y \circ \varphi(z) = \wp'(\lambda z, \lambda\mathfrak{f}).$$

Let s be an idele of K, and σ an automorphism of $\mathbf{C}$ over K such that σ restricted to K^{ab} is (s, K). We then have the curve A^σ defined by the equation

$$y_\sigma^2 = 4x_\sigma^3 - g_2^\sigma x_\sigma - g_3^\sigma.$$

One form of Shimura's reciprocity law states that there is a unique parametrization

$$\psi : \mathbf{C}/s^{-1}\mathfrak{f} \to A_{\mathbf{C}}^\sigma$$

such that the following diagram is commutative:

$$\begin{array}{ccc} K/\mathfrak{f} & \xrightarrow{\varphi} & A_{\mathrm{tor}} \\ {\scriptstyle s^{-1}}\downarrow & & \downarrow{\scriptstyle \sigma} \\ K/s^{-1}\mathfrak{f} & \xrightarrow[\psi]{} & A_{\mathrm{tor}}^\sigma \end{array}$$

Then there exists a unique λ' such that

$$x_\sigma \circ \psi(z) = \wp(\lambda' z, \lambda' s^{-1}\mathfrak{f}) \quad \text{and} \quad y_\sigma \circ \psi(z) = \wp'(\lambda' z, \lambda' s^{-1}\mathfrak{f}).$$

Thus the data φ, s, σ determine λ, ψ, λ'.

We fix an ideal $\mathfrak{c}$ prime to $N(\mathfrak{f})$, and we let s be an idele such that $s_\mathfrak{p} = 1$ if $\mathfrak{p} \nmid \mathfrak{c}$. If $\mathfrak{p} | \mathfrak{c}$, we let $s_\mathfrak{p}$ have the same order at $\mathfrak{p}$ as $\mathfrak{c}$. Then $s^{-1}\mathfrak{f} = \mathfrak{c}^{-1}\mathfrak{f}$. Also note that $s_p = 1$ if $p | N(\mathfrak{f})$. We fix σ such that σ restricted to K^{ab} is equal to (s, K). The following lemmas relate to this choice of data.

We now apply this to various modular forms. The first two lemmas deal with expressions considered in Chapter 10, §2.

If $L' \supset L$ are two lattices, and $(L' : L)$ is odd, we recall the formula of Chapter 10, §2:

$$\mathfrak{k}(z, L'/L) = \prod_t (\wp(z, L) - \wp(t, L))^{-1},$$

where the product is taken for $t \in (L'/L)/\pm 1$, $t \neq 0$. We use this formula when $L = \mathfrak{f}$ and $L' = \mathfrak{a}^{-1}\mathfrak{f}$.

Lemma 2.2. *Let* $\mathfrak{a}$ *be an odd ideal. Then*

$$\mathfrak{k}(\lambda, \lambda\mathfrak{a}^{-1}\mathfrak{f}/\lambda\mathfrak{f})^\sigma = \mathfrak{k}(\lambda', \lambda'\mathfrak{a}^{-1}\mathfrak{c}^{-1}\mathfrak{f}/\lambda'\mathfrak{c}^{-1}\mathfrak{f}).$$

Proof. The proof is entirely similar to the proof of Theorem 4.2 of Chapter 11, but it is even easier since all references to the delta function can be omitted. Thus we let

$$F(x) = \prod_t (x - x \circ \varphi(t))^{-1},$$

and we proceed as before using the Shimura reciprocity law. The desired relation falls out.

Lemma 2.3. *Let* $\mathfrak{a}_1, \mathfrak{a}_2$ *be ideals such that* $\mathfrak{a}_1^{-1}/\mathfrak{o}$ *and* $\mathfrak{a}_2^{-1}/\mathfrak{o}$ *have the same points of order* 2 *(as subgroups of* $K/\mathfrak{o}$*). Then the product*

$$h(\lambda, \mathfrak{f}) = \mathfrak{k}(\lambda, \lambda\mathfrak{a}_1^{-1}\mathfrak{f}/\lambda\mathfrak{f})\mathfrak{k}(\lambda, \lambda\mathfrak{a}_2^{-1}\mathfrak{f}/\lambda\mathfrak{f})$$

satisfies

$$h(\lambda, \mathfrak{f})^\sigma = h(\lambda', \mathfrak{c}^{-1}\mathfrak{f}).$$

Proof. The proof is similar to the previous ones of the same type, but we have to use Theorem 2.3 of Chapter 10, and the corresponding rational function of x which will be a product of three terms, corresponding to the three terms in the expression of Theorem 2.3. We leave the details to the reader.

Remark. In Lemma 2.3, note that $\mathfrak{a}_1^{-1}/\mathfrak{o}$ and $\mathfrak{a}_2^{-1}/\mathfrak{o}$ have the same points of order 2 if and only if $(\mathfrak{a}_1, 2) = (\mathfrak{a}_2, 2)$.

The next two lemmas deal with powers of a single Klein form. First we have a purely function theoretic property. Let τ denote the variable in the upper half plane. If $a \in \mathbf{Z}^2/N\mathbf{Z}^2$, $a \neq 0$ we use the notation $a(\tau) = a_1\tau + a_2$.

Lemma 2.4. *Let N be an integer > 1, and let $e \in \mathbf{Z}^2/N\mathbf{Z}^2$, $e \neq 0$. Then there exists a family of elements a, $b \in \mathbf{Z}^2/N\mathbf{Z}^2$, a, $b \neq 0$, and integers $n(a, b)$ such that*

$$\mathfrak{k}^{2N}\left(\frac{e(\tau)}{N}, [\tau, 1]\right) = \prod_{(a,b)} \left[\wp\left(\frac{a(\tau)}{N}, [\tau, 1]\right) - \wp\left(\frac{b(\tau)}{N}, [\tau, 1]\right)\right]^{n(a,b)}$$

and such that for $\alpha \in \mathrm{GL}_2(\mathbf{Z}/N\mathbf{Z})$,

$$\mathfrak{k}^{2N}\left(\frac{(e\alpha)(\tau)}{N}, [\tau, 1]\right) = \prod_{(a,b)} \left[\wp\left(\frac{(a\alpha)(\tau)}{N}, [\tau, 1]\right) - \wp\left(\frac{(b\alpha)(\tau)}{N}, [\tau, 1]\right)\right]^{n(a,b)}.$$

Proof. By the theory of quadratic relations, Theorem 2.7 of Chapter 3, there exist elements $a, b \in \mathbf{Z}^2/N\mathbf{Z}^2$ and integers $n(a, b)$ such that

$$2N(e) = \sum_{(a,b)} n(a, b)p(a, b)$$

where $p(a, b)$ is the parallelogram as in Chapter 3. This implies that

$$2N(e\alpha) = \sum n(a, b)p(a\alpha, b\alpha).$$

The theorem follows at once by "applying $\mathfrak{k}$" as in Chapter 3, §3 since

$$\wp_a - \wp_b = -\frac{\mathfrak{k}_{a+b}\mathfrak{k}_{a-b}}{\mathfrak{k}_a^2\mathfrak{k}_b^2}.$$

This concludes the proof.

Remark. From the degree of homogeneity in the lemma, we note that

$$2N = \sum_{(a,b)} -2n(a, b).$$

Lemma 2.5. *Let $N = N(\mathfrak{f})$. We have*

$$\mathfrak{k}^{2N}(\lambda, \lambda\mathfrak{f})^\sigma = \mathfrak{k}^{2N}(\lambda', \lambda'\mathfrak{f}\mathfrak{c}^{-1}).$$

Proof. Let $\mathfrak{f}/N = [\tau, 1]$ with $\tau \in K$. By Lemma 2.4 with $e = (0, 1)$ we have

$$\mathfrak{k}^{2N}\left(\frac{\lambda}{N}, \lambda[\tau, 1]\right) = \prod_{(a,b)} \left[\wp\left(\lambda \frac{a(\tau)}{N}, \lambda[\tau, 1]\right) - \wp\left(\lambda \frac{b(\tau)}{N}, \lambda[\tau, 1]\right)\right]^{n(a,b)}.$$

Let

$$s^{-1}\begin{pmatrix}\tau\\1\end{pmatrix} = u\beta\begin{pmatrix}\tau\\1\end{pmatrix} \quad \text{with } u \in \prod_p \mathrm{GL}_2(\mathbf{Z}_p),\ \beta \in \mathrm{GL}_2^+(\mathbf{Q})$$

and such that β^{-1} is integral (i.e. has coefficients in $\mathbf{Z}$). This is possible since $\mathfrak{c}$ is integral. Note that if $p \mid N$ then $u_p = \beta^{-1}$ since $s_p = 1$. We let

$$j_\beta(\tau) = c\tau + d \quad \text{if} \quad \beta = \begin{pmatrix} * & * \\ c & d\end{pmatrix}.$$

For convenience of notation, we shall write

$$W_\tau = \begin{pmatrix}\tau\\1\end{pmatrix}.$$

If $W = \begin{pmatrix}\omega_1\\\omega_2\end{pmatrix}$ is a column vector, we also write $\wp(z, W) = \wp(z, [W])$, where $[W]$ is the lattice generated by ω_1 and ω_2.

By Shimura's reciprocity law, we find:

$$\begin{aligned}
&\mathfrak{k}^{2N}\left(\frac{\lambda}{N}, \lambda[\tau, 1]\right)^{\sigma} \\
&\quad = \prod_{(a,b)} \left[\wp\left(\lambda' \frac{a(\tau)}{N} s^{-1}, s^{-1}[\tau, 1]\right) - \wp\left(\lambda' \frac{b(\tau)}{N} s^{-1}, \lambda' s^{-1}[\tau, 1]\right)\right]^{n(a,b)} \\
&\quad = \prod_{(a,b)} \left[\wp\left(\lambda' \frac{au}{N} \beta W_\tau, \lambda'\beta W_\tau\right) - \wp\left(\lambda' \frac{bu}{N} \beta W_\tau, \lambda'\beta W_\tau\right)\right]^{n(a,b)} \\
&\quad = \prod_{(a,b)} \left[\wp\left(\lambda' \frac{a\beta^{-1}}{N} \beta W_\tau, \lambda'\beta W_\tau\right) - \wp\left(\lambda' \frac{b\beta^{-1}}{N} \beta W_\tau, \lambda'\beta W_\tau\right)\right]^{n(a,b)} \\
&\quad = \prod_{(a,b)} (c\tau + d)^{-2n(a,b)} \left[\wp\left(\lambda' \frac{a}{N} \beta^{-1} W_{\beta(\tau)}, \lambda' W_{\beta(\tau)}\right)\right. \\
&\qquad\qquad\qquad \left. - \wp\left(\lambda' \frac{b}{N} \beta^{-1} W_{\beta(\tau)}, \lambda' W_{\beta(\tau)}\right)\right]^{n(a,b)} \\
&\quad = (c\tau + d)^{2N} \mathfrak{k}^{2N}\left(\lambda' \frac{e}{N} \beta^{-1} W_{\beta(\tau)}, \lambda' W_{\beta(\tau)}\right) \quad \text{(by Lemma 2.4)} \\
&\quad = (c\tau + d)^{2N} \mathfrak{k}^{2N}(\lambda' j_{\beta^{-1}}(\beta(\tau)), \lambda'[\beta(\tau), 1]).
\end{aligned}$$

On the other hand, we note that

$$j_\beta(\tau)^{-1} = j_{\beta^{-1}}(\beta(\tau)),$$

or in other words,

$$(c\tau + d)^{-1} = j_{\beta^{-1}}(\beta(\tau));$$

this is a special case of the cocycle relation satisfied by $j_\beta(\tau)$. Hence

$$\begin{aligned}
\mathfrak{k}^{2N}\left(\frac{1}{N}, \mathfrak{c}^{-1}\frac{\mathfrak{f}}{N}\right) &= \mathfrak{k}^{2N}\left(\frac{1}{N}, \beta W_\tau\right) \\
&= \mathfrak{k}^{2N}\left(\frac{1}{N}, (c\tau + d)[\beta(\tau), 1]\right) \\
&= (c\tau + d)^{2N}\mathfrak{k}^{2N}\left(\frac{(c\tau + d)^{-1}}{N}, [\beta(\tau), 1]\right) \\
&= (c\tau + d)^{2N}\mathfrak{k}^{2N}\left(\frac{1}{N} j_{\beta^{-1}}(\beta(\tau)), [\beta(\tau), 1]\right).
\end{aligned}$$

This proves the lemma.

The main theorem at the beginning of this section is formulated for elements of degree 0. As the two lemmas indicate, we can also give a non-homogeneous formulation as follows. Define

$$\mathfrak{k}_{\mathfrak{f},\lambda}(\mathbf{a}; \mathfrak{c}) = \prod \mathfrak{k}(\lambda, \lambda\mathfrak{f}\mathfrak{a}^{-1}\mathfrak{c}^{-1})^{n(\mathfrak{a})} = \lambda^{d(\mathbf{a})}\mathfrak{k}_\mathfrak{f}(\mathbf{a}; \mathfrak{c}).$$

Lemma 2.6. *Let* $\mathbf{a} \in \mathbf{I}_2(\mathfrak{f})$ *be of even degree, and suppose we are in case* **I 3(i)**. *Then*

$$\mathfrak{k}_{\mathfrak{f},\lambda}(\mathbf{a}; \mathfrak{o})^\sigma = \mathfrak{k}_{\mathfrak{f},\lambda'}(\mathbf{a}; \mathfrak{c}).$$

Proof. By assumption, $\mathbf{Na}/N(\mathfrak{f})$ and all the $d_\mathfrak{d}(\mathbf{a})$ are even. Let us write

$$\begin{aligned}
\mathfrak{k}_{\mathfrak{f},\lambda}(\mathbf{a}; \mathfrak{o}) &= \prod_{\mathfrak{a}} \mathfrak{k}(\lambda, \lambda\mathfrak{f}\mathfrak{a}^{-1})^{n(\mathfrak{a})} \\
&= \mathfrak{k}(\lambda, \lambda\mathfrak{f})^{\mathbf{Na}} \cdot \prod_{\mathfrak{d}|2} \prod_{(\mathfrak{a},2)=\mathfrak{d}} \left(\frac{\mathfrak{k}(\lambda, \lambda\mathfrak{f}\mathfrak{a}^{-1})}{\mathfrak{k}(\lambda, \lambda\mathfrak{f})^{\mathbf{Na}}}\right)^{n(\mathfrak{a})}. \qquad (1)
\end{aligned}$$

But if $(\mathfrak{a}_1, 2) = (\mathfrak{a}_2, 2)$ then

$$\lambda\mathfrak{f}\mathfrak{a}_1^{-1} \quad \text{and} \quad \lambda\mathfrak{f}\mathfrak{a}_2^{-1}$$

contain the same points of order 2 with respect to $\lambda\mathfrak{f}$. Since for each $\mathfrak{d}|(2)$, $d_{\mathfrak{d}}(\mathbf{a})$ is even, we may apply Lemma 2.3 to conclude that the factor

$$\prod_{(\mathfrak{a},2)=\mathfrak{d}} \mathfrak{k}(\lambda, \lambda\mathfrak{f}\mathfrak{a}^{-1}/\lambda\mathfrak{f})^{-n(\mathfrak{a})}$$

which occurs in (1) transforms to

$$\prod_{(\mathfrak{a},2)=\mathfrak{d}} \mathfrak{k}(\lambda', \lambda'\mathfrak{f}\mathfrak{a}^{-1}\mathfrak{c}^{-1}/\lambda\mathfrak{f}\mathfrak{c}^{-1})^{-n(\mathfrak{a})}$$

under σ.

Furthermore, we have $2N(\mathfrak{f})|\mathbf{Na}$, so by Lemma 2.5, $\mathfrak{k}(\lambda, \lambda\mathfrak{f})^{\mathbf{Na}}$ transforms to $\mathfrak{k}(\lambda', \lambda'\mathfrak{f}\mathfrak{c}^{-1})^{\mathbf{Na}}$. This concludes the proof of the lemma.

We now revert to elements of total degree 0, and to the proof proper of the main theorem. For simplicity, let us first deal with $\mathfrak{k}_{\mathfrak{f}}(\mathbf{a}, C)$. If $\mathbf{a} \in \mathbf{I}_0(\mathfrak{f})$ and satisfies **I 3(i)**. then Lemma 2.6 shows that we have the formula

$$\mathfrak{k}_{\mathfrak{f}}(\mathbf{a}; C_0)^{\sigma} = \mathfrak{k}_{\mathfrak{f}}(\mathbf{a}; C).$$

This implies that $\mathfrak{k}_{\mathfrak{f}}(\mathbf{a}; C)$ lies in $K(\mathfrak{f})$, and satisfies the desired transformation law. In particular, we have proved Theorem 2.1 in case $\mathbf{a}$ satisfies condition **I 3(i)**

The remaining cases **I 3(ii)** through **(iv)** can be reduced to the preceding case as follows. We apply Remark 3 after the definition of $\mathbf{I}(\mathfrak{f})$. By the first part of the proof, we then obtain

$$\mathfrak{k}_{\mathfrak{f}}(\mathfrak{a}; \mathfrak{c}) = \mathfrak{k}_{2\mathfrak{f}}((2)\mathbf{a}; \mathfrak{c}) \in K(2\mathfrak{f}),$$

and for $\mathfrak{c}'$ prime to $2\mathfrak{f}$;

$$\begin{aligned} \mathfrak{k}_{\mathfrak{f}}(\mathbf{a}; \mathfrak{c})^{(\mathfrak{c}',K)} &= \mathfrak{k}_{2\mathfrak{f}}((2)\mathbf{a}; \mathfrak{c})^{(\mathfrak{c}',K)} \\ &= \mathfrak{k}_{2\mathfrak{f}}((2)\mathbf{a}; \mathfrak{c}\mathfrak{c}') \\ &= \mathfrak{k}_{\mathfrak{f}}(\mathbf{a}; \mathfrak{c}\mathfrak{c}'). \end{aligned}$$

Since $\mathfrak{k}_{\mathfrak{f}}(\mathbf{a}; \mathfrak{c})$ and $\mathfrak{k}_{\mathfrak{f}}(\mathbf{a}; \mathfrak{c}\mathfrak{c}')$ depend only on the classes of $\mathfrak{c}$, $\mathfrak{c}'$ in $\mathrm{Cl}(\mathfrak{f})$ by Proposition 1.1(ii), it follows that $\mathfrak{k}_{\mathfrak{f}}(\mathbf{a}; \mathfrak{c}) \in K(\mathfrak{f})$, and the transformation under the Galois group is as claimed. This completes the proof of Theorem 2.1 for $\mathfrak{k}_{\mathfrak{f}}(\mathbf{a}, C)$.

The proof with the extra delta term, for $\beta_{\mathfrak{f}}(\mathbf{a}, \mathbf{b}; C)$ is entirely similar, since we have

$$\Delta(\lambda\mathfrak{b}^{-1})^{\sigma} = \Delta(\lambda'\mathfrak{b}^{-1}\mathfrak{c}^{-1}).$$

As before, we get the result first when **a** satisfies **I 3(i)** and has degree divisible by 12 directly from Lemma 2.6. The remaining cases are reduced to this one in the same way as before since we merely replaced $\mathfrak{f}$ by $2\mathfrak{f}$ and **a** by $(2)\mathbf{a}$. This does not affect the degree of **a**, nor does it affect **b**. Thus we conclude the proof of Theorem 2.1.

§3. Modular Units in $K(1)$ as Klein Units

In Chapter 9, §5, we had considered the group

$$\Delta_{K(1)}(c_0, w_{K(1)})^{1/12wh},$$

which had the "right" index in the group of all units. We shall now give a more natural definition of this group without the extraneous construction of taking roots. We give it directly in terms of power products of values of Klein forms.

The construction of §1, §2 does not immediately give elements of $K(1)$ since we took $\mathfrak{f} \neq 0$. To obtain such elements, we take the norm down to $K(1)$ of invariants at level $\mathfrak{p}$ for various primes $\mathfrak{p}$. That we recover units constructed with the delta function arises essentially from the distribution relations.

Since we shall deal with forms which are not of degree 0, rather than taking the norm, it will give us some extra flexibility if we take products over certain representatives. Thus for each prime $\mathfrak{p}$ we let $R(\mathfrak{p})$ be a set of representatives in $\mathfrak{o}$ for $\mathfrak{o}(\mathfrak{p})^*/\mathfrak{o}^*$, which is isomorphic to the Galois group

$$\mathrm{Gal}(K(\mathfrak{p})/K(1)).$$

We define the **norm with respect to the set of representatives**:

$$\mathbf{N}_{R(\mathfrak{p})}\mathfrak{k}_{\mathfrak{p}}(\mathbf{a}; \mathfrak{c}) = \prod_{\alpha \in R(\mathfrak{p})} \mathfrak{k}_{\mathfrak{p}}(\mathbf{a}; \mathfrak{c}\alpha),$$

where $\mathbf{a} \in \mathbf{I}(\mathfrak{p})$. If **a** has degree 0, then $\mathfrak{k}_{\mathfrak{p}}(\mathbf{a}; \mathfrak{c}\alpha) \in K(\mathfrak{p})$ by Theorem 2.1, and we do have

$$\mathbf{N}_{R(\mathfrak{p})}\mathfrak{k}_{\mathfrak{p}}(\mathbf{a}; \mathfrak{c}) = \mathbf{N}_{K(\mathfrak{p})/K(1)}\mathfrak{k}_{\mathfrak{p}}(\mathbf{a}; \mathfrak{c}).$$

otherwise, the product may depend on the choice of representatives. Let:

S = a finite set of primes $\mathfrak{p}$ relatively prime to $w_{K(1)}$;
$R(\mathfrak{p})$ = a set of representatives for $\mathfrak{o}(\mathfrak{p})^*/\mathfrak{o}^*$;
R_S = family of representatives $\{R(\mathfrak{p})\}$ for $\mathfrak{p} \in S$;
$\mathbf{a}_{\mathfrak{p}} \in \mathbf{I}_w(\mathfrak{p})$; we let $\mathbf{a}_S$ denote the family $\{\mathbf{a}_{\mathfrak{p}}\}$ for $\mathfrak{p} \in S$;
$\mathfrak{c}$ = an ideal prime to all $\mathfrak{p} \in S$.

We also assume that

$$\sum_{\mathfrak{p} \in S} \frac{\mathbf{N}\mathfrak{p} - 1}{w} d(\mathbf{a}_{\mathfrak{p}}) = 0.$$

Putting

$$\mathbf{a} = \sum_{\mathfrak{p} \in S} \sum_{\alpha \in R(\mathfrak{p})} (\alpha)\mathbf{a}_{\mathfrak{p}},$$

this condition means that *the total degree is* 0, that is $d(\mathbf{a}) = 0$. Given such data, we define the **Klein invariants**

$$\begin{aligned} \mathfrak{k}(\mathbf{a}_S, R_S; \mathfrak{c}) &= \prod_{\mathfrak{p} \in S} \prod_{\alpha \in R(\mathfrak{p})} \mathfrak{k}_{\mathfrak{p}}(\mathbf{a}_{\mathfrak{p}}; \mathfrak{c}\alpha) \\ &= \prod_{\mathfrak{p} \in S} \mathbf{N}_{R(\mathfrak{p})} \mathfrak{k}_{\mathfrak{p}}(\mathbf{a}_{\mathfrak{p}}; \mathfrak{c}). \end{aligned}$$

The dependence on the various data, especially the set of representatives, will be investigated below. If we fix all the variables except $\mathfrak{c}$, then we also write

$$\beta(\mathfrak{c}) = \mathfrak{k}(\mathbf{a}_S, R_S; \mathfrak{c}).$$

It will be proved below that $\beta(\mathfrak{c})$ depends only on the ordinary ideal class of $\mathfrak{c}$, and lies in $K(1)$. We may therefore replace $\mathfrak{c}$ by $C \in \mathrm{Cl}(1)$. We then define the group of **Klein numbers** in $K(1)$ to be:

$\mathfrak{K}(\mathfrak{o})$ = group consisting of all elements $\mathfrak{k}(\mathbf{a}_S, \mathrm{R}_S; \mathfrak{c})$
for all choices of S, $R(\mathfrak{p})$, $\mathbf{a}_{\mathfrak{p}}$ and $\mathfrak{c}$ as above.

Since

$$\mathfrak{k}_{\mathfrak{p}}(\mathbf{a}_{\mathfrak{p}}; \mathfrak{c}) = \mathfrak{k}_{\mathfrak{p}}(\mathbf{a}_{\mathfrak{p}}\mathfrak{c}; \mathfrak{o}),$$

we would get the same group $\mathfrak{K}(\mathfrak{o})$ if we assumed $\mathfrak{c} = \mathfrak{o}$ in the above definition. However, we retain the general $\mathfrak{c}$ in order to be able to write the reciprocity law in the form

$$\beta(\mathfrak{c})^{\sigma(\mathfrak{c}')} = \beta(\mathfrak{c}\mathfrak{c}')$$

as before. Of course, once we have proved this formula with ideals $\mathfrak{c}, \mathfrak{c}'$, we get the analogous formula with ideal classes.

Remark. Let X be a finite set of primes containing all those dividing $w_{K(1)}$. We define $\mathfrak{K}_X(\mathfrak{o})$ just as we defined $\mathfrak{K}(\mathfrak{o})$, except that we require the sets S not to contain any prime of X. It will be shown later that

$$\boldsymbol{\mu}_{K(1)}\mathfrak{K}_X(\mathfrak{o}) = \boldsymbol{\mu}_{K(1)}\mathfrak{K}(\mathfrak{o}).$$

Similarly, we may perform the above constructions with the more general invariant involving the delta function. Given S, $\mathbf{a}_S, R_S$ and $\mathfrak{c}$ as before, but assuming only that

$$\sum_{\mathfrak{p} \in S} \frac{\mathbf{N}\mathfrak{p} - 1}{w} d(\mathbf{a}_\mathfrak{p}) \equiv 0 \bmod 12,$$

we let $\mathbf{b}$ be such that

$$12d(\mathbf{b}) = \sum_{\mathfrak{p} \in S} \frac{\mathbf{N}\mathfrak{p} - 1}{w} d(\mathbf{a}_\mathfrak{p}),$$

and we defined the **modular invariant**

$$\beta(\mathbf{a}_S, R_S, \mathbf{b}; \mathfrak{c}) = \mathfrak{k}(\mathbf{a}_S, R_S; \mathfrak{c})\,\Delta(\mathbf{b}; \mathfrak{c}).$$

Again we denote it by $\beta(\mathfrak{c})$. It will be proved below that it depends only on the ordinary ideal class of $\mathfrak{c}$, and lies in $K(1)$. We then define the group of **modular numbers** in $K(1)$ to be:

$\mathcal{M}(\mathfrak{o})$ = group consisting of all elements $\beta(\mathbf{a}_S, R_S, \mathbf{b}; \mathfrak{c})$ for all choices of S, $R(\mathfrak{p})$, $\mathbf{a}_\mathfrak{p}$, $\mathbf{b}$, $\mathfrak{c}$ *as above.*

If all the sets S are further required not to contain any prime in a given set X, then we define the group $\mathcal{M}_X(\mathfrak{o})$.

We define the group of **modular units** in $K(1)$ to be

$$E_{\mathrm{mod}}(K(1)) = \boldsymbol{\mu}_{K(1)}(\mathcal{M}(\mathfrak{o}) \cap E),$$

where E is the group of all units. In other words, it is the group generated by the roots of unity in $K(1)$, and by those elements of $\mathcal{M}(\mathfrak{o})$ which are units.

If H is an unramified abelian extension of K, we define

$$E_{\mathrm{mod}}(H) = \boldsymbol{\mu}_H \mathbf{N}_{K(1)/H} E_{\mathrm{mod}}(K(1)).$$

Using the notation of Chapter 9, we shall prove:

Theorem 3.1. $E_{\mathrm{mod}}(H) = \Delta_H(c_0, ww_H)^{1/12wh}$,

where the right hand side consists of those elements of H whose $12wh$-power lies in $\Delta_H(c_0, ww_H)$.

The rest of this section carries out the above program. First note that by Lemma 5.5 of Chapter 9, the proof of Theorem 3.1 can be limited to the case when $H = K(1)$. We now investigate the dependence of our products on the choice of representatives.

The reciprocity law mapping establishes a natural isomorphism

$$(\mathfrak{o}/\mathfrak{f})^*/\mathfrak{o}^* \approx \mathrm{Gal}(K(\mathfrak{f})/K(1)).$$

If $\mathfrak{g}|\mathfrak{f}$, then $\mathrm{Gal}(K(\mathfrak{f})/K(\mathfrak{g}))$ is then isomorphic to a subgroup of $(\mathfrak{o}/\mathfrak{f})^*/\mathfrak{o}^*$.

Lemma 3.2. *Let* $\mathfrak{g}|\mathfrak{f}$. *Let* R *and* $R' \subset \mathfrak{o}$ *be two sets of representatives for the class group isomorphic to* $\mathrm{Gal}(K(\mathfrak{f})/K(\mathfrak{g}))$. *Let* $\mathbf{a} \in \mathbf{I}(\mathfrak{f})$. *Then for any ideal* $\mathfrak{c}$ *prime to* $\mathfrak{f}$, *we have*:

$$\frac{\prod_{\alpha \in R} \mathfrak{k}_{\mathfrak{f}}(\mathbf{a}; \mathfrak{c}\alpha)}{\prod_{\alpha' \in R'} \mathfrak{k}_{\mathfrak{f}}(\mathbf{a}; \mathfrak{c}\alpha')} = \varepsilon^{d(\mathbf{a})} \left(\frac{\prod_{\alpha' \in R'} \alpha'}{\prod_{\alpha \in R} \alpha} \right)^{d(\mathbf{a})}$$

where ε *is a* $w(\mathfrak{g})$-th *root of unity. In particular, if* $\mathbf{a} \in \mathbf{I}_{w(\mathfrak{g})}(\mathfrak{f})$, *then* $\varepsilon^{d(\mathbf{a})} = 1$.

Proof. Each $\alpha \in R$ is of the form

$$\alpha = \alpha' \varepsilon_\alpha \gamma_\alpha$$

for some $\alpha' \in R'$ $\varepsilon_\alpha \in \boldsymbol{\mu}_K$, $\varepsilon_\alpha \equiv 1 \bmod \mathfrak{g}$, and $\gamma_\alpha \equiv 1 \bmod^* \mathfrak{f}$. Choose $\gamma \in \mathfrak{o}$, $\gamma \equiv 1 \bmod \mathfrak{f}$, such that all $\gamma\gamma_\alpha \in \mathfrak{o}$. Then

$$\begin{aligned}\prod_{\alpha \in R} \mathfrak{k}_{\mathfrak{f}}(\mathbf{a}; \mathfrak{c}\alpha) &= \gamma^{d(\mathbf{a})(\mathbf{N}\mathfrak{p} - 1)} \prod_{\alpha \in R} \mathfrak{k}_{\mathfrak{f}}(\mathbf{a}; \mathfrak{c}\alpha\gamma) \quad [\text{by Proposition 1.1}] \\ &= \gamma^{d(\mathbf{a})(\mathbf{N}\mathfrak{p} - 1)} \prod_{\alpha \in R} \mathfrak{k}_{\mathfrak{f}}(\mathbf{a}; \mathfrak{c}\alpha'\varepsilon_\alpha\gamma_\alpha\gamma) \\ &= \prod_{\alpha} \gamma_\alpha^{-d(\mathbf{a})} \prod_{\alpha \in R} \mathfrak{k}_{\mathfrak{f}}(\mathbf{a}; \mathfrak{c}\alpha')\end{aligned}$$

[by Proposition 1.1 and the fact that $\mathfrak{c}\varepsilon_\alpha = \mathfrak{c}$]

$$= \prod_{\alpha} \varepsilon_\alpha^{d(\mathbf{a})} \prod_{\alpha} (\alpha'/\alpha)^{d(\mathbf{a})} \prod_{\alpha' \in R'} \mathfrak{k}_{\mathfrak{f}}(\mathbf{a}; \mathfrak{c}\alpha')$$

by definitions. This proves the lemma.

Corollary 3.3. *Let* X *be a finite set of primes containing all those dividing* $w_{K(1)}$. *Then*

$$K^{*w} \subset \mathfrak{K}_X(\mathfrak{o}).$$

Proof. Suppose $\alpha \in \mathfrak{o}$ and $\alpha \equiv 1 \bmod \mathfrak{p}$ for some $\mathfrak{p} \notin X$. Choose $\mathbf{a} \in \mathbf{I}_w(\mathfrak{p})$ with $d(\mathbf{a}) = w$, and let R be a set of representatives for $\mathfrak{o}(\mathfrak{p})^*/\mathfrak{o}^*$ with $\alpha \in R$. Let R' be the same set R with α replaced by 1. Then the product of Lemma 3.2 becomes α^w since $\varepsilon^{d(\mathbf{a})} = 1$. Thus for any such α, we get $\alpha^w \in \mathfrak{K}_X(\mathfrak{o})$. On the other hand, if $\alpha \in K^*$ and $\alpha = \alpha_1/\alpha_2$ with $\alpha_1, \alpha_2 \in \mathfrak{o}$, we select two primes $\mathfrak{p}_1, \mathfrak{p}_2$ relatively prime to $w_{K(1)}\alpha_1\alpha_2$, not in X, and $\lambda \in \mathfrak{o}$ such that

$$\lambda\alpha_1 \equiv 1 \bmod \mathfrak{p}_1 \quad \text{and} \quad \lambda\alpha_2 \equiv 1 \bmod \mathfrak{p}_2.$$

Then $\lambda\alpha_1$ and $\lambda\alpha_2$ satisfy the conditions in the first part of the proof; and $\alpha = \lambda\alpha_1/\lambda\alpha_2$ so $\alpha^w \in \mathfrak{K}_X(\mathfrak{o})$. This proves the corollary.

Theorem 3.4. *Let* $\beta(\mathfrak{c}) = \mathfrak{k}(\mathbf{a}_S, R_S; \mathfrak{c})\Delta(\mathbf{b}; \mathfrak{c})$. *Then* $\beta(\mathfrak{c}) \in K(1)$, *and depends only on the ideal class of* $\mathfrak{c}$. *It satisfies*

$$\beta(C)^{\sigma(C')} = \beta(CC').$$

Proof. Let

$$\mathfrak{f} = 2 \prod_{\mathfrak{p} \in S} \mathfrak{p},$$

and let

$$\mathbf{a} = \sum_{\mathfrak{p} \in S} \sum_{\alpha \in R(\mathfrak{p})} \mathbf{a}_{\mathfrak{p}} \cdot \mathfrak{f}\mathfrak{p}^{-1}(\alpha).$$

Then $\mathbf{a} \in \mathbf{I}_{12}(\mathfrak{f})$ by Remark 2 of §1 [applied with $\mathfrak{f} = \mathfrak{p}$, $\mathfrak{c} = \mathfrak{f}\mathfrak{p}^{-1}$, and $\mathfrak{b} = (\alpha)$], and we have

$$\beta(\mathfrak{c}) = \mathfrak{k}_{\mathfrak{f}}(\mathbf{a}; \mathfrak{c})\Delta(\mathbf{b}; \mathfrak{c}).$$

Therefore $\beta(\mathfrak{c}) \in K(\mathfrak{f})$, and for $\mathfrak{c}'$ prime to $\mathfrak{f}$ we get

$$\beta(\mathfrak{c})^{\sigma(\mathfrak{c}')} = \beta(\mathfrak{c}\mathfrak{c}').$$

Suppose now that $\mathfrak{c}' = (\gamma)$ is principal, and prime to $\mathfrak{f}$. We show that $\beta(\mathfrak{c}\gamma) = \beta(\mathfrak{c})$. For this we use Lemma 3.2. For each $\mathfrak{p}$, let

$$R'(\mathfrak{p}) = \{\gamma\alpha : \alpha \in R(\mathfrak{p})\}.$$

Then $R'(\mathfrak{p})$ is another set of representatives for $\mathfrak{o}(\mathfrak{p})^*/\mathfrak{o}^*$, and

$$\begin{aligned}\frac{\beta(\mathfrak{c}\gamma)}{\beta(\mathfrak{c})} &= \frac{\prod_{\mathfrak{p} \in S} \mathbf{N}_{R'}(\mathfrak{k}_{\mathfrak{p}}(\mathbf{a}_{\mathfrak{p}}; \mathfrak{c}))}{\prod_{\mathfrak{p} \in S} \mathbf{N}_R(\mathfrak{k}_{\mathfrak{p}}(\mathbf{a}_{\mathfrak{p}}; \mathfrak{c}))} \frac{\Delta(\mathbf{b}, \mathfrak{c}\gamma)}{\Delta(\mathbf{b}, \mathfrak{c})} \\ &= \prod_{\mathfrak{p} \in S} \gamma^{-d(\mathbf{a}_{\mathfrak{p}})(\mathbf{N}\mathfrak{p} - 1)/w} \cdot \gamma^{12d(\mathbf{b})}\end{aligned}$$

by Lemma 3.2, and the fact that the power of ε is equal to 1 since w divides $d(\mathbf{a}_\mathfrak{p})$ by assumption. This final expression is now seen to be equal to 1 since the total degree is 0 by assumption, thus concluding the proof.

Thus we have proved the formula $\beta(\mathfrak{c})^{\sigma(\mathfrak{c}')} = \beta(\mathfrak{c}\mathfrak{c}')$ when $\mathfrak{c}$, $\mathfrak{c}'$ are prime to $\mathfrak{f}$, and we know that it depends only on the classes of $\mathfrak{c}$, $\mathfrak{c}'$ up to multiplication by principal ideals γ prime to $\mathfrak{f}$. It follows by class field theory that $\beta(\mathfrak{c})$ lies in $K(1)$, and the formula holds for classes C, C' instead of $\mathfrak{c}$, $\mathfrak{c}'$. This concludes the proof of Theorem 3.4.

Our next step is to give an expression for the $12w$-th power of the type of product that enters in the definition of $\mathfrak{K}(\mathfrak{o})$.

Lemma 3.5. *Let* $\mathbf{a} \in \mathbf{I}(\mathfrak{p})$ *with* $\mathfrak{p}$ *prime to* $w_{K(1)}$. *Let* $R = R(\mathfrak{p})$ *be a set of representatives for* $\mathfrak{o}(\mathfrak{p})^*/\mathfrak{o}^*$. *Then for any* $\mathfrak{c}$ *prime to* $\mathfrak{p}$, *we have*

$$\prod_{\alpha \in R} \mathfrak{k}_\mathfrak{p}(\mathbf{a}; \mathfrak{c}\alpha)^{12w} = \lambda \prod_{\mathfrak{a}} \left(\frac{\Delta(\mathfrak{a}^{-1}\mathfrak{c}^{-1})}{\Delta(\mathfrak{p}\mathfrak{a}^{-1}\mathfrak{c}^{-1})^{\mathrm{N}\mathfrak{p}}} \right)^{n(\mathfrak{a})}$$

where

$$\lambda = \prod_{\alpha \in R} \alpha^{12wd(\mathbf{a})}.$$

Proof. It will suffice to prove that the $N(\mathfrak{p})$-th power of the left hand side equals the $N(\mathfrak{p})$-th power of the right hand side, because the quotient l.h.s./r.h.s lies in $K(1)$ by Theorem 3.4. We have

$$\begin{aligned} \prod_{\alpha \in R} \mathfrak{k}_\mathfrak{p}(\mathbf{a}; \mathfrak{c}\alpha)^{12wN(\mathfrak{p})} &= \prod_{\alpha \in R} \prod_{\mathfrak{a}} \mathfrak{k}(1, \mathfrak{p}\mathfrak{a}^{-1}\mathfrak{c}^{-1})^{12wN(\mathfrak{p})n(\mathfrak{a})} \\ &= \mathrm{Num}/\mathrm{Den} \end{aligned}$$

where the numerator and denominator are given in terms of g and Δ by:

$$\begin{aligned} \mathrm{Num} &= \prod_{\mathfrak{a}} \prod_{\alpha \in R} g_\mathfrak{p}(C(\mathfrak{a}\mathfrak{c}\alpha))^{wn(\mathfrak{a})} \\ \mathrm{Den} &= \prod_{\mathfrak{a},\alpha} \Delta(\mathfrak{p}\mathfrak{a}^{-1}\mathfrak{c}^{-1}\alpha^{-1})^{wN(\mathfrak{p})n(\mathfrak{a})}. \end{aligned}$$

By the distribution relations of Theorem 1.4 in Chapter 11, we see that

$$\mathrm{Num} = \prod_{\mathfrak{a}} (\Delta(\mathfrak{a}^{-1}\mathfrak{c}^{-1})/\Delta(\mathfrak{p}\mathfrak{a}^{-1}\mathfrak{c}^{-1}))^{N(\mathfrak{p})n(\mathfrak{a})}.$$

Furthermore,

$$\begin{aligned}\text{Den} &= \prod_{\alpha,\mathfrak{a}} (\Delta(\mathfrak{p}\mathfrak{a}^{-1}\mathfrak{c}^{-1})\alpha^{12})^{wN(\mathfrak{p})n(\mathfrak{a})}\\ &= \prod_{\mathfrak{a}} \Delta(\mathfrak{p}\mathfrak{a}^{-1}\mathfrak{c}^{-1})^{(\mathbf{N}\mathfrak{p}-1)N(\mathfrak{p})n(\mathfrak{a})}\cdot \lambda^{N(\mathfrak{p})},\end{aligned}$$

where

$$\lambda = \prod_{\alpha} \alpha^{12wd(\mathbf{a})}.$$

Thus combining like terms, we get

$$\prod_{\alpha\in R} \mathfrak{k}_{\mathfrak{p}}(\mathbf{a}; c\alpha)^{12wN(\mathfrak{p})} = \prod_{\mathfrak{a}} \left(\frac{\Delta(\mathfrak{a}^{-1}\mathfrak{c}^{-1})}{\Delta(\mathfrak{p}\mathfrak{a}^{-1}\mathfrak{c}^{-1})^{\mathbf{N}\mathfrak{p}}}\right)^{n(\mathfrak{a})N(\mathfrak{p})} \cdot \lambda^{N(\mathfrak{p})}.$$

This concludes the proof of Lemma 3.5.

For an individual term of degree 0, the formula of Lemma 3.5 can be rewritten in terms of the norm $\mathbf{N}_{\mathfrak{p},\mathfrak{o}}$ from $K(\mathfrak{p})$ to $K(1)$, as follows.

Theorem 3.6. *Let* $\mathbf{a} \in \mathbf{I}_0(\mathfrak{p})$ *with* $\mathfrak{p}$ *prime to* $w_{K(1)}$. *Then for any* $\mathfrak{c}$ *prime to* $\mathfrak{p}$, *we have*

$$\mathbf{N}_{\mathfrak{p},\mathfrak{o}}\mathfrak{k}_{\mathfrak{p}}(\mathbf{a};\mathfrak{c})^{12w} = \left(\prod_{\mathfrak{a}} \Delta(\mathfrak{p}\mathfrak{a}^{-1}\mathfrak{c}^{-1})^{n(\mathfrak{a})}\right)^{\sigma(\mathfrak{p})-\mathbf{N}\mathfrak{p}}.$$

Proof. Since $\mathbf{a}$ has degree 0, both $\mathfrak{k}_{\mathfrak{p}}(\mathbf{a};\mathfrak{c})$ and the product on the right lie in $K(1)$. In Lemma 3.5, taking the $N(\mathfrak{p})$-th root introduces a p-th root of unity ε, which thus lies in $K(1)$, and must be equal to 1. This proves Theorem 3.6.

Theorem 3.7. *If* $\mathbf{a} = \sum n(\mathfrak{a})\mathfrak{a} \in \mathbf{I}_0(\mathfrak{p})$, *then*

$$\mathbf{N}_{\mathfrak{p},\mathfrak{o}}(\mathfrak{k}_{\mathfrak{p}}(\mathbf{a};\mathfrak{c})) \sim \left(\prod_{\mathfrak{a}} \mathfrak{a}^{n(\mathfrak{a})}\right)^{(1-\mathbf{N}\mathfrak{p})/w}.$$

The sign $\sim$ *means that the right hand side is the ideal factorization of the left hand side.*

Proof. This follows immediately from Theorem 3.6 and Theorem 3.1 of Chapter 11, giving the factorization for $\Delta(\mathfrak{a})/\Delta(\mathfrak{o})$.

Next we need a simple lemma.

Lemma 3.8. *Let n be an integer prime to the discriminant of K over* $\mathbf{Q}$. *Then*

$$K(\boldsymbol{\mu}_n) \cap K(1) = K.$$

Equivalently, any ideal class of K contains an ideal $\mathfrak{a}$ *satisfying*

$$\mathbf{N}\mathfrak{a} \equiv 1 \pmod{n}.$$

Proof. The equivalence of the two conditions follows from the fact that for any ideal $\mathfrak{a}$ prime to n, the action of $(\mathfrak{a}, K)$ on $K(\boldsymbol{\mu}_n)$ is given by $\zeta \mapsto \zeta^{\mathbf{N}\mathfrak{a}}$. The fact that the intersection of $K(\boldsymbol{\mu}_n)$ with $K(1)$ is K follows from the observation that $K(\boldsymbol{\mu}_n)$ is a compositum of fields $K(\boldsymbol{\mu}_{p^\nu})$ with p prime dividing n, and that these latter extensions are totally ramified above each prime of K lying above p, so $\mathrm{Gal}(K(\boldsymbol{\mu}_n)/K)$ is generated by all inertia groups, whence the required intersection follows.

Theorem 3.1 will follow as a corollary from the following result, which describes the group $\mathfrak{K}(\mathfrak{o})^{12w}$. We recall the **admissibility** conditions which we have already met in Chapter 9, §5, concerning an ideal $\mathfrak{a}$ prime to w_H with H now equal to $K(1)$:

$$\begin{cases} \text{If } 4 \nmid w_{K(1)}, & \text{then } \mathbf{N}\mathfrak{a} \equiv 1 \bmod 4 \\ \text{If } 4 \mid w_{K(1)}, & \text{then } \mathbf{N}\mathfrak{a} \equiv 1 \text{ or } 3 \bmod 8. \end{cases}$$

These conditions are meant to apply only when $K \neq \mathbf{Q}(\sqrt{-1}), \mathbf{Q}(\sqrt{-3})$.

Theorem 3.9. *Assume* $K \neq \mathbf{Q}(\sqrt{-1}), \mathbf{Q}(\sqrt{-3})$. *Then*

$$\mathfrak{K}_X(\mathfrak{o})^{12w} = \mathscr{M}(\mathfrak{o})^{12w},$$

and this group is also equal to the set of all products

$$\Delta(\mathbf{a}) = \prod \Delta(\mathfrak{a}^{-1})^{n(\mathfrak{a})} \quad \textit{with} \quad \mathbf{a} = \sum n(\mathfrak{a})\mathfrak{a} \in \mathbf{I}$$

satisfying the conditions:

(1) $d(\mathbf{a}) = 0$;

(2w) $\prod \mathfrak{a}^{n(\mathfrak{a})}$ *is the w-*th *power of a fractional ideal of K*;

(3) $\sum n(\mathfrak{a})\mathbf{N}\mathfrak{a}' \equiv 0 \bmod ww_{K(1)}$, *where* $\mathfrak{a}'$ *is any K(1)-admissible ideal in the same class as* $\mathfrak{a}$.

Proof. By Lemma 3.5 any element of $\mathcal{M}(\mathfrak{o})^{12W}$ is of the form

$$\xi^{12w^2} \prod_{\mathfrak{p} \in S} \prod_{\mathfrak{a}} \left(\frac{\Delta(\mathfrak{a}^{-1}\mathfrak{c}^{-1})}{\Delta(\mathfrak{p}\mathfrak{a}^{-1}\mathfrak{c}^{-1})^{\mathbf{N}\mathfrak{p}}} \right)^{n_{\mathfrak{p}}(\mathfrak{a})} \cdot \Delta(\mathbf{b}; \mathfrak{c})^{12w}$$

with $\xi \in K$, with $d(\mathbf{a}_{\mathfrak{p}}) = \sum n_{\mathfrak{p}}(\mathfrak{a})$ divisible by w for each $\mathfrak{p}$, and with total degree 0. Note that we may write

$$\xi^{-12w} = \Delta(\xi^w \mathfrak{o})/\Delta(\mathfrak{o}).$$

Thus (1) is satisfied.

We may choose $\gamma \in \mathfrak{o}$ such that $\gamma\xi^w$, $\gamma\mathfrak{p}^{-1}$, $\gamma\mathfrak{b}$ are integral for all $\mathfrak{p} \in S$ and all $\mathfrak{b}$ occurring in $\mathbf{b}$. Since the product has degree 0, we can then multiply by γ^{-1} throughout to see that the above product is equal to

$$\left(\frac{\Delta(\gamma\xi^w\mathfrak{o})}{\Delta(\gamma\mathfrak{o})} \right)^{-w} \prod_{\mathfrak{p} \in S} \prod_{\mathfrak{a}} \left(\frac{\Delta(\gamma^{-1}\mathfrak{a}^{-1}\mathfrak{c}^{-1})}{\Delta(\mathfrak{p}\gamma^{-1}\mathfrak{a}^{-1}\mathfrak{c}^{-1})^{\mathbf{N}\mathfrak{p}}} \right)^{n_{\mathfrak{p}}(\mathfrak{a})} \cdot \prod_{\mathfrak{b}} \Delta(\gamma^{-1}\mathfrak{b}^{-1}\mathfrak{c}^{-1})^{12wm(\mathfrak{b})}$$

with inverses of integral ideals. Then

$$\prod_{\mathfrak{p} \in S} \prod_{\mathfrak{a}} ((\gamma^{-1}\mathfrak{a}^{-1}\mathfrak{c}^{-1})/(\mathfrak{p}\gamma^{-1}\mathfrak{a}^{-1}\mathfrak{c}^{-1})^{\mathbf{N}\mathfrak{p}})^{n_{\mathfrak{p}}(\mathfrak{a})}$$

$$= \prod_{\mathfrak{p} \in S} \left[\mathfrak{p}^{-\mathbf{N}\mathfrak{p} \cdot d(\mathbf{a}_{\mathfrak{p}})} \prod_{\mathfrak{a}} ((\gamma\mathfrak{a}\mathfrak{c})^{\mathbf{N}\mathfrak{p}-1})^{n_{\mathfrak{p}}(\mathfrak{a})} \right],$$

and both $d(\mathbf{a}_{\mathfrak{p}})$ and $\mathbf{N}\mathfrak{p} - 1$ are divisible by w. Since the fractional ideal associated with $\Delta(\mathbf{b}; \mathfrak{c})^{12w}$ is a w-th power, this shows that condition ($2w$) is satisfied.

Finally, for (3), we note that in each factor

$$\prod_{\mathfrak{a}} ((\Delta(\gamma^{-1}\mathfrak{a}^{-1}\mathfrak{c}^{-1})/\Delta(\mathfrak{p}\gamma^{-1}\mathfrak{a}^{-1}\mathfrak{c}^{-1})^{\mathbf{N}\mathfrak{p}})^{n_{\mathfrak{p}}(\mathfrak{a})}$$

condition (3) is satisfied. Indeed, choose $K(1)$- admissible ideals $\mathfrak{a}'$ in the class of $\gamma\mathfrak{a}\mathfrak{c}$ and $\mathfrak{p}'$ in the class of $\mathfrak{p}^{-1}$. Then the sum to consider is

$$\sum_{\mathfrak{a}} (\mathbf{N}\mathfrak{a}' - \mathbf{N}\mathfrak{p}\mathbf{N}\mathfrak{p}'\mathbf{N}\mathfrak{a}') = \sum_{\mathfrak{a}} \mathbf{N}\mathfrak{a}'(1 - \mathbf{N}\mathfrak{p}\mathbf{N}\mathfrak{p}').$$

Here

$$\sum_{\mathfrak{a}} \mathbf{N}\mathfrak{a}' \equiv d(\mathbf{a}_{\mathfrak{p}}) \equiv 0 \bmod w \quad \text{and} \quad 1 - \mathbf{N}\mathfrak{p}\mathbf{N}\mathfrak{p}' \equiv 0 \bmod w_{K(1)},$$

so the sum is divisible by $ww_{K(1)}$.

Since $w_{K(1)}$ divides 12, the factor $\Delta(\mathbf{b}; \mathfrak{c}\gamma)^{12w}$ obviously satisfies (3). The condition is also obvious for the ξ-factor, since

$$\mathbf{N}\xi_1^w - 1 \equiv 0 \bmod w_{K(1)},$$

where ξ_1 is an algebraic integer whose ideal is in the same class as (ξ) and is $K(1)$-admissible. This proves (3).

Thus $\mathscr{M}(\mathfrak{o})^{12w}$ is contained in the group determined by conditions (1), (2w), (3).

We now show that the group determined by these three conditions is contained in $\mathfrak{K}_X(\mathfrak{o})^{12w}$, which completes the proof.

Let

$$\Delta(\mathbf{a}) = \prod_{\mathfrak{a}} \left(\frac{\Delta(\mathfrak{a}^{-1})}{\Delta(\mathfrak{o})}\right)^{n(\mathfrak{a})}$$

where $\mathbf{a}$ satisfies (1), (2w), (3). We shall first prove that without loss of generality, we may assume that $\mathbf{a}$ satisfies the stronger condition

$$\prod \mathfrak{a}^{n(\mathfrak{a})} = \mathfrak{o} \tag{2}$$

instead of (2w). To see this, suppose

$$\prod \mathfrak{a}^{n(\mathfrak{a})} = \mathfrak{b}^w$$

with some fractional ideal $\mathfrak{b}$. Choose primes $\mathfrak{p}$ not dividing numerator or denominator of $\mathfrak{b}$, prime to $w_{K(1)}$, not in X, and integers $m(\mathfrak{p})$ such that

$$\sum m(\mathfrak{p})N(\mathfrak{p})(1 - \mathbf{N}\mathfrak{p}) = w.$$

This can be done, since by the root of unity Lemma 4.2 of Chapter 9 the g.c.d. of all numbers $N(\mathfrak{p})(1 - \mathbf{N}\mathfrak{p})$ is w.

Let $\lambda \in \mathfrak{o}$ be prime to $w_{K(1)}$ and all of the $\mathfrak{p}$, and let $R(\mathfrak{p})$ be any set of representatives for $\mathfrak{o}(\mathfrak{p})^*/\mathfrak{o}^*$. Define.

$$\beta = \lambda^w \prod_{\mathfrak{p}} \mathbf{N}_{R(\mathfrak{p})}\mathfrak{k}_{\mathfrak{p}}(N(\mathfrak{p}) \cdot (\mathfrak{b} - (\lambda^w)); \mathfrak{o})^{m(\mathfrak{p})}.$$

Then $\beta \in \mathfrak{K}_X(\mathfrak{o})$ since both the $\mathfrak{p}$-product and λ^w lie in $\mathfrak{K}_X(\mathfrak{o})$ (by Corollary 3.3). We show that $\Delta(\mathbf{a})\beta^{12w}$ satisfies (2). Indeed, by Theorem 3.6,

$$\beta^{12w} = \left(\frac{\Delta(\lambda)}{\Delta(\mathfrak{o})}\right)^{w^2} \prod_{\mathfrak{p}} \left(\frac{\Delta(\mathfrak{b}^{-1})}{\Delta(\mathfrak{p}\mathfrak{b}^{-1})^{\mathbf{N}\mathfrak{p}}}\right)^{m(\mathfrak{p})N(\mathfrak{p})} \prod_{\mathfrak{p}} \left(\frac{\Delta((\lambda))^{-w}}{\Delta(\mathfrak{p}\lambda^{-w})^{\mathbf{N}\mathfrak{p}}}\right)^{-m(\mathfrak{p})N(\mathfrak{p})},$$

so the product associated to this expression by $(2w)$ is immediately computed to be $\mathfrak{b}^{-w}$. This proves that we could assume $\mathbf{a}$ to satisfy the stronger condition (2).

Recall that conditions (1), (2), (3) on $\mathbf{a}$ define the group which we denoted by $\mathbf{I}_0(\mathfrak{o}, ww_H)$, where $H = K(1)$ in the present instance. In Chapter 9, Lemma 5.3, we saw that this group had generators

$$\mathbf{a} = n((\mathfrak{a}_1) + (\mathfrak{a}_2) - (\mathfrak{a}_1\mathfrak{a}_2) - (\mathfrak{o}))$$

where n is any integer such that

$$n(\mathbf{N}\mathfrak{a}_1 - 1)(\mathbf{N}\mathfrak{a}_2 - 1) \equiv 0 \bmod ww_H.$$

It will therefore suffice to prove that $\Delta(\mathbf{a}) \in \mathfrak{K}_X(\mathfrak{o})^{12w}$ for such generators $\mathbf{a}$. We show that $\Delta(\mathbf{a})^2$ and $\Delta(\mathbf{a})^3$ lie in $\mathfrak{K}_X(\mathfrak{o})^{12w}$, which will complete the proof.

For $\Delta(\mathbf{a})^2$: If $3 \mid (w_{K(1)}/w)$, then we may assume without loss of generality that $3 \mid n(\mathbf{N}\mathfrak{a}_2 - 1)$. Choose principal ideals $\mathfrak{p} \notin X$ prime to the discriminant of K over $\mathbf{Q}$ and to $w_{K(1)}$, and integers $m(\mathfrak{p})$ such that

$$\sum m(\mathfrak{p})(1 - \mathbf{N}\mathfrak{p}) = w_{K(1)}.$$

This can be done by Lemma 4.2 of Chapter 9. Also choose a prime $\mathfrak{q} \notin X$ in the ideal class of $\mathfrak{a}_2$, prime to the discriminant of K over $\mathbf{Q}$ and to $w_{K(1)}$. Finally, choose an odd ideal $\mathfrak{a}$ in the ideal class of $\mathfrak{a}_1$ such that

$$\mathbf{N}\mathfrak{a} \equiv 1 \bmod N(\mathfrak{q}) \quad \text{and mod each } N(\mathfrak{p}),$$

which can be done by Lemma 3.8. Then $(\mathfrak{a}) - (\mathfrak{o})$ lies in $\mathbf{I}_0(\mathfrak{q})$ and $\mathbf{I}_0(\mathfrak{p})$.

We can write

$$\begin{aligned}\Delta(\mathbf{a})^2 &= (\Delta(\mathfrak{a}_1^{-1})\Delta(\mathfrak{a}_2^{-1})/\Delta(\mathfrak{a}_1^{-1}\mathfrak{a}_2^{-1})\Delta(\mathfrak{o}))^{2n} \\ &= \left(\frac{\Delta(\mathfrak{a}^{-1})/\Delta(\mathfrak{o})}{\Delta(\mathfrak{q}^{-1}\mathfrak{a}^{-1})/\Delta(\mathfrak{q}^{-1})}\right)^{2n} \\ &= \left(\frac{(\Delta(\mathfrak{a}^{-1})/\Delta(\mathfrak{o}))^{1-\mathbf{N}\mathfrak{q}}}{(\Delta(\mathfrak{a}^{-1})/\Delta(\mathfrak{o}))^{\sigma(\mathfrak{q})-\mathbf{N}\mathfrak{q}}}\right)^{2n}.\end{aligned}$$

By Theorem 3.6 we see that the denominator of this expression lies in $\mathfrak{K}_X(\mathfrak{o})^{12w}$, because $(\mathfrak{a}) - (\mathfrak{o})$ lies in $\mathbf{I}_0(\mathfrak{q})$ and

$$(\Delta(\mathfrak{a}^{-1})/\Delta(\mathfrak{o}))^{\sigma(\mathfrak{q})-\mathbf{N}\mathfrak{q}} = \mathbf{N}_{\mathfrak{q},\mathfrak{o}}\mathfrak{k}_{\mathfrak{q}}(\mathfrak{a} - \mathfrak{o}; \mathfrak{o})^{12w}.$$

For the numerator, we see that

$$w_{K(1)} \mid 2n(\mathbf{N}\mathfrak{q} - 1)$$

since $w_{K(1)}\,|\,12$ by Lemma 4.3 of Chapter 9; $2\,|\,(\mathbf{N}\mathfrak{q}-1)$; and by assumption, if $3\,|\,(w_{K(1)}/w)$ then $3\,|\,n(\mathbf{N}\mathfrak{q}-1)$. Say

$$2n(\mathbf{N}\mathfrak{q}-1) = mw_{K(1)}.$$

Then

$$\begin{aligned}(\Delta(\mathfrak{a}^{-1})/\Delta(\mathfrak{o}))^{(\mathbf{N}\mathfrak{q}-1)2n} &= (\Delta(\mathfrak{a}^{-1})/\Delta(\mathfrak{o}))^{\Sigma m(\mathfrak{p})(1-\mathbf{N}\mathfrak{p})m}\\ &= \prod ((\Delta(\mathfrak{a}^{-1})/\Delta(\mathfrak{o}))^{\sigma(\mathfrak{p})-\mathbf{N}\mathfrak{p}})^{m(\mathfrak{p})m}\end{aligned}$$

since all $\mathfrak{p}$ are principal. As before,

$$\left(\frac{\Delta(\mathfrak{a}^{-1})}{\Delta(\mathfrak{o})}\right)^{\sigma(\mathfrak{p})-\mathbf{N}\mathfrak{p}} = \mathbf{N}_{\mathfrak{p},\mathfrak{o}}\mathfrak{k}_{\mathfrak{p}}(\mathfrak{a}-\mathfrak{o};\mathfrak{o})^{12w},$$

and $(\mathfrak{a})-(\mathfrak{o}) \in \mathbf{I}_0(\mathfrak{p})$. Thus $\Delta(\mathbf{a})^2 \in \mathfrak{R}_X(\mathfrak{o})^{12w}$.

The argument for $\Delta(\mathbf{a})^3$ is similar. This time, if $2\,|\,(w_{K(1)}/w)$, we assume $4\,|\,n(\mathbf{N}\mathfrak{a}_2-1)$, so that $w_{K(1)}\,|\,3n(\mathbf{N}\mathfrak{a}_2-1)$, and then proceed as above. This completes the proof.

Corollary 3.10. *We have*

$$\boldsymbol{\mu}_{K(1)}\mathfrak{R}_X(\mathfrak{o}) = \boldsymbol{\mu}_{K(1)}\mathfrak{R}(\mathfrak{o}) = \boldsymbol{\mu}_{K(1)}\mathcal{M}(\mathfrak{o}),$$

and $E_{\mathrm{mod}}(K(1)) = \boldsymbol{\mu}_{K(1)}(\mathfrak{R}_X(\mathfrak{o}) \cap E)$.

Proof. Obvious.

Of course, the question arises as to which roots of unity are actually contained in $\mathfrak{R}(\mathfrak{o})$ or $\mathcal{M}(\mathfrak{o})$. We do not deal with this question here.

From the preceding theorem, we obtain a result proved by Deuring in his monograph, where he refers to Hasse [Has 1] and Fricke, who had obtained similar results earlier. Cf. [Deu] p. 41.

Corollary 3.11. *For any ideal* $\mathfrak{a}$ *the number* $\Delta(\mathfrak{a}^2)/\Delta(\mathfrak{o})$ *is a* 24-th *power in* $K(1)$

Proof. This is obvious if $K = \mathbf{Q}(i)$ or $\mathbf{Q}(\sqrt{-3})$, so we may suppose $w = 2$. Then $\Delta(\mathfrak{a}^2)/\Delta(\mathfrak{o})$ satisfies conditions (1), $(2w)$, (3). Indeed, (1) and $(2w)$ are clear. For (3), we choose $\mathfrak{a}'$ in the class of $\mathfrak{a}$ to be $K(1)$-admissible and prime to 6. Then $\mathbf{N}\mathfrak{a}'^2 \equiv 1 \bmod 24$, so a fortiori $\equiv 1 \bmod ww_{K(1)}$. This concludes the proof.

As already remarked by Fricke, Hasse and Deuring, a result like the above gives an explicit realization of the principal ideal theorem.

Corollary 3.12. *If* $\mathfrak{b}$ *is any ideal then there exists* $\beta \in \mathfrak{K}(\mathfrak{o})$ *which generates* $\mathfrak{b}\mathfrak{o}_{K(1)}$.

Proof. As in Corollary 3.11 let $\Delta(\mathfrak{b}^2)/\Delta(\mathfrak{o}) = \beta^{-24}$ with $\beta \in K(\mathfrak{o})$. By the ideal factorization of Δ, this gives $\mathfrak{b}^{-24} \sim \beta^{-24}$, so $\mathfrak{b} \sim \beta$ as desired. A refinement is given in the next corollary.

Corollary 3.13. *Let* $\mathfrak{a}$ *again be prime to* $w_{K(1)}$, *and let* r *be the* g.c.d. *of* $\mathbf{N}\mathfrak{a}' - 1$ *and* $ww_{K(1)}$ *for any* $K(1)$*-admissible ideal* $\mathfrak{a}'$ *in the same class as* $\mathfrak{a}$. *Let* $e = ww_{K(1)}/r$. *If* $\mathfrak{a}^e$ *is the* w-th *power of an ideal in* K, *then* $(\Delta(\mathfrak{a})/\Delta(\mathfrak{o}))^e$ *is a* 24-th *power in* $K(1)$. *If* $\mathfrak{a}^e$ *is not* w-th *power of an ideal in* K, *then* $(\Delta(\mathfrak{a})/\Delta(\mathfrak{o}))^{2e}$ *is a* 24-th *power in* $K(1)$.

Finally, we prove Theorem 3.1. We have $\boldsymbol{\mu}_{K(1)} \subset E_{\text{mod}}(K(1))$ by definition, and there is equality when $K = \mathbf{Q}(\sqrt{-1})$ or $\mathbf{Q}(\sqrt{-3})$, in which cases the theorem is immediate because $\Delta_{K(1)}(c_0, ww_{K(1)}) = 1$ and $E_{\text{mod}}(K(1)) = \boldsymbol{\mu}_{K(1)}$. Hence we assume that we are not in these special cases. It suffices to show that

$$E_{\text{mod}}(K(1))^{12wh} = \Delta_{K(1)}(c_0, ww_{K(1)}).$$

By Theorem 3.9, $E_{\text{mod}}(K(1))^{12wh}$ is the group of all products

$$\prod \Delta(\mathfrak{a}^{-1})^{n(\mathfrak{a})h},$$

satisfying (1), (2w), (3), and the further condition that the product be a unit. By the factorization of $\Delta(\mathfrak{a})/\Delta(\mathfrak{o})$, this last condition reads

$$\prod \mathfrak{a}^{n(\mathfrak{a})h} = \mathfrak{o}, \quad \text{or equivalently,} \quad \prod \mathfrak{a}^{n(\mathfrak{a})} = \mathfrak{o}.$$

This is just condition (2). For any $\mathfrak{a}$, let $\alpha(\mathfrak{a})$ be a generator for $\mathfrak{a}^h$. Then for a product satisfying (1), (2), (3), we have

$$\begin{aligned}\prod \Delta(\mathfrak{a}^{-1})^{n(\mathfrak{a})h} &= \prod \left(\frac{\alpha(\mathfrak{a})^{12}\,\Delta^h(\mathfrak{a}^{-1})}{\Delta^h(\mathfrak{o})}\right)^{n(\mathfrak{a})} \\ &= \prod \delta(c(\mathfrak{a}))^{n(\mathfrak{a})}.\end{aligned}$$

Furthermore, (2) and (3) translate into the defining conditions of $\Delta_{K(1)}(c_0, ww_{K(1)})$. This completes the proof of Theorem 3.1.

§4. Modular Units in $K(\mathfrak{f})$ as Klein Units

In this section we define what we shall call the group of modular units in the ray class field $K(\mathfrak{f})$. One's first inclination might be to define this as the group of all units of the form $\mathfrak{k}_{\mathfrak{f}}(\mathbf{a}; C)$ with $\mathbf{a} \in \mathbf{I}_0(\mathfrak{f})$. However, this group is not, in general, large enough. For instance it is not always of finite index in the group of all units of $K(\mathfrak{f})$. The reason for this is, essentially, that we do not get enough unramified units by this construction. To remedy this defect, we use the same trick as in §3, namely we introduce extra primes $\mathfrak{p}$ and symmetrize with respect to the Galois group $\mathrm{Gal}(K(\mathfrak{p})/K(\mathfrak{o}))$.

Let:

$\mathbf{a} \in \mathbf{I}_w(\mathfrak{f})$;
S = a finite set of primes $\mathfrak{p}$, $\mathfrak{p} \nmid \mathfrak{f}$, and $\mathfrak{p}$ prime to $w_{K(1)}$;
$R(\mathfrak{p})$ = a set of representatives for $(\mathfrak{o}/\mathfrak{p})^*/\mathfrak{o}^*$, as in §3;
R_S = family of representatives $\{R(\mathfrak{p})\}$ for $\mathfrak{p} \in S$;
$\mathfrak{c}$ = an ideal prime to $\mathfrak{f}$ and all $\mathfrak{p} \in S$;
we assume that

$$d(\mathbf{a}) + \sum_{\mathfrak{p} \in S} \frac{\mathbf{N}\mathfrak{p} - 1}{w} d(\mathbf{a}_{\mathfrak{p}}) = 0.$$

The expression on the left hand side will be called the **total degree** of $(\mathbf{a}, \mathbf{a}_S)$, and we thus require the total degree to be 0. Corresponding to such data, we define an element

$$\mathfrak{k}_{\mathfrak{f}}(\mathbf{a}, \mathbf{a}_S, R_S; \mathfrak{c}) = \mathfrak{k}_{\mathfrak{f}}(\mathbf{a}; \mathfrak{c}) \prod_{\mathfrak{p} \in S} \prod_{\alpha \in R(\mathfrak{p})} \mathfrak{k}_{\mathfrak{p}}(\mathbf{a}_{\mathfrak{p}}; \mathfrak{c}\alpha),$$

again denoted by $\beta(\mathfrak{c})$ if all the other variables are fixed. Note that the product over α on the right hand side is like a norm considered in §3, but is not actually a norm. It depends on the choice of representatives, because this product by itself is not of degree 0.

Remark. As in the last section, we could also define

$$\beta(\mathfrak{c}) = \mathfrak{k}_{\mathfrak{f}}(\mathbf{a}, \mathbf{a}_S, R_S; \mathfrak{c})\, \Delta(\mathfrak{b}; \mathfrak{c}),$$

with total degree 0. Then the next theorem would also be valid for these, as usual. For simplicity of notation, we carry the proof out without the delta factor.

Theorem 4.1. *The number $\beta(\mathfrak{c})$ lies in $K(\mathfrak{f})$ and depends only on the class of $\mathfrak{c}$ in $\mathrm{Cl}(\mathfrak{f})$. We have*

$$\beta(\mathfrak{c})^{\sigma(\mathfrak{c}')} = \beta(\mathfrak{c}\mathfrak{c}').$$

Proof. The proof proceeds exactly along the lines of Theorem 3.4, which is actually a special case of the present result. Let

$$\mathfrak{f}' = 2\mathfrak{f} \prod_{\mathfrak{p} \in S} \mathfrak{p}$$

$$\mathbf{a}' = \mathbf{a} \cdot \mathfrak{f}'\mathfrak{f}^{-1} + \sum_{\mathfrak{p}} \sum_{\alpha \in R(\mathfrak{p})} \mathbf{a}_{\mathfrak{p}} \cdot \mathfrak{f}'\mathfrak{p}^{-1}\alpha.$$

By Remark 2 of §1, we know that $\mathbf{a}' \in \mathbf{I}_0(\mathfrak{f}')$, and

$$\beta(\mathfrak{c}) = \mathfrak{k}_{\mathfrak{f}'}(\mathbf{a}'; \mathfrak{c}),$$

so $\beta(\mathfrak{c}) \in K(\mathfrak{f}')$ and transforms properly relative to $\mathfrak{f}'$.

Now suppose $\mathfrak{c}' = (\gamma)$ and $\gamma \equiv 1 \bmod \mathfrak{f}$, γ prime to $\mathfrak{f}'$. We want to show that $\beta(\mathfrak{c}\gamma) = \beta(\mathfrak{c})$. For this, we again use Lemma 3.2, as well as Proposition 1.1. Let

$$R'(\mathfrak{p}) = \{\gamma\alpha : \alpha \in R(\mathfrak{p})\}.$$

Then

$$\begin{aligned}\frac{\beta(\mathfrak{c}\gamma)}{\beta(\mathfrak{c})} &= \frac{\mathfrak{k}_{\mathfrak{f}}(\mathbf{a}; \mathfrak{c}\gamma)}{\mathfrak{k}_{\mathfrak{f}}(\mathbf{a}; \mathfrak{c})} \prod_{\mathfrak{p} \in S} \frac{\mathbf{N}_{R'}(\mathfrak{k}_{\mathfrak{p}}(\mathbf{a}_{\mathfrak{p}}; \mathfrak{c}))}{\mathbf{N}_{R}(\mathfrak{k}_{\mathfrak{p}}(\mathbf{a}_{\mathfrak{p}}; \mathfrak{c}))} \\ &= \gamma^{-d(\mathbf{a})} \prod_{\mathfrak{p} \in S} \gamma^{-d(\mathbf{a}_{\mathfrak{p}})(\mathbf{N}\mathfrak{p} - 1)/w} \\ &= \gamma^{-d(\mathbf{a}')}.\end{aligned}$$

Here we have used Proposition 1.1 on the $\mathfrak{f}$-factor (since $\gamma \equiv 1 \bmod \mathfrak{f}$), and Lemma 3.2 on the $\mathfrak{p}$-factor. Since $d(\mathbf{a}') = 0$, we have proved $\beta(\mathfrak{c}\gamma) = \beta(\mathfrak{c})$. This shows that $\beta(\mathfrak{c}) \in K(\mathfrak{f})$ because of the formula

$$\beta(\mathfrak{c})^{\sigma(\mathfrak{c}')} = \beta(\mathfrak{c}\mathfrak{c}'),$$

relative to $K(\mathfrak{f}')$ over K. The theorem follows.

We define the group of **Klein numbers** in $K(\mathfrak{f})$ to be:

$\mathfrak{K}(\mathfrak{f})$ = group generated by all elements $\mathfrak{k}_{\mathfrak{f}}(\mathbf{a}, \mathbf{a}_S, R_S; \mathfrak{c})$ for all choices of $\mathbf{a}$, S, $\mathbf{a}_S$, R_S, $\mathfrak{c}$ as above.

Remark. We could also define the group $\mathscr{M}(\mathfrak{f})$ by using more general invariants of the form

$$\beta(\mathfrak{c}) = \mathfrak{k}_{\mathfrak{f}}(\mathbf{a}, \mathbf{a}_S, R_S; \mathfrak{c})\,\Delta(\mathbf{b}; \mathfrak{c}).$$

However, one can find a finite set of primes T, $\mathbf{a}_T$, R_T such that

$$\mathfrak{k}(\mathbf{a}_T, R_T; \mathfrak{c})\,\Delta(\mathbf{b}; \mathfrak{c}) \in \mathscr{M}(\mathfrak{o}) \subset K(1),$$

and Corollary 3.10 shows that

$$\boldsymbol{\mu}_{K(1)}\mathscr{M}(\mathfrak{f}) = \boldsymbol{\mu}_{K(1)}\mathfrak{K}(\mathfrak{f}).$$

We define the group of **modular units of** $K(\mathfrak{f})$ to be

$$E_{\text{mod}}(K(\mathfrak{f})) = \boldsymbol{\mu}_{K(\mathfrak{f})}(\mathfrak{K}(\mathfrak{f}) \cap E)$$

where E is the group of all units. In other words, $E_{\text{mod}}(K(\mathfrak{f}))$ is the group generated by the roots of unity in $K(\mathfrak{f})$ and those elements of $\mathfrak{K}(\mathfrak{f})$ which are units. The above remark shows that using $\mathscr{M}(\mathfrak{f})$ instead of $\mathfrak{K}(\mathfrak{f})$ would not yield a larger group of units.

If $\mathfrak{g} | \mathfrak{f}$ then $\mathfrak{K}(\mathfrak{g}) \subset \mathfrak{K}(\mathfrak{f})$. We also note that our definition of $\mathfrak{K}(\mathfrak{f})$ includes the definition of $\mathfrak{K}(\mathfrak{o})$ as a special case. From a strictly logical point of view, there was no need to treat the cases $\mathfrak{K}(\mathfrak{f})$ and $\mathfrak{K}(\mathfrak{o})$ separately, but we have done so to introduce the main ideas of the construction before giving the full complications.

Remark. Let X be a finite set of primes containing all those dividing $w_{K(1)}$. We may then define $\mathfrak{K}_X(\mathfrak{f})$ just as we defined $\mathfrak{K}(\mathfrak{f})$, except for the additional requirement that the sets S should be also disjoint from X. We shall see below that

$$\boldsymbol{\mu}_{K(\mathfrak{f})}\mathfrak{K}_X(\mathfrak{f}) = \boldsymbol{\mu}_{K(\mathfrak{f})}\mathfrak{K}(\mathfrak{f}).$$

It follows that the group of modular units could be defined by using $\mathfrak{K}_X(\mathfrak{f})$ instead of $\mathfrak{K}(\mathfrak{f})$.

In §5 we shall discuss the power $E_{\text{mod}}(K(\mathfrak{f}))^{12hwN(\mathfrak{f})}$. When $\mathfrak{f}$ is a prime power, this calculation will be needed in Chapter 13. For now, we content ourselves with showing that the elements $u_{\mathfrak{a}}(C_0)$ and the groups $\mathscr{R}_{\mathfrak{f}}^*$, $\Phi_{\mathfrak{f}}(w_{K(\mathfrak{f})})$ are included in the present construction.

Lemma 4.2. *Let $\mathfrak{a}$ be prime to w. Then*

$$u_{\mathfrak{a}}^{(\mathfrak{f})}(C_0)^w \in \mathfrak{K}_X(\mathfrak{f})^{12w}.$$

Proof. Note that if $\mathbf{a} = \mathbf{N}\mathfrak{a}(\mathfrak{o}) - (\mathfrak{a})$, then $\mathbf{a} \in \mathbf{I}_w(\mathfrak{g})$ for any $\mathfrak{g} \nmid \mathfrak{a}$. Pick a set S of primes $\mathfrak{p} \nmid \mathfrak{a}$, $\mathfrak{p} | p$, p prime to $\mathfrak{f}$, $\mathfrak{p} \notin X$, and integers $m(\mathfrak{p})$ such that

$$\sum_{\mathfrak{p} \in S} m(\mathfrak{p})(\mathbf{N}\mathfrak{p} - 1) = w.$$

Let $\mathbf{a}_\mathfrak{p} = -m(\mathfrak{p})\mathbf{a}$ for all $\mathfrak{p} \in S$, and let $R(\mathfrak{p})$ be any set of representatives. Consider

$$\begin{aligned}\beta &= \mathfrak{k}_\mathfrak{f}(\mathbf{a}, \mathbf{a}_S, R_S; \mathfrak{o}) \\ &= \mathfrak{k}_\mathfrak{f}(\mathbf{N}\mathfrak{a}(\mathfrak{o}) - (\mathfrak{a}); \mathfrak{o}) \prod_{\mathfrak{p} \in S} \mathbf{N}_{R(\mathfrak{p})}\mathfrak{k}_\mathfrak{p}(\mathbf{N}\mathfrak{a}(\mathfrak{o}) - (\mathfrak{a}); \mathfrak{o})^{-m(\mathfrak{p})}.\end{aligned}$$

We easily check that $\beta \in \mathfrak{R}_X(\mathfrak{f})$, namely we have already noted that $\mathbf{a} \in \mathbf{I}_w(\mathfrak{f})$, $\mathbf{a}_\mathfrak{p} \in \mathbf{I}_w(\mathfrak{p})$, and the total degree is

$$(\mathbf{N}\mathfrak{a} - 1) - \sum_\mathfrak{p} m(\mathfrak{p}) \frac{\mathbf{N}\mathfrak{p} - 1}{w} (\mathbf{N}\mathfrak{a} - 1) = 0$$

by our choice of $m(\mathfrak{p})$. We can then apply Theorem 4.1.

Now consider

$$u_\mathfrak{a}(C_0)^w \beta^{-12w}.$$

Recall that

$$u_\mathfrak{a}(C_0) = \frac{g^{12}(1, \mathfrak{f})^{\mathbf{N}\mathfrak{a}}}{g^{12}(1, \mathfrak{f}\mathfrak{a}^{-1})} = \left(\frac{\mathfrak{k}(1, \mathfrak{f})^{\mathbf{N}\mathfrak{a}}}{\mathfrak{k}(1, \mathfrak{f}\mathfrak{a}^{-1})}\right)^{12} \frac{\Delta(\mathfrak{f})^{\mathbf{N}\mathfrak{a}}}{\Delta(\mathfrak{f}\mathfrak{a}^{-1})}.$$

Also, using Lemma 3.5,

$$\beta^{12w} = \left(\frac{\mathfrak{k}(1, \mathfrak{f})^{\mathbf{N}\mathfrak{a}}}{\mathfrak{k}(1, \mathfrak{f}\mathfrak{a}^{-1})}\right)^w \prod_\mathfrak{p} \left(\gamma(\mathfrak{p}) \frac{\Delta(\mathfrak{o})^{\mathbf{N}\mathfrak{a}}\, \Delta(\mathfrak{p}\mathfrak{a}^{-1})^{\mathbf{N}\mathfrak{p}}}{\Delta(\mathfrak{p})^{\mathbf{N}(\mathfrak{a}\mathfrak{p})}\, \Delta(\mathfrak{a}^{-1})}\right)^{-m(\mathfrak{p})}$$

where

$$\gamma(\mathfrak{p}) = \prod_{\alpha \in R(\mathfrak{p})} \alpha^{12wd(\mathbf{a}_\mathfrak{p})}.$$

In particular, since $w \mid d(\mathbf{a}_\mathfrak{p})$, we get

$$\gamma(\mathfrak{p})^{m(\mathfrak{p})} \in K^{*12w^2} \subset \mathfrak{R}(\mathfrak{o})^{12w} \subset \mathfrak{R}_X(\mathfrak{f})^{12w},$$

by Corollary 3.3. Thus we get the following expression:

$$(u_\mathfrak{a}(C_0)\beta^{-12})^w = \gamma\left[\left(\frac{\Delta(\mathfrak{f})^{\mathbf{N}\mathfrak{a}}}{\Delta(\mathfrak{f}\mathfrak{a}^{-1})}\right)^w \prod_\mathfrak{p} \left(\frac{\Delta(\mathfrak{p})^{\mathbf{N}(\mathfrak{a}\mathfrak{p})}\, \Delta(\mathfrak{a}^{-1})}{\Delta(\mathfrak{o})^{\mathbf{N}\mathfrak{a}}\, \Delta(\mathfrak{p}\mathfrak{a}^{-1})^{\mathbf{N}\mathfrak{p}}}\right)^{-m(\mathfrak{p})}\right]$$

where $\gamma \in \mathfrak{R}(\mathfrak{o})^{12w}$.

If $K = \mathbf{Q}(\sqrt{-1})$ or $\mathbf{Q}(\sqrt{-3})$, then every ideal of K is principal, and the product in brackets on the right hand side is a $12w$-th power in K, so we are done. Assume therefore that $K \neq \mathbf{Q}(\sqrt{-1}), \mathbf{Q}(\sqrt{-3})$.

We claim that the product in brackets also lies in $\mathfrak{K}(\mathfrak{o})^{12w}$. We simply need to check the conditions (1), (2w), (3) of Theorem 3.9.

(1) The total degree is $(\mathbf{N}\mathfrak{a} - 1) - \sum m(\mathfrak{p})(\mathbf{N}\mathfrak{a} - 1)(\mathbf{N}\mathfrak{p} - 1) = 0$. Given this, choose λ in $\mathfrak{o}$ so that $\lambda\mathfrak{f}^{-1}$ and $\lambda\mathfrak{p}^{-1}$ for all $\mathfrak{p}$ are integral ideals, and rewrite the product in the form of Theorem 3.9 by multiplying inside each Δ by λ^{-1}.

(2w) The corresponding product reduces to

$$\left[\mathfrak{f}^w \prod_{\mathfrak{p}} \mathfrak{p}^{-m(\mathfrak{p})\mathbf{N}\mathfrak{p}}\right]^{(\mathbf{N}\mathfrak{a}-1)}$$

which is a w-th power, since $w \mid (\mathbf{N}\mathfrak{a} - 1)$.

(3) Choose $\lambda_1, \mathfrak{f}_1, \mathfrak{p}_1, \mathfrak{a}_1$ in the ideal classes of $\lambda, \mathfrak{f}^{-1}, \mathfrak{p}^{-1}, \mathfrak{a}$ respectively such that they are all $K(1)$-admissible. Then the sum in (3) becomes

$$\mathbf{N}\lambda_1(\mathbf{N}\mathfrak{a} - \mathbf{N}\mathfrak{a}_1)[w\mathbf{N}\mathfrak{f}_1 - \sum m(\mathfrak{p})(\mathbf{N}(\mathfrak{p}\mathfrak{p}_1) - 1)].$$

Then $(\mathbf{N}\mathfrak{a} - \mathbf{N}\mathfrak{a}_1)$ is divisible by $w_{K(1)}$, while $(\mathbf{N}(\mathfrak{p}\mathfrak{p}_1) - 1)$ is divisible by w, so the expression in brackets is divisible by w, and the product is divisible by $ww_{K(1)}$, as desired.

Theorem 4.3.

(i) *If* $\mathfrak{g} \mid \mathfrak{f}$, *then* $\mathscr{R}_{\mathfrak{g}}^* \subset E_{\text{mod}}(K(\mathfrak{f}))^{12}$.

(ii) *If* $\mathfrak{g} \mid \mathfrak{f}$, *then* $\Phi_{\mathfrak{g}}(w_{K(\mathfrak{g})}) \subset E_{\text{mod}}(K(\mathfrak{f}))^{12N(\mathfrak{g})}$.

Proof. (i) Since $E_{\text{mod}}(K(\mathfrak{g})) \subset E_{\text{mod}}(K(\mathfrak{f}))$, it suffices to prove this when $\mathfrak{f} = \mathfrak{g}$. Since $\mathscr{R}_{\mathfrak{f}}^* \subset K(\mathfrak{f})$ and $E_{\text{mod}}(K(\mathfrak{f}))$ contains all $w_{K(\mathfrak{f})}$ roots of unity, it suffices to show that

$$\mathscr{R}_{\mathfrak{f}}^{*w} \subset E_{\text{mod}}(K(\mathfrak{f}))^{12w},$$

and this is immediate from the lemma.

(ii) The second part follows immediately from (i), via Theorem 4.3 of Chapter 11.

We shall not go into any extensive discussion of the relations between the Klein units at different levels. Here we merely indicate briefly a compatibility property.

Suppose H is an abelian extension of K of conductor $\mathfrak{f}$. How should we define the group of modular units of H? It should contain at least

$$\mathbf{N}_{K(\mathfrak{f})/H}(E_{\text{mod}}(K(\mathfrak{f}))).$$

Also if $H' \subset H$ then we should have

$$E_{\text{mod}}(H') \subset E_{\text{mod}}(H).$$

With these two prerequisites in mind, we define:

$E_{\text{mod}}(H) =$ group generated by $\boldsymbol{\mu}_H$ and the subgroups

$$\mathbf{N}_{K(\mathfrak{g})/H_{\mathfrak{g}}}(E_{\text{mod}}(K(\mathfrak{g})))$$

for all $\mathfrak{g} \mid \mathfrak{f}$. We denote $H_{\mathfrak{g}} = H \cap K(\mathfrak{g})$, just as $H_{(1)} = H \cap K(1)$.

Lemma 4.4 *If $H' \subset H$ then $E_{\text{mod}}(H') \subset E_{\text{mod}}(H)$.*

Proof. Let $\mathfrak{f}' =$ conductor of H', and suppose $\mathfrak{g} \mid \mathfrak{f}'$. Then

$$\mathbf{N}_{K(\mathfrak{g})/H'_{\mathfrak{g}}}(E_{\text{mod}}(K(\mathfrak{g}))) = \mathbf{N}_{H_{\mathfrak{g}}/H'_{\mathfrak{g}}} \circ \mathbf{N}_{K(\mathfrak{g})/H_{\mathfrak{g}}}(E_{\text{mod}}(K(\mathfrak{g}))) \subset E_{\text{mod}}(H).$$

Thus $E_{\text{mod}}(H') \subset E_{\text{mod}}(H)$.

When $\mathfrak{f}$ is a prime power, we refer to §4 of the next chapter for a more precise description of the norm group from one level to a lower level. The situation appears both complicated and interesting.

§5. A Description of $E_{\text{mod}}(K(\mathfrak{f}))^{12hwN(\mathfrak{f})}$

This section is due to Kersey. It describes the $12hwN(\mathfrak{f})$ power of the modular units in a way which is suited for the index computation of the next chapter. As we know from the very beginning of the theory, only a power of the Klein forms, suitably multiplied by the delta function, satisfies distribution relations. Thus the purpose of the present section is to express powers of modular units in terms of those functions satisfying the distribution relations. The section is valid still for arbitrary conductors $\mathfrak{f} \neq (1)$. The application of the next section will be to the case when $\mathfrak{f} = \mathfrak{p}^n$ is a prime power, $\mathfrak{p} \nmid w(H_{(1)})$. For the general case, cf. [Ke 2].

In §4, we defined an element

$$\beta = \mathfrak{k}_{\mathfrak{f}}(\mathbf{a}, \mathbf{a}_S, R_S, \mathfrak{d}) = \mathfrak{k}_{\mathfrak{f}}(\mathbf{a}; \mathfrak{d}) \prod_{\mathfrak{p} \in S} \prod_{\alpha \in R(\mathfrak{p})} \mathfrak{k}_{\mathfrak{p}}(\mathbf{a}_{\mathfrak{p}}; \alpha\mathfrak{d}).$$

where

$$\mathbf{a} = \sum n(\mathfrak{a})\mathfrak{a} \quad \text{and} \quad \mathbf{a}_{\mathfrak{p}} = \sum n_{\mathfrak{p}}(\mathfrak{a})\mathfrak{a}$$

are subject to appropriate conditions. Here we let $\mathfrak{c} = \mathfrak{o}$ for convenience. When taking the $12hwN(\mathfrak{f})$ power of this element, we shall find a decomposition corresponding to the $\mathfrak{k}_{\mathfrak{f}}(\mathbf{a}; \mathfrak{o})$ factor, and to the other factors. Thus it will be convenient to define numbers as follows. The first two numbers will take care of the $\mathfrak{k}_{\mathfrak{f}}(\mathbf{a}; \mathfrak{o})$ factor.

$$\gamma_1(\mathbf{a}, \mathfrak{f}) = \prod_{\mathfrak{a}} g_{[\mathfrak{f},\mathfrak{a}]}(C([\mathfrak{a}, \mathfrak{f}]))^{n(\mathfrak{a})hwN(\mathfrak{f})/N([\mathfrak{f},\mathfrak{a}])}$$

where $[\mathfrak{f}, \mathfrak{a}] = \mathfrak{f}/(\mathfrak{a}, \mathfrak{f})$ and $[\mathfrak{a}, \mathfrak{f}] = \mathfrak{a}/(\mathfrak{a}, \mathfrak{f})$;

$$\gamma_2(\mathbf{a}, \mathfrak{f}) = \prod_{\mathfrak{a}} \Delta(\mathfrak{f}\mathfrak{a}^{-1})^{-n(\mathfrak{a})hwN(\mathfrak{f})}$$

$$\gamma_3(\mathbf{a}, \mathfrak{f}) = \prod_{\mathfrak{p}} \prod_{\mathfrak{a}} \left(\frac{\Delta(\mathfrak{a}^{-1})}{\Delta(\mathfrak{p}\mathfrak{a}^{-1})^{\mathbf{N}\mathfrak{p}}} \right)^{n_{\mathfrak{p}}(\mathfrak{a})hN(\mathfrak{f})}$$

$$\gamma_4(\mathbf{a}_S, R_S) = \prod_{\mathfrak{p}} \prod_{\alpha \in R(\mathfrak{p})} \alpha^{12hwN(\mathfrak{f})d(\mathbf{a}_{\mathfrak{p}})}.$$

For simplicity, we shall also write $\gamma_1, \gamma_2, \gamma_3, \gamma_4$ without the reference to $\mathbf{a}$, $\mathfrak{f}$, $\mathbf{a}_S$, R_S.

Lemma 5.1. *We have* $\beta^{12hwN(\mathfrak{f})} = \gamma_1\gamma_2\gamma_3\gamma_4$.

Proof. This is straightforward. By definition, we have

$$\mathfrak{k}(1, \mathfrak{f}\mathfrak{a}^{-1})^{12N(\mathfrak{f})} = g^{12N(\mathfrak{f})}(1, \mathfrak{f}\mathfrak{a}^{-1})\,\Delta(\mathfrak{f}\mathfrak{a}^{-1})^{-N(\mathfrak{f})}$$

and

$$g^{12N(\mathfrak{f})}(1, \mathfrak{f}\mathfrak{a}^{-1}) = g_{\mathfrak{f}/(\mathfrak{a},\mathfrak{f})}(C(\mathfrak{a}/(\mathfrak{a}, \mathfrak{f})))^{N(\mathfrak{f})/N([\mathfrak{f},\mathfrak{a}])}.$$

Thus

$$\mathfrak{k}_{\mathfrak{f}}(\mathbf{a}, \mathfrak{o})^{12hwN(\mathfrak{f})} = \gamma_1\gamma_2.$$

By Lemma 3.5, the rest of the product for $\beta^{12hwN(\mathfrak{f})}$ is $\gamma_3\gamma_4$, thus proving Lemma 5.1.

Remarks

(i) We note that $\gamma_2\gamma_3 \in K(1)$, since the total number of Δ-factors is

$$\begin{aligned}
&\sum_{\mathfrak{a}} -n(\mathfrak{a})hwN(\mathfrak{f}) + \sum_{\mathfrak{p}} \sum_{\mathfrak{a}} n_{\mathfrak{p}}(\mathfrak{a})(1 - \mathbf{N}\mathfrak{p})hN(\mathfrak{f}) \\
&\quad = -\left(d(\mathbf{a}) + \sum_{\mathfrak{p}} \frac{\mathbf{N}\mathfrak{p} - 1}{w} d(\mathbf{a}_{\mathfrak{p}}) \right) hwN(\mathfrak{f}) \\
&\quad = 0.
\end{aligned}$$

(ii) We have an ideal factorization

$$(\gamma_2\gamma_3\gamma_4) = \mathfrak{b}^{12hwN(\mathfrak{f})},$$

where $\mathfrak{b}$ is an ideal of $\mathfrak{o}_K$. Indeed,

$$\begin{aligned}(\gamma_2\gamma_3) &= \prod_{\mathfrak{a}} (\mathfrak{f}^{-1}\mathfrak{a})^{-12n(\mathfrak{a})hwN(\mathfrak{f})} \prod_{\mathfrak{p}} \prod_{\mathfrak{a}} \left(\frac{\mathfrak{a}}{(\mathfrak{p}^{-1}\mathfrak{a})^{\mathbf{N}\mathfrak{p}}}\right)^{12n_{\mathfrak{p}}(\mathfrak{a})hN(\mathfrak{f})} \\ &= \left[\mathfrak{f}^{d(\mathbf{a})} \prod_{\mathfrak{a}} \mathfrak{a}^{n(\mathfrak{a})} \prod_{\mathfrak{p}} \mathfrak{p}^{\mathbf{N}\mathfrak{p}\cdot d(\mathbf{a}_{\mathfrak{p}})/w} \prod_{\mathfrak{a}} \mathfrak{a}^{n_{\mathfrak{p}}(\mathfrak{a})(1-\mathbf{N}\mathfrak{p})/w}\right]^{12hwN(\mathfrak{f})}\end{aligned}$$

while γ_4 by definition is the $12hwN(\mathfrak{f})$-power of an element of K.

Now we assume that β is a unit, that is $\beta \in E_{\text{mod}}(K(\mathfrak{f}))$, and we use the remarks to rewrite the product of Lemma 5.1 in a more useful form for the applications to the next chapter, §4.

Since β is a unit, we have

$$\gamma_1 \sim (\gamma_2\gamma_3\gamma_4)^{-1},$$

so by Remark (ii) above, we get

$$\gamma_1 \sim \mathfrak{b}^{-12hwN(\mathfrak{f})}$$

with some ideal $\mathfrak{b}$ of $\mathfrak{o}_K$. On the other hand, using Theorem 3.2, we have

$$\gamma_1 \sim \Bigg(\prod_{\substack{\mathfrak{q}^m \mid \mathfrak{f} \\ \mathfrak{q}\text{ prime}}} \prod_{\mathfrak{a}:[\mathfrak{f},\mathfrak{a}]=\mathfrak{q}^m} \mathfrak{q}^{n(\mathfrak{a})/\phi(\mathfrak{q}^m)}\Bigg)^{12hwN(\mathfrak{f})}.$$

The first product is taken over all prime powers $\mathfrak{q}^m$ dividing $\mathfrak{f}$, $m \geqq 1$. Thus for each prime $\mathfrak{q} \mid \mathfrak{f}$, the sum

$$k_{\mathfrak{q}} = \sum_m \sum_{\mathfrak{a}:[\mathfrak{f},\mathfrak{a}]=\mathfrak{q}^m} n(\mathfrak{a})/\phi(\mathfrak{q}^m)$$

must be an *integer*.

Suppose $\mathfrak{q}^{m(\mathfrak{q})} \| \mathfrak{f}$. By Theorem 1.4,

$$\prod_{\substack{C \in \mathrm{Cl}(\mathfrak{q}^{m(\mathfrak{q})}) \\ C \mid c_0 \text{ in } \mathrm{Cl}(1)}} g_{\mathfrak{q}^{m(\mathfrak{q})}}(C)^{w/w(\mathfrak{q}^{m(\mathfrak{q})})} = \left(\frac{\Delta(\mathfrak{o})}{\Delta(\mathfrak{q})}\right)^{N(\mathfrak{q}^{m(\mathfrak{q})})} \sim \mathfrak{q}^{12N(\mathfrak{q}^{m(\mathfrak{q})})},$$

so if we let

(1) $$\gamma_1' = \gamma_1 \prod_{\mathfrak{q}^{m(\mathfrak{q})}\|\mathfrak{f}} \prod_{C|c_0} g_{\mathfrak{q}^{m(\mathfrak{q})}}(C)^{-k_{\mathfrak{q}}hwwN(\mathfrak{f})/w(\mathfrak{q}^{m(\mathfrak{q})})N(\mathfrak{q}^{m(\mathfrak{q})})},$$

and

$$\gamma_2' = \gamma_2\left(\frac{\Delta(\mathfrak{o})}{\Delta(\mathfrak{q})}\right)^{k_{\mathfrak{q}}hwN(\mathfrak{f})}$$

then

$$\gamma_1'\gamma_2' = \gamma_1\gamma_2, \quad \text{and} \quad \gamma_1' \text{ is a unit.}$$

Next, we can write the factor $\gamma_2'\gamma_3$ in terms of $g_{(1)}$ rather than Δ, recalling that

(2) $$g_{(1)}(c) = \alpha^{12}\frac{\Delta^h(\mathfrak{a})}{\Delta^h(\mathfrak{o})} \quad \text{if } \mathfrak{a} \in c^{-1},\ \mathfrak{a}^h = (\alpha).$$

Let

(3) $$\gamma_2'' = \prod_{\mathfrak{a}} g_{(1)}(c(\mathfrak{f}^{-1}\mathfrak{a}))^{-n(\mathfrak{a})wN(\mathfrak{f})} \cdot \prod_{\mathfrak{q}^m\|\mathfrak{f}} g_{(1)}(c(\mathfrak{q}^{-1}))^{-k_{\mathfrak{q}}wN(\mathfrak{f})} \cdot \prod_{\mathfrak{p}}\prod_{\mathfrak{a}}\left(\frac{g_{(1)}(c(\mathfrak{a}))}{g_{(1)}(c(\mathfrak{a}\mathfrak{p}^{-1}))^{\mathbf{N}\mathfrak{p}}}\right)^{n_{\mathfrak{p}}(\mathfrak{a})N(\mathfrak{f})}$$

Then

$$\gamma_2'\gamma_3/\gamma_2'' \in K^{12N(\mathfrak{f})} \quad \text{and} \quad \gamma_2'' \text{ is a unit.}$$

But

$$\gamma_1\gamma_2\gamma_3\gamma_4 = \gamma_1'\gamma_2''\left(\frac{\gamma_2'\gamma_3}{\gamma_2''}\gamma_4\right)$$

and $\gamma_1\gamma_2\gamma_3\gamma_4$, γ_1', γ_2'' are all units. Since $(\gamma_2'\gamma_3/\gamma_2'')\gamma_4 \in K^{12}$ and is a unit, it follows that

(4) $$(\gamma_2'\gamma_3/\gamma_2'')\gamma_4 = 1,$$

so

(5) $$\beta^{12hwN(\mathfrak{f})} = \gamma_1'\gamma_2''.$$

Finally, by the definition of γ_2'' and the relation between $g_{(1)}$ and Δ, we get the ideal factorization

$$(\gamma_2'\gamma_3/\gamma_2'') = \left[\prod_{\mathfrak{a}} (\mathfrak{f}\mathfrak{a}^{-1})^{-n(\mathfrak{a})} \prod_{\mathfrak{q}^m \| \mathfrak{f}} \mathfrak{q}^{-k_{\mathfrak{q}}} \prod_{\mathfrak{p}} \mathfrak{p}^{-\mathbf{N}\mathfrak{p} d_{\mathfrak{p}}(\mathfrak{a})/w} \prod_{\mathfrak{p}} \prod_{\mathfrak{a}} \mathfrak{a}^{n\,(\mathfrak{a})(\mathbf{N}\mathfrak{p}-1)/w}\right]^{12hN(\mathfrak{f})}$$

By (4), γ_4^{-1} generates this same ideal. But

$$(\gamma_4^{-1}) = \prod_{\mathfrak{p}} \prod_{\alpha} (\alpha)^{-d(\mathfrak{a}_{\mathfrak{p}})12hwN(\mathfrak{f})}$$

is the $12hwN(\mathfrak{f})$-power of a principal ideal. Hence the ideal in brackets above must be principal. In order words, if we gather together like factors in the definition of γ_2'' to write

$$\gamma_2'' = \prod_{c \in \mathrm{Cl}(1)} g_{(1)}(c)^{n_{(1)}(c)N(\mathfrak{f})},$$

then

(6) $$\prod c^{n_{(1)}(c)} = c_0,$$

which we recognize as condition 2 in the discussion of $E_{\mathrm{mod}}(K(1))$.

It will be convenient to summarize the results of this section using a different normalization of the Siegel units g, in order to eliminate some exponents.

Let H be an abelian extension of K of conductor $\mathfrak{f}$, and let $\mathfrak{g} | \mathfrak{f}$, $C \in \mathrm{Cl}(\mathfrak{g})$. Let $H_{\mathfrak{g}} = H \cap K(\mathfrak{g})$ and **define** the **Kersey invariant**:

$$\boxed{\psi_H(\mathfrak{g}, C) = \mathbf{N}_{K(\mathfrak{g})/H_{\mathfrak{g}}}(g_{\mathfrak{g}}(C))^{\eta(\mathfrak{g})wN(\mathfrak{f})/w(\mathfrak{g})N(\mathfrak{g})}}$$

where $\eta(\mathfrak{g}) = h$ if $\mathfrak{g} \neq (1)$ and $\eta(1) = 1$.

The distribution relations satisfied by the ψ contain no exponents and read as follows.

Lemma 5.2. *Let $\mathfrak{g}' | \mathfrak{f}$, $\mathfrak{g}' = \mathfrak{p}\mathfrak{g}$ with some prime $\mathfrak{p}$. Let $C \in \mathrm{Cl}(H_{\mathfrak{g}})$. If $\mathfrak{p} \nmid \mathfrak{g}$, let $C(\mathfrak{p})$ be the class of $\mathfrak{p}$ in $\mathrm{Cl}(H_{\mathfrak{g}})$. Let $\mathfrak{p}^h = (\gamma)$. Then:*

$$\prod_{\substack{C' \in \mathrm{Cl}(H_{\mathfrak{g}}) \\ C' | C}} \psi_H(\mathfrak{g}, C') = \begin{cases} \psi_H(\mathfrak{g}, C) & \text{if } \mathfrak{p} | \mathfrak{g} \\[2ex] \dfrac{\psi_H(\mathfrak{g}, C)}{\psi_H(\mathfrak{g}, CC(\mathfrak{p})^{-1})} & \text{if } \mathfrak{p} \nmid \mathfrak{g} \text{ and } \mathfrak{g} \neq (1) \\[2ex] \gamma^{12[K(1):H_{(1)}]} \dfrac{\psi_H((1), C)}{\psi_H((1), CC(\mathfrak{p})^{-1})} & \text{if } \mathfrak{g} = (1). \end{cases}$$

Proof. This follows easily from the distribution relations Theorems 1.3 and 1.4 of Chapter 11.

We now describe our elements $\beta^{12hwN(\mathfrak{f})} \in E_{\text{mod}}(K(\mathfrak{f}))^{12hwN(\mathfrak{f})}$ in terms of the ψ.

Theorem 5.3. *An element of* $E_{\text{mod}}(K(\mathfrak{f}))^{12hwN(\mathfrak{f})}$ *can be written in the form*

$$\beta^{12hwN(\mathfrak{f})} = \delta_1\delta_2\delta_3$$

where:

$$\delta_1 = \prod_{\substack{\mathfrak{g}|\mathfrak{f} \\ \mathfrak{g}\neq(1)}} \prod_{C\in\text{Cl}(\mathfrak{g})} \left(\frac{\psi_{K(\mathfrak{f})}(\mathfrak{g}, C)^{w(\mathfrak{g})}}{\psi_{K(\mathfrak{f})}((1), c(\mathfrak{g}^{-1}C))^{w}}\right)^{n_\mathfrak{g}(C)}$$

$$\delta_2 = \prod_{\substack{\mathfrak{q}^{m(\mathfrak{q})}||\mathfrak{f} \\ \mathfrak{q}\text{ prime}}} \left[\frac{\prod_{\substack{C\in\text{Cl}(\mathfrak{q}^{m(\mathfrak{q})}) \\ C|c_0}} \psi_{K(\mathfrak{f})}(\mathfrak{q}^{m(\mathfrak{q})}, C)}{\psi_{K(\mathfrak{f})}((1), c_0)/\psi_{K(\mathfrak{f})}((1), c(\mathfrak{q})^{-1})}\right]^{wk_\mathfrak{q}}$$

$$\delta_3 = \prod_{c\in\text{Cl}(1)} \prod_{\mathfrak{p}} \left(\frac{\psi_{K(\mathfrak{f})}((1), c)}{\psi_{K(\mathfrak{f})}((1), cc(\mathfrak{p})^{-1})^{\mathbf{N}\mathfrak{p}}}\right)^{n_\mathfrak{p}(c)}.$$

If we rewrite this product separating ramified levels ($\mathfrak{g} \neq (1)$) from unramified levels ($\mathfrak{g} = (1)$) as

$$\prod_{\substack{\mathfrak{g}|\mathfrak{f} \\ \mathfrak{g}\neq(1)}} \prod_{C\in\text{Cl}(\mathfrak{g})} \psi_{K(\mathfrak{f})}(\mathfrak{g}, C)^{m_\mathfrak{g}(C)} \cdot \prod_{c\in\text{Cl}(1)} \psi_{K(\mathfrak{f})}((1), c)^{m_{(1)}(c)},$$

then the following two conditions hold:

5.3.1 $\prod c^{m_{(1)}(c)} = c_0$

5.3.2 *If* $\mathfrak{f}$ *is prime to* $w(K(1))$, *then*

$$\sum_{\substack{\mathfrak{g}|\mathfrak{f} \\ \mathfrak{g}\neq(1)}} \mathbf{N}(\mathfrak{f}/\mathfrak{g}) \sum_{C\in\text{Cl}(\mathfrak{g})} m_\mathfrak{g}(C)\mathbf{N}\mathfrak{a}(C) \equiv 0 \bmod w_{K(\mathfrak{f})}/w_{K(1)}.$$

Proof. We regroup the factors in the expression

$$\beta^{12hwN(\mathfrak{f})} = \gamma_1'\gamma_2''.$$

Let $\delta_1 = \gamma_1$ times the first product in the definition (3) of γ_2''. If we unite the factors

$$g_{[\mathfrak{f},\mathfrak{a}]}(C([\mathfrak{a},\mathfrak{f}]))^{n(\mathfrak{a})hwN(\mathfrak{f})/N([\mathfrak{f},\mathfrak{a}])}$$

and

$$g_{(1)}(c(\mathfrak{f}^{-1}\mathfrak{a}))^{-n(\mathfrak{a})wN(\mathfrak{f})},$$

write such a product in terms of the ψ, and gather like factors, then we get a product of the desired form for δ_1.

Similarly let $\delta_2 = \gamma_1'/\gamma_1$ (see (1)) times the second product in the definition of γ_2''.

Let $\delta_3 =$ third product in the definition of γ_2''.

Condition 5.3.1 is the same as (6).

Finally, suppose $\mathfrak{f}$ is prime to $w_{K(1)}$, and consider condition 5.3.2. Since $\mathbf{a} \in \mathbf{I}(\mathfrak{f})$ to begin with, we know that

$$\mathbf{Na} = \sum n(\mathfrak{a})\mathbf{N}\mathfrak{a} \equiv 0 \bmod N(\mathfrak{f}).$$

On the other hand, by construction and by Lemma 4.4 of Chapter 9, we have

$$\sum n(\mathfrak{a})\mathbf{N}\mathfrak{a} \equiv \sum_{\substack{\mathfrak{g}|\mathfrak{f}\\ \mathfrak{g}\neq(1)}} \mathbf{N}(\mathfrak{f}/\mathfrak{g}) \sum_{C\in \mathrm{Cl}(\mathfrak{g})} n_{\mathfrak{g}}(C)\mathbf{N}\mathfrak{a}(C) \bmod w_{K(\mathfrak{f})}/w_{K(1)}.$$

Since $w(K(\mathfrak{f}))/w(K(1))$ divides $N(\mathfrak{f})$ [in fact,

$$\frac{w(K(\mathfrak{f}))}{w(K(1))} = \text{g.c.d.}(w(K(\mathfrak{f})), N(\mathfrak{f}))$$

because $N(\mathfrak{f})$ is prime to $w(K(1))$], it follows that

$$\sum_{\substack{\mathfrak{g}|\mathfrak{f}\\ \mathfrak{g}\neq(1)}} \mathbf{N}(\mathfrak{f}/\mathfrak{g}) \sum_{C\in \mathrm{Cl}(\mathfrak{g})} n_{\mathfrak{g}}(C)\mathbf{N}\mathfrak{a}(C) \equiv 0 \bmod w_{K(\mathfrak{f})}/w_{K(1)}.$$

To establish 5.3.2 it then suffices to see that for $\mathfrak{q}^{m(\mathfrak{q})}\|\mathfrak{f}$, we have

$$\mathbf{N}(\mathfrak{f}\mathfrak{q}^{-m(\mathfrak{q})}) \sum_{\substack{C\in \mathrm{Cl}(\mathfrak{q}^{m(\mathfrak{q})})\\ C|c_0}} \mathbf{N}\mathfrak{a}(C) \equiv 0 \bmod w_{K(\mathfrak{f})}/w_{K(1)}.$$

By arguments similar to those of Chapter 9, §4 we see that

$$\frac{w(K(\mathfrak{f}))}{w(K(\mathfrak{q}^{m(\mathfrak{q})}))} \quad \text{divides} \quad \mathbf{N}(\mathfrak{f}\mathfrak{q}^{-m(\mathfrak{q})}).$$

Also, let ζ be a primitive $w(K(\mathfrak{q}^{m(\mathfrak{q})}))$-th root of unity. Then

$$\zeta^{\Sigma \mathbf{N}\mathfrak{a}(C)} = \mathbf{N}_{K(\mathfrak{q}^{m(\mathfrak{q})})/K(1)}(\zeta)$$

is a $w(K(1))$-th root of unity, so

$$w(K(\mathfrak{q}^{m(\mathfrak{q})}))/w(K(1)) \quad \text{divides} \quad \sum \mathbf{N}\mathfrak{a}(C).$$

This completes the proof of Theorem 5.3.

CHAPTER 13

Computation of a Unit Index

This chapter is due to Kersey. We determine the index of the group of Klein units in the group of all units in case the conductor is the power of a prime ideal $\mathfrak{p}^n$. For the composite case, we refer to [Ke 2]. The treatment differs from, say, Robert's [Ro 1] in several ways. First we deal with the somewhat larger group of Klein units rather than Robert units; and second, we introduce intermediate groups other than those of Robert, more in line with the Sinnott computation of the index in the composite case for cyclotomic fields.

We continue to use the following notation:

H = abelian extension of K of conductor $\mathfrak{p}^n = \mathfrak{f}$.
$K(1)$ = Hilbert class field of K.
$H_\mathfrak{g} = H \cap K(\mathfrak{g})$ for $\mathfrak{g} \mid \mathfrak{f}$.
$G = G(H/K) = \mathrm{Gal}(H/K)$.
$\mathrm{Cl}(H) = \mathrm{Cl}(H/K)$ = ideal class group corresponding to H.
$G(H_{(1)}/K)$ and $\mathrm{Cl}(H_{(1)})$ are the similar objects for $H_{(1)}$.
$E(H)$ = full group of units of H.
$W = \boldsymbol{\mu}_H$ = group of roots of unity of H, having order $w(H) = w_H$.
$h = h_K$ = class number of K.

§1. The Regulator Map and the Inertia Group

In Chapter 12, §5 we replaced the invariants $g_\mathfrak{f}(C)$ by appropriate powers so as to eliminate the exponents arising in the distribution relations. We repeat this now for $\mathfrak{f}$ equal to the prime power $\mathfrak{p}^n$ for convenience of notation in what follows. Thus for $C \in \mathrm{Cl}(H)$ we choose $C' \in \mathrm{Cl}(\mathfrak{p}^n)$ lying above C, and define

$$\psi_H(\mathfrak{p}^n, C) = \mathbf{N}_{K(\mathfrak{p}^n)/H}(g_{\mathfrak{p}^n}(C'))^{hw/w(\mathfrak{p}^n)}.$$

Similarly, if $c \in \mathrm{Cl}(H_{(1)})$, choose $c' \in \mathrm{Cl}(1)$ lying above c; let $g_{(1)}(c') = \delta(c')$ as in Chapter 9, §1; and define

$$\psi_H((1), c) = \mathbf{N}_{K(1)/H_{(1)}}(g_{(1)}(c'))^{N(\mathfrak{p}^n)}.$$

With these definitions, the distribution relations satisfied by the ψ_H are given in the following lemma.

Lemma 1.1. *For any $c \in \mathrm{Cl}(H_{(1)})$ we have*

$$\prod_{\substack{C \in \mathrm{Cl}(H) \\ C|c}} \psi_H(\mathfrak{p}^n, C) = \gamma^{12} \frac{\psi_H((1), c)}{\psi_H((1), cc(\mathfrak{p})^{-1})},$$

where $c(\mathfrak{p})$ is the class of $\mathfrak{p}$ in $\mathrm{Cl}(1)$, and γ is any generator of the fractional ideal $\mathfrak{p}^v$, with $v = -[HK(1) : H]N(\mathfrak{p}^n)$.

Proof. For the convenience of the reader, we give the proof passing directly from level $\mathfrak{p}^n$ to level (1). By Theorems 1.3 and 1.4 of Chapter 11, we have:

$$\begin{aligned}
\prod_{\substack{C \in \mathrm{Cl}(H) \\ C|c}} \psi_H(\mathfrak{p}^n, C) &= \prod_{\substack{C \in \mathrm{Cl}(\mathfrak{p}^n) \\ C|c}} g_{\mathfrak{p}^n}(C)^{hw/w(\mathfrak{p}^n)} \\
&= \prod_{\substack{c' \in \mathrm{Cl}(1) \\ c'|c}} \prod_{\substack{C \in \mathrm{Cl}(\mathfrak{p}^n) \\ C|c'}} g_{\mathfrak{p}^n}(C)^{hw/w(\mathfrak{p}^n)} \\
&= \prod_{\substack{c' \in \mathrm{Cl}(1) \\ c'|c}} (\Delta(\mathfrak{c}'^{-1})/\Delta(\mathfrak{p}\mathfrak{c}'^{-1}))^{N(\mathfrak{p}^n)h}
\end{aligned}$$

[for any ideal $\mathfrak{c}' \in c'$, by Theorem 1.4 of Chapter 11]

$$= \prod_{\substack{c' \in \mathrm{Cl}(1) \\ c'|c}} (\alpha^{-12}\delta(c')/\delta(c'c(\mathfrak{p})^{-1}))^{N(\mathfrak{p}^n)}$$

[where $(\alpha) = \mathfrak{p}^h$, by the definition of Chapter 9, §1]

$$\begin{aligned}
&= \mathbf{N}_{K(1)/H_{(1)}}(\alpha^{-12}\delta(c')/\delta(c'c(\mathfrak{p})^{-1})^{N(\mathfrak{p}^n)}) \\
&= (\alpha^{-12[K(1):H_{(1)}]}\mathbf{N}_{K(1)/H_{(1)}}(g_{(1)}(c')/g_{(1)}(c'c(\mathfrak{p})^{-1})))^{N(\mathfrak{p}^n)} \\
&= \gamma^{12}\psi_H((1), c)/\psi_H((1), cc(\mathfrak{p})^{-1})
\end{aligned}$$

as desired.

For any non-trivial character χ of $\mathrm{Cl}(H) \approx G$, we let

$$S_{\mathfrak{p}^n}(\chi, \psi) = \sum_{C \in \mathrm{Cl}(H)} \chi(C) \log |\psi_H(\mathfrak{p}^n, C)|.$$

If χ factors through $\mathrm{Cl}(H_{(1)})$, we also let

$$S_{(1)}(\chi, \psi) = \sum_{C \in \mathrm{Cl}(H)} \chi(C) \log |\psi_H((1), C)|.$$

The relation between these numbers and the $S_{\mathfrak{f}}(\chi, g_{\mathfrak{f}})$ previously defined in Chapter 11, §2 is given by the following lemma.

Lemma 1.2. *If* $\mathfrak{f} | \mathfrak{p}^n$ *and* χ *is a character of* $\mathrm{Cl}(H)$, *which, viewed as a character of* $\mathrm{Cl}(\mathfrak{p}^n)$, *factors through* $\mathrm{Cl}(\mathfrak{f})$, *then*:

$$S_{\mathfrak{p}^n}(\chi, \psi) = \frac{hwN(\mathfrak{p}^n)}{w(\mathfrak{f})N(\mathfrak{f})} S_{\mathfrak{f}}(\chi, g_{\mathfrak{f}}) \quad \text{if } \mathfrak{f} \neq (1)$$

$$S_{(1)}(\chi, \psi) = N(\mathfrak{p}^n) S_{(1)}(\chi, g_{(1)}) \quad \text{if } \mathfrak{f} = (1).$$

Proof. This is a routine calculation. Suppose $\mathfrak{f} \neq (1)$. Then:

$$\begin{aligned} S_{\mathfrak{p}^n}(\chi, \psi) &= \sum_{C \in \mathrm{Cl}(H)} \chi(C) \log |\mathbf{N}_{K(\mathfrak{p}^n)/H}(g_{\mathfrak{p}^n}(C'))^{hw/w(\mathfrak{p}^n)}| \\ &= \frac{hw}{w(\mathfrak{p}^n)} \sum_{C \in \mathrm{Cl}(H)} \chi(C) \log |g_{\mathfrak{p}^n}(C')| \\ &= \frac{hw}{w(\mathfrak{p}^n)} S_{\mathfrak{p}^n}(\chi, g_{\mathfrak{p}^n}) \\ &= \frac{hw}{w(\mathfrak{p}^n)} \frac{w(\mathfrak{p}^n)N(\mathfrak{p}^n)}{w(\mathfrak{f})N(\mathfrak{f})} S_{\mathfrak{f}}(\chi, g_{\mathfrak{f}}) \end{aligned}$$

[by Theorem 2.1 of Chapter 11]

$$= \frac{hwN(\mathfrak{p}^n)}{w(\mathfrak{f})N(\mathfrak{f})} S_{\mathfrak{f}}(\chi, g_{\mathfrak{f}}).$$

The proof for $\mathfrak{f} = (1)$ is similar.

If χ is a character of $\mathrm{Cl}(H)$, viewed as a character of $\mathrm{Cl}(\mathfrak{p}^n)$, let $\mathfrak{f}_\chi$ denote its conductor, and **define**:

$$[\mathfrak{f}_\chi] = \begin{cases} \mathfrak{p}^n & \text{if } \mathfrak{f}_\chi \neq (1) \\ (1) & \text{if } \mathfrak{f}_\chi = (1). \end{cases}$$

Since $S_{\mathfrak{f}}(\chi, g) \neq 0$ for $\chi \neq 1$, Lemma 1.2 shows that

$$S_{[\mathfrak{f}_\chi]}(\chi, \psi) \neq 0.$$

Translating Theorem 2.1 of Chapter 11 into the present situation yields:

Lemma 1.3. *If* χ *is a character of* $\mathrm{Cl}(H)$, *then*

$$S_{\mathfrak{p}^n}(\chi, \psi) = (1 - \chi(\mathfrak{p}))S_{[\mathfrak{f}_\chi]}(\chi, \psi).$$

Let Ψ_E be the multiplicative subgroup of H generated by the units

$$\psi_H(\mathfrak{f}, C)/\psi_H(\mathfrak{f}, C')$$

where $\mathfrak{f}$ is either $\mathfrak{p}^n$ or (1), and $C, C' \in \mathrm{Cl}(H_{\mathfrak{f}})$. Our first goal will be to determine the index of this group Ψ_E in the group of all units of H. We can conduct our investiation of Ψ_E in the group ring $\mathbf{R}[G]$ via the regulator map

$$\rho : H^* \to \mathbf{R}[G].$$

If, as agreed, we let $E = E(H)$ be the group of all units, and $W = \boldsymbol{\mu}_H$ be the group of roots of unity in H, then

$$(\mathrm{Ker}\ \rho) \cap E = W,$$

since the kernel of ρ consists of elements of H^* with absolute value 1 at all archimedean absolute values. Thus

$$E/W\Psi_E \approx \rho(E)/\rho(\Psi_E).$$

We shall determine the order of this factor group by introducing a certain $\mathbf{Z}[G]$-submodule U of $\mathbf{R}[H]$, analogous to the module U employed by Iwasawa and Sinnott in the study of cyclotomic extensions of the rationals. Before defining U, we must fix some notation.

If $A \subset G$ we let

$$s(A) = \sum_{\sigma \in A} \sigma \in \mathbf{Z}[G].$$

For any character χ, we let

$$e_\chi = \frac{1}{|G|} \sum_{\sigma \in G} \chi(\sigma)\sigma^{-1}$$

be the usual idempotent of $\mathbf{C}[G]$ associated with χ.

Let $T_{\mathfrak{p}}$ be the inertia group of $\mathfrak{p}$ in G.

$\lambda_{\mathfrak{p}}$ = any Frobenius automorphism for $\mathfrak{p}$, well defined mod $T_{\mathfrak{p}}$.

For any $\mathbf{Z}[G]$-module B, we let B_0 denote the submodule of elements annihilated by $s(G)$, and:

$B^{T_{\mathfrak{p}}}$ = subset of elements of B fixed under the action of $T_{\mathfrak{p}}$. Let

$$\alpha(1) = 1 - \lambda_{\mathfrak{p}}^{-1}\frac{s(T_{\mathfrak{p}})}{|T_{\mathfrak{p}}|} \quad \text{and} \quad \alpha(\mathfrak{p}^n) = s(T_{\mathfrak{p}}).$$

We are now ready to define

$$U = \mathbf{Z}[G]_0\alpha(1) + \mathbf{Z}[G]_0\alpha(\mathfrak{p}^n).$$

The relation between U and the group Ψ_E is given by the following proposition, cf. Sinnott [Sin], p. 114.

Proposition 1.4. *If we put*

$$\omega = \sum_{\chi \neq 1} S_{[\mathfrak{f}_\chi]}(\bar{\chi}, \psi)e_\chi,$$

then

$$\rho(\Psi_E) = \omega U.$$

Proof. For $\mathfrak{f} = \mathfrak{p}^n$ or (1), let

$$\eta(\mathfrak{f}) = \sum_{C \in \mathrm{Cl}(H)} \log|\psi_H(\mathfrak{f}, C)|\,\sigma(C)^{-1}.$$

We first show that

$$\rho(\Psi_E) = \mathbf{Z}[G]_0\eta(\mathfrak{p}^n) + \mathbf{Z}[G]_0\eta(1). \tag{1}$$

To see this, notice that if $C_0 \in \mathrm{Cl}(H/K)$ is the unit class, then

$$\eta(\mathfrak{p}^n) = \rho(\psi_H(\mathfrak{p}^n, C_0)),$$

and for any $C \in \mathrm{Cl}(H)$,

$$\rho(\psi_H(\mathfrak{p}^n, C)) = \sigma(C)\eta(\mathfrak{p}^n).$$

This follows from the fact that the Galois action on the $\psi_H(\mathfrak{p}^n, C)$ is given by

$$\psi_H(\mathfrak{p}^n, C)^{\sigma(C')} = \psi_H(\mathfrak{p}^n, CC')$$

for C, $C' \in \mathrm{Cl}(H)$, which in turn follows from the similar rule for the Galois action on the numbers $g_H(C)$.

For $\mathfrak{f} = (1)$, the situation is similar, except that here the Galois action is

$$\psi_H((1), C)^{\sigma(C')} = \psi_H((1), CC')/\psi_H((1), C').$$

Noting this, we find that for $C \in \mathrm{Cl}(H_{(1)})$,

$$(2) \qquad \rho(\psi_H((1), C)) = (\sigma(C) - 1)\eta(1).$$

In either case, we have

$$\rho(\psi_H(\mathfrak{f}, C)/\psi_H(\mathfrak{f}, C')) = (\sigma(C) - \sigma(C'))\eta(\mathfrak{f})$$

for $\mathfrak{f} = (1)$ or $\mathfrak{p}^n$ and C, $C' \in \mathrm{Cl}(H)$. Since these elements generate $\rho(\Psi_E)$ (as $\mathbf{Z}$-module), we see that (1) holds.

Thus to see that $\omega U = \rho(\Psi_E)$, it will suffice to show that

$$\omega\alpha(\mathfrak{f}) = (1 - e_1)\eta(\mathfrak{f}) \quad \text{for } \mathfrak{f} = (1) \text{ or } \mathfrak{p}^n.$$

Indeed, since $\mathbf{Z}[G]_0 e_1 = 0$, we have

$$\mathbf{Z}[G]_0(1 - e_1)\eta(\mathfrak{f}) = \mathbf{Z}[G]_0\eta(\mathfrak{f}).$$

Given a character χ of G, we may extend χ to a ring homomorphism $\mathbf{C}[G] \to \mathbf{C}$, for which we use the same symbol χ. An element $\beta \in \mathbf{C}[G]$ is determined by the values $\chi(\beta)$ for all χ.

Suppose first that $\chi \neq 1$, and consider $\chi(\eta(\mathfrak{f}))$ for $\mathfrak{f} = (1)$ or $\mathfrak{p}^n$. If $\mathfrak{f}_\chi \nmid \mathfrak{f}$, that is if $\mathfrak{f} = (1)$ and $\mathfrak{f}_\chi \neq (1)$, then there is some $C \in \mathrm{Cl}(H)$ lying above the unit class $c_0 \in \mathrm{Cl}(H_{(1)})$ for which $\chi(C) \neq 1$. But for such C, we have $\psi_H((1), C) = 1$, so by (2) above

$$\sigma(C)\eta(1) = \eta(1).$$

Thus $\chi(\sigma(C))\chi(\eta(1)) = \chi(\eta(1))$. Since $\chi(\sigma(C)) \neq 1$, we see that $\chi(\eta(1)) = 0$.

On the other hand, suppose $\mathfrak{f}_\chi | \mathfrak{f}$. Then

$$\chi(\eta(\mathfrak{f})) = \sum_{C \in \mathrm{Cl}(H)} \log|\psi_H(\mathfrak{f}, C)|\bar{\chi}(C).$$

If $\mathfrak{f} = \mathfrak{p}^n$, this is simply $S_{\mathfrak{p}^n}(\bar{\chi}, \psi)$, which by Lemma 1.3 gives

$$\chi(\eta(\mathfrak{f})) = (1 - \bar{\chi}(\mathfrak{p}))S_{[\mathfrak{f}_\chi]}(\bar{\chi}, \psi).$$

If $\mathfrak{f} = (1)$, then χ factors through $\mathrm{Cl}(H_{(1)})$), and

$$\chi(\eta(\mathfrak{f})) = [H : H_{(1)}] \sum_{c \in \mathrm{Cl}(H_{(1)})} \log|\psi_H((1), c)|\,\bar{\chi}(c).$$

The degree $[H : H_{(1)}]$ is exactly $|T_\mathfrak{p}|$, so this last expression is equal to $|T_\mathfrak{p}|S_{(1)}(\bar{\chi}, \psi)$.

Also, $\chi(\omega) = S_{[\mathfrak{f}_\chi]}(\bar{\chi}, \psi)$, and:

$$\chi(\alpha(\mathfrak{p}^n)) = \begin{cases} 0 & \text{if } \chi \text{ is non-trivial on } T_\mathfrak{p} \\ |T_\mathfrak{p}| & \text{if } \chi \text{ is trivial on } T_\mathfrak{p} \end{cases}$$

$$\chi(\alpha(1)) = \chi\left(1 - \lambda_\mathfrak{p}^{-1}\frac{s(T_\mathfrak{p})}{T_\mathfrak{p}}\right) = \begin{cases} 1 - \bar{\chi}(\mathfrak{p}) & \text{if } \chi = 1 \text{ on } T_\mathfrak{p} \\ 1 & \text{if } \chi \neq 1 \text{ on } T_\mathfrak{p}. \end{cases}$$

Since χ is trivial on $T_\mathfrak{p}$ if and only if χ factors through $\mathrm{Cl}(H_{(1)})$, i.e. if and only if $\mathfrak{f}_\chi = (1)$; and since $\bar{\chi}(\mathfrak{p})$ is by definition 0 if $\mathfrak{f}_\chi \neq (1)$, we see that in any case

$$\chi(\omega\alpha(\mathfrak{f})) = \chi(\eta(\mathfrak{f})).$$

This is true for $\chi \neq 1$. Since $\chi(\omega) = \chi(1 - e_1) = 0$ for $\chi = 1$, we have for *all* χ,

$$\chi(\omega\alpha(\mathfrak{f})) = \chi((1 - e_1)\eta(\mathfrak{f})),$$

and so

$$\omega\alpha(\mathfrak{f}) = (1 - e_1)\eta(\mathfrak{f})$$

as claimed. This completes the proof.

§2. An Index Computation

In this section, we prove:

Theorem 2.1. $(E : W\Psi_E) = \dfrac{h(H)/w(H)}{h/w}(12hwN(\mathfrak{p}^n))^r$

where $r = |G| - 1$, *as usual.*

To do this, we compute the index $(\rho(E):\rho(\Psi_E))$ by splitting it into three pieces:

$$(\rho(E):\rho(\Psi_E)) = (\rho(E):\mathbf{Z}[G]_0)(\mathbf{Z}[G]_0:U)(U:\rho(\Psi_E)),$$

and computing each piece separately.

The indices are to be interpreted as in [Sin], p. 110. Thus if L and M are lattices (discrete subgroups of maximal rank) in a $\mathbf{R}$-space V, we define $(L:M)$ to be $|\det A|$, where A is any linear transformation of V mapping L to M. If $M \subset L$ then $(L:M)$ is the usual group index. Also, if N is another lattice in V, then

$$(L:M) = (L:N)(N:M).$$

We now compute the three indices.

Lemma 2.2. $(\rho(E):\mathbf{Z}[G]_0) = 2^r/R(H)$.

where $R(H)$ is the regulator of H.

Proof. We may take the elements $\sigma - 1$, with $\sigma \in G$, $\sigma \neq 1$ as a basis for $\mathbf{Z}[G]_0$, which of course is a lattice in $\mathbf{R}[G]_0$. Dirichlet's unit theorem says that $\rho(E)$ is also a lattice, so this index is defined. Let $\varepsilon_1, \ldots, \varepsilon_r$ be a system of fundamental units for H. Then in terms of the above basis for $\mathbf{Z}[G]_0$, we have

$$\rho(\varepsilon_i) = \sum_{\sigma \in G} \log|\varepsilon_i^\sigma|(\sigma^{-1} - 1).$$

The absolute value of the determinant

$$\det(\log|\varepsilon_i^\sigma|), \qquad i = 1, \ldots, r \text{ and } \sigma \neq 1$$

is by definition $R(H)/2^r$. This gives the first index.

Lemma 2.3. $(\mathbf{Z}[G]_0:U) = 1$.

Proof. Let

$$e(\mathfrak{p}) = s(T_\mathfrak{p})/|T_\mathfrak{p}|$$

be the idempotent of $\mathbf{C}[G]$ associated with the subgroup $T_\mathfrak{p}$ of G, and consider the linear transformation A of $\mathbf{Q}[G]_0$ given by multiplication with $1 - e(\mathfrak{p})$. Note that the kernel of A consists of those elements of $\mathbf{Q}[G]_0$ on which $T_\mathfrak{p}$ acts trivially. Let $U_\mathbf{Q}$ be the $\mathbf{Q}$-vector space spanned by U. Then we have the following result.

Lemma 2.4.

(i)
$$(1 - e(\mathfrak{p}))\mathbf{Z}[G]_0 = (1 - e(\mathfrak{p}))U$$
$$\textit{and} \quad (1 - e(\mathfrak{p}))\mathbf{Q}[G]_0 = (1 - e(\mathfrak{p}))U_{\mathbf{Q}}.$$

(ii)
$$U^{T_\mathfrak{p}} = \mathbf{Z}[G]_0^{T_\mathfrak{p}} \quad \textit{and} \quad U_{\mathbf{Q}}^{T_\mathfrak{p}} = \mathbf{Q}[G]_0^{T_\mathfrak{p}}.$$

Proof. For (i), simply note that

$$(1 - e(\mathfrak{p}))\alpha(\mathfrak{p}^n) = (1 - e(\mathfrak{p}))s(T_\mathfrak{p}) = 0$$

and

$$(1 - e(\mathfrak{p}))\alpha(1) = (1 - e(\mathfrak{p}))(1 - \lambda_\mathfrak{p}^{-1}e(\mathfrak{p})) = 1 - e(\mathfrak{p})$$

since $e(\mathfrak{p})$ is idempotent. Since

$$U = \mathbf{Z}[G]_0\alpha(1) + \mathbf{Z}[G]_0\alpha(\mathfrak{p}^n),$$

both claims in (i) are obvious.

Let us prove (ii). To see that $U^{T_\mathfrak{p}} \subset \mathbf{Z}[G]_0^{T_\mathfrak{p}}$, note that if $z(1), z(\mathfrak{p}^n) \in \mathbf{Z}[G]_0$, then

$$z(1)\alpha(1) + z(\mathfrak{p}^n)\alpha(\mathfrak{p}^n) = z(1)(1 - \lambda_\mathfrak{p}^{-1}e(\mathfrak{p})) + z(\mathfrak{p}^n)s(T_\mathfrak{p})$$

is fixed under the action of $T_\mathfrak{p}$ if and only if $z(1)$ is fixed under the action of $T_\mathfrak{p}$, i.e., if and only if $z(1) \in \mathbf{Z}[G]_0^{T_\mathfrak{p}}$. But in this case, $z(1)e(\mathfrak{p}) = z(1)$, so

$$z(1)\alpha(1) + z(\mathfrak{p}^n)\alpha(\mathfrak{p}^n) = z(1)(1 - \lambda_\mathfrak{p}^{-1}) + z(\mathfrak{p}^n)s(T_\mathfrak{p}),$$

and this clearly lies in $\mathbf{Z}[G]_0^{T_\mathfrak{p}}$. This proves $U^{T_\mathfrak{p}} \subset \mathbf{Z}[G]_0^{T_\mathfrak{p}}$.

Conversely, it is clear that

$$\mathbf{Z}[G]_0^{T_\mathfrak{p}} = s(T_\mathfrak{p})\mathbf{Z}[G]_0 \subset U,$$

and so $\mathbf{Z}[G]_0^{T_\mathfrak{p}} \subset U^{T_\mathfrak{p}}$, and the first part of (ii) is proved. The second part of (ii) follows from the same calculation, thus proving the lemma.

In particular, we see that the $\mathbf{Q}$-dimension of $U_{\mathbf{Q}}$ is equal to that of $\mathbf{Q}[G]_0$ (namely r), since

$$\begin{aligned}
\dim U_{\mathbf{Q}} &= \dim A(U_{\mathbf{Q}}) + \dim(\operatorname{Ker} A \cap U_{\mathbf{Q}}) \\
&= \dim(1 - e(\mathfrak{p}))U_{\mathbf{Q}} + \dim U_{\mathbf{Q}}^{T_\mathfrak{p}} \\
&= \dim(1 - e(\mathfrak{p}))\mathbf{Q}[G]_0 + \dim \mathbf{Q}[G]_0^{T_\mathfrak{p}} \\
&= \dim \mathbf{Q}[G]_0
\end{aligned}$$

by the above lemma. Thus the **Q**-rank of U is maximal in $\mathbf{Q}[G]_0$, and the index

$$(\mathbf{Z}[G]_0 : U)$$

is defined. We are thus in a position to apply a standard lemma of elementary algebra, as in Sinnott, Lemma 6.1.

Let B, B' be discrete subgroups of $\mathbf{Q}[G]$, such that $(B : B')$ is defined. Let $\beta \in \mathbf{Q}[G]$, and let B_β denote the kernel in B of multiplication by β. Define B'_β similarly. Then both $(B_\beta : B'_\beta)$ and $(\beta B : \beta B')$ are defined, and

$$(B : B') = (B_\beta : B'_\beta)(\beta B : \beta B').$$

For the proof, one may assume without loss of generality that $B' \subset B$, and the lemma can then be proved as an elementary application of the Noether homomorphism theorems, cf. [L 4], Chapter II, §3.

Applying this with $B = \mathbf{Z}[G]_0$, $B' = U$, and $\beta = 1 - e(\mathfrak{p})$, and using Lemma 2.4, we see that

$$(\mathbf{Z}[G]_0 : U) = 1,$$

thus proving Lemma 2.3.

Lemma 2.5. $(U : \rho(\Psi_E)) = \prod_{\chi \neq 1} S_{[\mathfrak{f}_\chi]}(\bar{\chi}, \psi)$.

Proof. We have seen that U has maximal rank. Consider the linear transformation Λ of $\mathbf{R}[G]_0$ given by multiplication with the element ω defined in Proposition 1.4. By this proposition, Λ maps U to $\omega U = \rho(\Psi_E))$. Since

$$\omega = \prod_{\chi \neq 1} S_{[\mathfrak{f}_\chi]}(\bar{\chi}, \psi) e_\chi,$$

we have

$$|\det \Lambda| = \prod_{\chi \neq 1} |S_{[\mathfrak{f}_\chi]}(\bar{\chi}, \psi)|,$$

and this is non-zero by the remark following Lemma 1.2. This proves Lemma 2.5.

Putting our three indices together yields

$$(E : W\Psi_E) = \frac{2^r}{R(H)} \prod_{\chi \neq 1} |S_{[\mathfrak{f}_\chi]}(\bar{\chi}, \psi)|.$$

Using the analytic class number formula **LF 3** of Chapter 11, §2, the evaluation **LF 1** of $L(1, \chi)$ in terms of the numbers $S_{\mathfrak{f}}(\bar{\chi}, g_{\mathfrak{f}})$, and the relation between these numbers and $S_{[\mathfrak{f}_\chi]}(\bar{\chi}, \psi)$, we find precisely the formula stated in Theorem 2.1.

§3. Freeness Results

In this section we show that the only relations satisfied by the numbers ψ_H are those imposed by the distribution relations. Let:

F = the free abelian group on the disjoint union

$$\mathrm{Cl}(H) \cup \mathrm{Cl}(H_{(1)}) \cup \{\gamma\}.$$

D = the subgroup generated by the distribution relations

DR(H, c) $$\sum_{\substack{C \in \mathrm{Cl}(H) \\ C | c}} (C) - (\gamma) - ((c) - (cc(\mathfrak{p})^{-1}))$$

for all $c \in \mathrm{Cl}(H_{(1)})$, and by (C_0).

Ψ = the group generated by the numbers $\psi_H(\mathfrak{p}^n, C)$ and $\psi_H((1), c)$ for all classes C and c.

We can map $\mathbf{F} \to \Psi$ by

$$(C) \mapsto \psi_H(\mathfrak{p}^n, C)$$
$$(c) \mapsto \psi_H((1), c)$$
$$\gamma \mapsto \gamma_{\mathfrak{p}}^{12}.$$

This factors to give a homomorphism

$$\mathbf{F}/\mathbf{D} \to \Psi,$$

since Ψ satisfies the relations **D**.

We shall show that this is an isomorphism. Furthermore, we shall show that Ψ is a free abelian group, and we shall give a free set of generators.

To this end, let M be the subgroup of $\mathrm{Cl}(H_{(1)})$ generated by $c(\mathfrak{p})$, and let m be the index,

$$m = (\mathrm{Cl}(H_{(1)}) : M).$$

Let Y_1 be a set of representatives for the $m - 1$ cosets of M other than M itself. Let Z be a subset of $\mathrm{Cl}(H)$ consisting of C_0, and exactly one class lying above each element of Y_1, and let $Y_2 = \mathrm{Cl}(H) - Z$. Let

$$Y = Y_1 \cup Y_2.$$

Theorem 3.1.

(i) *The map* $\mathbf{F}/\mathbf{D} \to \Psi$ *defined above is an isomorphism.*
(ii) $\mathbf{F}/\mathbf{D}$ *is a free abelian group of rank* $|\mathrm{Cl}(H)| - 1$. *A free set of generators is given by*

$$\{\gamma\} \cup \{(C) : C \in Y\}.$$

Proof. From §2, we know that

$$\operatorname{rank} \Psi_E = |\mathrm{Cl}(H)| - 1.$$

It follows from the factorization of the values ψ that

$$\Psi/\Psi_E \approx \mathbf{Z},$$

so $\operatorname{rank} \Psi = |\mathrm{Cl}(H)|$. Thus it suffices to show that $\mathbf{F}/\mathbf{D}$ is generated by the elements

$$\gamma \quad \text{and} \quad (C), \qquad C \in Y.$$

This follows easily from the distribution relations. Indeed, let the group generated by these elements be F'. If $C \in Z$, then we see that $(C) \in F'$ because, if we add the relations $\mathbf{DR}(H, c)$ for all c of the form CC', $C' \in M$, $C|c$, we get

$$\sum_{C'|MC} (C') = |M| \cdot \gamma.$$

The only summand on the left which does not automatically lie in F' is (C). Thus $(C) \in F'$.

Now suppose $c \in \mathrm{Cl}(H_{(1)})$. The distribution relations $\mathbf{DR}(H, c)$ now show that if $(c) \in F'$, then $(cc(\mathfrak{p})^{-1}) \in F'$. Since $0 = (c_0) \in F'$, and every other coset of M has a representative $c \in Y_1$ with $(c) \in F'$, the result then follows. This completes the proof.

Thus the only relations satisfied by the numbers

$$\psi_H(\mathfrak{p}^n, C), \qquad \psi_H((1), c), \quad \text{and} \quad \gamma_{\mathfrak{p}}^{12}$$

are those which are consequences of the relations $\mathbf{DR}(H, c)$. Similarly, we see that the only relations satisfied by the numbers

$$\psi_H(\mathfrak{p}^n, C) \quad \text{and} \quad \psi_H((1), c)$$

are those which do not involve $\gamma_{\mathfrak{p}}^{12}$, and so are consequences of relations of the form

$$\mathbf{DR}(H, c_1, c_2) = \mathbf{DR}(H, c_1) - \mathbf{DR}(H, c_2)$$

obtained by taking the differences of two relations $\mathbf{DR}(H', c)$.

§4. The Index $(E_H : E_{\mathrm{mod}}(H))$

Theorem 4.1. *Assume that $\mathfrak{p}$ is a prime not dividing $w(H_{(1)})$. Let H be an abelian extension of K of conductor $\mathfrak{f} = \mathfrak{p}^n$ with $n \geqq 1$, $\mathfrak{p}$ prime. Then*

$$(E(H) : E_{\mathrm{mod}}(H)) = h_H/[K(1) : H_{(1)}].$$

This section is devoted to the proof.
Since we already know by Theorem 2.1 that

$$(E_H : W\Psi_E) = \frac{h(H)/w(H)}{h/w}(12hwN(\mathfrak{f}))^r,$$

and

$$(E_{\mathrm{mod}}(H) : WE_{\mathrm{mod}}(H)^{12hwN(\mathfrak{f})}) = (12hwN(\mathfrak{f}))^r,$$

it suffices to show that

$$(W\Psi_E\colon WE_{\mathrm{mod}}(H)^{12hwN(\mathfrak{f})}) = \frac{w(H)}{w}\frac{h}{[K(1) : H_{(1)}]}.$$

For this, we refine our description of $E_{\mathrm{mod}}(H)^{12hwN(\mathfrak{f})}$.

Theorem 4.2. *Let $\mathfrak{p}$ be a prime not dividing $w(H_{(1)})$. Then $E_{\mathrm{mod}}(H)^{12hwN(\mathfrak{f})}$ is the group of all products*

4.2.0 $$\prod \psi_H(\mathfrak{f}, C)^{n_{\mathfrak{f}}(C)} \cdot \prod \psi_H((1), c)^{n_{(1)}(c)}$$

satisfying the following five conditions (except that when $K = \mathbf{Q}(\sqrt{-1})$ *or* $\mathbf{Q}(\sqrt{-3})$, *only the first two are to be satisfied*):

4.2.1 $\sum n_{\mathfrak{f}}(C) = 0$

4.2.2 $\sum n_{\mathfrak{f}}(C)\mathbf{N}\mathfrak{a}(C) \equiv 0 \bmod w(H)/w(H_{(1)})$

4.2.3 $\sum n_{(1)}(c) = 0$

4.2.4 $\prod c^{n_{(1)}(c)} = c_0$

4.2.5 $\mathbf{N}\mathfrak{f} \sum n_{(1)}(c)\mathbf{N}\mathfrak{a}(c) \equiv w[K(\mathfrak{f}) : HK(1)] \sum n_{\mathfrak{f}}(C)\mathbf{N}\mathfrak{a}(C) \bmod ww\,(H_{(1)})$.

Here $\mathfrak{a}(c)$ is any $H_{(1)}$-admissible ideal in c, and as usual this is to apply only to the case when K is not one of the two exceptional fields. Furthermore $\mathfrak{a}(C)$ is any ideal in C prime to $w(H)$.

Since the proof of Lemma 4.2 is rather technical, we first show that the lemma implies Theorem 4.1.

Lemma 4.3. *The distribution relations* $\mathbf{DR}(H, c_1, c_2)$ *of* §3 *satisfy conditions* 4.2.1 *through* 4.2.5.

Proof. Consider the relation

$$\sum (C) - \sum (C') - ((c) - (c') - (cc(\mathfrak{p})^{-1}) + (c'c(\mathfrak{p})^{-1})).$$

Conditions 4.2.1 and 4.2.3 are obvious, as is 4.2.4 since

$$\frac{c \cdot c'c(\mathfrak{p})^{-1}}{c' \cdot cc(\mathfrak{p})^{-1}} = c_0.$$

As for 4.2.2, let ζ be a $w(H)$-th root of unity. Then letting $\zeta^x = \zeta[x]$ for typographical reasons, we see that

$$\zeta[\sum \mathbf{N}\mathfrak{a}(C) - \sum \mathbf{N}\mathfrak{a}(C')] = \mathbf{N}_{H/H_{(1)}}(\zeta[\mathbf{N}\mathfrak{a}(c) - \mathbf{N}\mathfrak{a}(c')])$$

is a $w(H_{(1)})$-th root of unity. Thus

$$\frac{w(H)}{w(H_{(1)})} \quad \text{divides} \quad \sum \mathbf{N}\mathfrak{a}(C) - \sum \mathbf{N}\mathfrak{a}(C'),$$

which is 4.2.2.

Finally, consider 4.2.5. Let $\mathfrak{p}' \in c(\mathfrak{p})^{-1}$ be $H_{(1)}$-admissible. The left hand side of the congruence in question is

$$\begin{aligned}&\mathbf{N}\mathfrak{f}(\mathbf{N}\mathfrak{a}(c) - \mathbf{N}\mathfrak{a}(c') - \mathbf{N}\mathfrak{a}(c)\mathbf{N}\mathfrak{p}' + \mathbf{N}\mathfrak{a}(c')\mathbf{N}\mathfrak{p}')\\ &\quad = \mathbf{N}\mathfrak{f}(\mathbf{N}\mathfrak{a}(c) - \mathbf{N}\mathfrak{a}(c'))(1 - \mathbf{N}\mathfrak{p}').\end{aligned}$$

The right hand side is

$$\begin{aligned}&w[K(\mathfrak{f}) : HK(1)](\textstyle\sum \mathbf{N}\mathfrak{a}(C) - \sum \mathbf{N}\mathfrak{a}(C'))\\ &\quad = w[K(\mathfrak{f}) : HK(1)](\mathbf{N}\mathfrak{a}(c) - \mathbf{N}\mathfrak{a}(c')) \displaystyle\sum_{C|C_0} \mathbf{N}\mathfrak{a}(C).\end{aligned}$$

Now

$$\begin{aligned}\sum_{C|c_0} \mathbf{N}\mathfrak{a}(C) &\equiv [H : H_{(1)}] \bmod w(H_{(1)})\\ &= [HK(1) : K(1)],\end{aligned}$$

so modulo $ww(H_{(1)})$ the right hand side becomes

$$\begin{aligned}&\equiv w[K(\mathfrak{f}) : HK(1)][HK(1) : K(1)](\mathbf{N}\mathfrak{a}(c) - \mathbf{N}\mathfrak{a}(c'))\\ &= w[K(\mathfrak{f}) : K(1)](\mathbf{N}\mathfrak{a}(c) - \mathbf{N}\mathfrak{a}(c'))\\ &= \mathbf{N}\mathfrak{p}^{n-1}(\mathbf{N}\mathfrak{p} - 1)(\mathbf{N}\mathfrak{a}(c) - \mathbf{N}\mathfrak{a}(c')).\end{aligned}$$

Since $\mathfrak{p}\mathfrak{p}'$ is in the unit class of $\mathrm{Cl}(H_{(1)})$, it follows that

$$1 - \mathbf{N}(\mathfrak{p}\mathfrak{p}') \equiv 0 \bmod w(H_{(1)}).$$

Also

$$\mathbf{N}\mathfrak{a}(c) - \mathbf{N}\mathfrak{a}(c') \equiv 0 \bmod w.$$

Thus condition 4.2.5 is also satisfied. This completes the proof of Lemma 4.3.

Lemma 4.4. $$(\Psi_E : E_{\mathrm{mod}}(H)^{12hwN(\mathfrak{f})}) = \frac{w(H)}{w}\frac{h}{[HK(1) : H]}.$$

Proof. This will follow routinely from Lemmas 4.2 and 4.3. When $K \neq \mathbf{Q}(\sqrt{-1}), \mathbf{Q}(\sqrt{-3})$, we define a homomorphism

$$\Psi_E \to \mathbf{Z}\Big/\frac{w(H)}{w(H_{(1)})}\mathbf{Z} \times \mathrm{Cl}(H_{(1)}) \times \mathbf{Z}/ww(H_{(1)})\mathbf{Z}$$

by mapping an element

$$\prod \psi_H(f, C)^{n_{\mathfrak{f}}(C)} \cdot \prod \psi_H((1), c)^{n_{(1)}(c)}$$

with $\sum n_{\mathfrak{f}}(C) = \sum n_{(1)}(c) = 0$, to the triple

$$(x, y, z)$$

where:

$$x = \sum n_{\mathfrak{f}}(C)\mathbf{N}\mathfrak{a}(C) \bmod w(H)/w(H_{(1)})$$

$$y = \prod c^{n_{(1)}(c)}$$

$$z = \mathbf{N}\mathfrak{f} \sum n_{(1)}(c)\mathbf{N}\mathfrak{a}(c) - w[K(\mathfrak{f}) : HK(1)] \sum n_{\mathfrak{f}}(C)\mathbf{N}\mathfrak{a}(C) \bmod ww(H_{(1)}).$$

This is well defined by Lemma 4.3 and Theorem 3.1; and its kernel is $E_{\text{mod}}(H)^{12hwN(\mathfrak{f})}$ by Lemma 4.2. It only remains to compute the size of its image.

We claim that the image is exactly

$$\mathbf{Z}\Big/\frac{w(H)}{w(H_{(1)})}\mathbf{Z} \times \mathrm{Cl}(H_{(1)}) \times w^2\mathbf{Z}/ww(H_{(1)})\mathbf{Z},$$

and so has cardinality

$$\frac{w(H)}{w(H_{(1)})} \cdot \frac{h}{[HK(1) : H]} \cdot \frac{ww(H_{(1)})}{w},$$

as desired. To see this, note first that since $w(H)/w(H_{(1)})$ is prime to w, we can certainly choose $n_{\mathfrak{f}}(C)$, $\mathfrak{a}(C)$ so that

$$\sum n_{\mathfrak{f}}(C) = 0 \quad \text{and} \quad \sum n_{\mathfrak{f}}(C)\mathbf{N}\mathfrak{a}(C) \equiv 1 \bmod w(H)/w(H_{(1)}).$$

Further, $w[K(\mathfrak{f}) : HK(1)] \sum n_{\mathfrak{f}}(C)\mathbf{N}\mathfrak{a}(C)$ is divisible by w^2, since

$$\sum n_{\mathfrak{f}}(C)\mathbf{N}\mathfrak{a}(C) \equiv \sum n_{\mathfrak{f}}(C) = 0 \bmod w.$$

It now follows as in the proof of Lemma 5.2 of Chapter 9 that the image of our map is as claimed.

When $K = \mathbf{Q}(\sqrt{-1})$ or $\mathbf{Q}(\sqrt{-3})$, then $w(H_{(1)}) = w$, one has to deal only with the first two conditions, and the proof is similar but easier, using a homomorphism

$$\Psi_E \to \mathbf{Z}/(w(H)/w)\mathbf{Z}.$$

We leave the details to the reader.

As noted at the beginning of this section, Lemma 4.4 provides the last piece in the computation of the index

$$(E_H : E_{\mathrm{mod}}(H)),$$

and thus completes the proof of Theorem 4.1.

The case when $\mathfrak{p}$ may divide $w(H_{(1)})$ and the composite case will be treated in [Ke 2].

For the rest of this chapter we assume $\mathfrak{p}$ does not divide $w(H_{(1)})$.

§5. More Roots of Unity Lemmas

For the proof of Theorem 4.2, we need more roots of unity lemmas. For convenience, we shall introduce the following notation. Let $F \supset F'$ be two abelian extensions of K, with class groups $\mathrm{Cl}(F)$ and $\mathrm{Cl}(F')$ respectively. Then we define

$$v(F, F') = \sum_{\substack{C \in \mathrm{Cl}(F) \\ C | C_0(F')}} \mathbf{N}\mathfrak{a}(C) \bmod w(F),$$

where $C_0(F')$ is the unit class of $\mathrm{Cl}(F')$, and $\mathfrak{a}(C)$ is any ideal in C, prime to $w(F)$.

It is immediate that the following **multiplicativity relation** holds for $F \supset F' \supset F''$:

$$v(F, F')v(F', F'') \equiv v(F, F'') \bmod w(F).$$

Of course, each factor on the left is defined only mod $w(F)$ and mod $w(F')$ respectively, but the relation is to be interpreted in the sense that if we substitute any integers satisfying these congruences, then the product will be well defined mod $w(F)$ and will satisfy the desired relation.

Remark. The integer $v(F, F')$ has the interpretation that if $\zeta \in \boldsymbol{\mu}(F)$ then

$$\zeta^{v(F,F')} = \mathbf{N}_{F/F'}(\zeta).$$

In particular, we have:

Lemma 5.1. *$w(F)/w(F')$ divides $v(F, F')$, and*

$$v(F, F') \equiv [F : F'] \bmod w(F').$$

Next we consider extensions corresponding to those used in the application to this chapter.

Lemma 5.2. *Let $1 \leqq m' \leqq m$ and let $\mathfrak{p}$ be a prime, $\mathfrak{p} \nmid w_{K(1)}$. Let $\mathfrak{q} = \mathfrak{p}^m$ and $\mathfrak{q}' = \mathfrak{p}^{m'}$. Then*

5.2.1 $$v(H_{\mathfrak{q}}, H_{\mathfrak{q}'}) \equiv [H_{\mathfrak{q}} : H_{\mathfrak{q}'}] \bmod w(H_{\mathfrak{q}})$$

5.2.2 $$[K(\mathfrak{q}) : H_{\mathfrak{q}}K(\mathfrak{q}')]v(H_{\mathfrak{q}}, H_{\mathfrak{q}'}) \equiv \mathbf{N}(\mathfrak{q}/\mathfrak{q}') \bmod w(H_{\mathfrak{q}}).$$

Proof. Since

$$[K(\mathfrak{q}) : H_{\mathfrak{q}}K(\mathfrak{q}')][H_{\mathfrak{q}} : H_{\mathfrak{q}'}] = [K(\mathfrak{q}) : K(\mathfrak{q}')] = \mathbf{N}(\mathfrak{q}/\mathfrak{q}'),$$

it follows that 5.2.1 implies 5.2.2. Also by the remark, 5.2.1 is equivalent to

5.2.1′ $$\mathbf{N}_{H_{\mathfrak{q}}/H_{\mathfrak{q}'}}(\zeta) = \zeta^{[H_{\mathfrak{q}} : H_{\mathfrak{q}'}]} \quad \text{for } \zeta \in \boldsymbol{\mu}(H_{\mathfrak{q}}).$$

Note that by the previous lemma, this is obvious if $\zeta \in \boldsymbol{\mu}(H_{\mathfrak{q}'})$. Also it suffices to prove 5.2.1′ when $m' = m - 1$, that is $\mathfrak{q} = \mathfrak{q}'\mathfrak{p}$, since the general case then follows by induction. Thus suppose $\mathfrak{q} = \mathfrak{p}\mathfrak{q}'$. Then $w(H_{\mathfrak{q}})/w(H_{\mathfrak{q}'})$ is either 1 or p, and in the former case we are done, so suppose it equals p.

Let

$$p^a \| w(H_{\mathfrak{q}'}) \quad \text{so} \quad p^{a+1} \| w(H_{\mathfrak{q}}).$$

We note that $a \geqq 1$, that is $\boldsymbol{\mu}_p \subset H_{\mathfrak{q}'}$. Indeed, we have the diagram

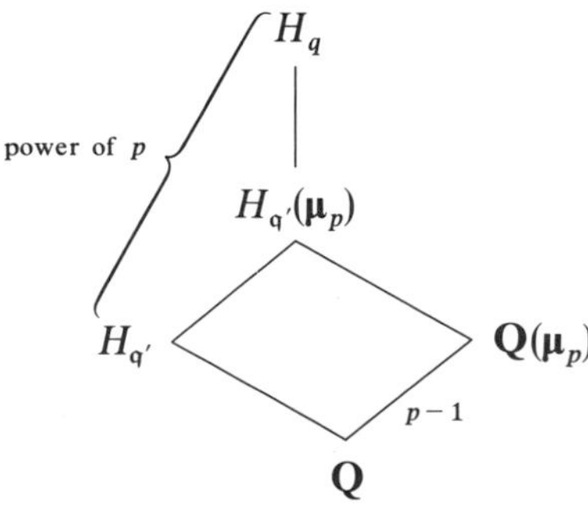

so $[H_{\mathfrak{q}'}(\boldsymbol{\mu}_p) : H_{\mathfrak{q}'}] = 1$, that is $\boldsymbol{\mu}_p \subset H_{\mathfrak{q}'}$.

Now consider the diagram

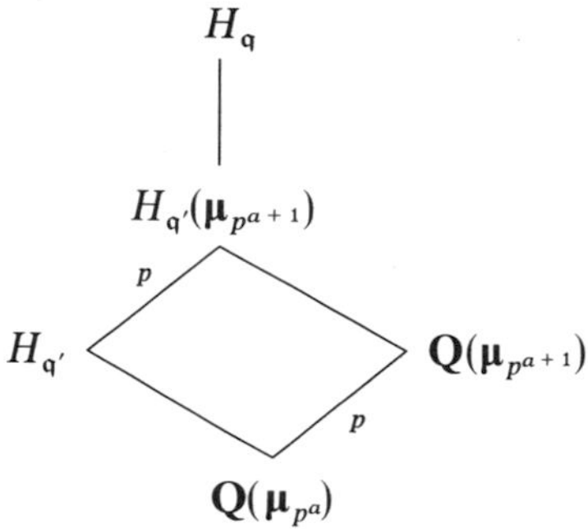

Then for $\zeta \in \boldsymbol{\mu}_{p^{a+1}}$ we have

$$\begin{aligned}\mathbf{N}_{H_{\mathfrak{q}}/H_{\mathfrak{q}'}}(\zeta) &= \mathbf{N}_{\mathbf{Q}(\boldsymbol{\mu}_{p^{a+1}})/\mathbf{Q}(\boldsymbol{\mu}_{p^a})}(\zeta^x) \quad \text{where } x = [H_{\mathfrak{q}} : H_{\mathfrak{q}'}(\boldsymbol{\mu}_{p^{a+1}})]\\ &= \zeta^{px}\\ &= \zeta^y \quad \text{where } y = [H_{\mathfrak{q}} : H_{\mathfrak{q}'}]\end{aligned}$$

as desired. Here we have used

$$\begin{aligned}\mathbf{N}_{\mathbf{Q}(\boldsymbol{\mu}_{p^{a+1}})/\mathbf{Q}(\boldsymbol{\mu}_{p^a})}(\zeta) &= \zeta^{1+(p^a+1)+\cdots+((p-1)p^a+1)}\\ &= \zeta^{p+p^a p(p-1)/2} = \zeta^p\end{aligned}$$

(p is assumed odd). This concludes the proof of Lemma 5.2.

§6. Proof of Theorem 4.2

For the convenience of the reader, we repeat the statement of Theorem 4.2, which occurred rather far back.

Theorem 4.2. *Assume that* $\mathfrak{p}$ *does not divide* $w(H_{(1)})$. *Then* $E_{\text{mod}}(H)^{12hwN(\mathfrak{f})}$ *is the group of all products*

4.2.0 $$\prod \psi_H(\mathfrak{f}, C)^{n_{\mathfrak{f}}(C)} \cdot \prod \psi_H((1), c)^{n_{(1)}(c)}$$

satisfying the following five conditions (except that when $K = \mathbf{Q}(\sqrt{-1})$ *or* $\mathbf{Q}(\sqrt{-3})$, *only the first two are to be satisfied):*

4.2.1 $$\sum n_{\mathfrak{f}}(C) = 0$$

4.2.2 $$\sum n_{\mathfrak{f}}(C)\mathbf{N}\mathfrak{a}(C) \equiv 0 \bmod w(H)/w(H_{(1)})$$

4.2.3 $$\sum n_{(1)}(c) = 0$$

4.2.4 $$\prod c^{n_{(1)}(c)} = c_0$$

4.2.5 $$\mathbf{N}\mathfrak{f} \sum n_{(1)}(c)\mathbf{N}\mathfrak{a}(c) \equiv w[K(\mathfrak{f}) : HK(1)] \sum n_{\mathfrak{f}}(C)\mathbf{N}\mathfrak{a}(C) \bmod ww(H_{(1)}).$$

Here $\mathfrak{a}(C)$ is any ideal in C prime to $w(H)$, and $\mathfrak{a}(c)$ is any $H_{(1)}$-admissible ideal in c. As usual, this last condition is to apply on to the case when K is not one of the two exceptional fields.

The rest of this section is devoted to the proof of Theorem 4.2. First we show that any element of $E_{\text{mod}}(H)^{12hwN(\mathfrak{f})}$ can be written in the desired form.

In Chapter 12, §4 we defined

$$E_{\text{mod}}(H) = \prod_{\mathfrak{g}|\mathfrak{f}} \mathbf{N}_{K(\mathfrak{g})/H_{\mathfrak{g}}} E_{\text{mod}}(K(\mathfrak{g})).$$

In theorem 3.1 of Chapter 12, we showed that

$$E_{\text{mod}}(H_{(1)})^{12hw} = \Delta_{H_{(1)}}(c_0, ww_{H_{(1)}}).$$

If we further raise this expression to the $N(\mathfrak{f})$-power, then conditions (1), (2), (3) in the definition of $\Delta_{H_{(1)}}(c_0, ww_{H_{(1)}})$ become conditions 4.2.3 through 4.2.5 of Theorem 4.2. Note that we use here $\mathbf{N}\mathfrak{f}$ prime to $w(H_{(1)})$ to assert the equivalence of (3) and 4.2.5. Conditions 4.2.1 and 4.2.2 are satisfied vacuously.

Now consider

$$\mathbf{N}_{K(\mathfrak{q})/H_{\mathfrak{q}}} E_{\text{mod}}(K(\mathfrak{q}))^{12hwN(\mathfrak{f})}$$

where $\mathfrak{q} = \mathfrak{p}^m$ and $1 \leqq m \leqq n$. We use the expression of Theorem 5.3 in Chapter 12, for the $12hwN(\mathfrak{q})$-power of an element $\beta \in E_{\text{mod}} K(\mathfrak{q})$. Note that for $\mathfrak{q}' | \mathfrak{q}$, $\mathfrak{q}' \neq (1)$, we have

$$\begin{aligned}\mathbf{N}_{K(\mathfrak{q})/H_{\mathfrak{q}}} \psi_{K(\mathfrak{q})}(\mathfrak{q}', C) &= \mathbf{N}_{K(\mathfrak{q}')/H_{\mathfrak{q}'}} \psi_{K(\mathfrak{q})}(\mathfrak{q}', C)^{[K(\mathfrak{q}) : H_{\mathfrak{q}} K(\mathfrak{q}')]} \\ &= \psi_{H_{\mathfrak{q}}}(\mathfrak{q}', C)^{[K(\mathfrak{q}) : H_{\mathfrak{q}} K(\mathfrak{q}')]}.\end{aligned}$$

The diagram of fields for the calculation is as follows.

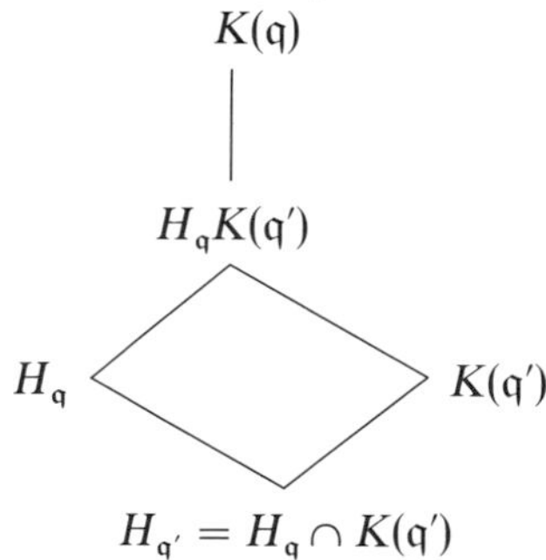

Also the distribution relations show that

$$\begin{aligned}\psi_{H_{\mathfrak{q}}}(\mathfrak{q}', C)^{N(\mathfrak{f})/N(\mathfrak{q})} &= \psi_H(\mathfrak{q}', C)\\ &= \prod_{\substack{C' \in \mathrm{Cl}(H)\\ C'|C}} \psi_H(\mathfrak{f}, C').\end{aligned}$$

Using these two facts, and the decomposition

$$\beta^{12hwN(\mathfrak{q})} = \delta_1\delta_2\delta_3$$

of Theorem 5.3 of Chapter 12, we find that

$$\mathbf{N}_{K(\mathfrak{q})/H_{\mathfrak{q}}}\beta^{12hwN(\mathfrak{f})} = \mathbf{N}_{K(\mathfrak{q})/H_{\mathfrak{q}}}(\delta_1\delta_2\delta_3)^{N(\mathfrak{f})/N(\mathfrak{q})}.$$

The right hand side is then a product over three factors, given by the following expressions.

$$\begin{aligned}&\mathbf{N}_{K(\mathfrak{q})/H_{\mathfrak{q}}}\delta_1^{N(\mathfrak{f})/N(\mathfrak{q})}\\ &= \mathbf{N}_{K(\mathfrak{q})/H_{\mathfrak{q}}} \prod_{\substack{\mathfrak{q}'|\mathfrak{q}\\ \mathfrak{q}'\neq(1)}} \prod_{C\in\mathrm{Cl}(\mathfrak{q}')} \left(\frac{\psi_{K(\mathfrak{q})}(\mathfrak{q}', C)}{\psi_{K(\mathfrak{q})}((1), c(\mathfrak{q}'^{-1}C))^w}\right)^{n_{\mathfrak{q}'}(C)N(\mathfrak{f})/N(\mathfrak{q})}\\ &= \prod_{\substack{\mathfrak{q}'|\mathfrak{q}\\ \mathfrak{q}'\neq(1)}} \prod_{C\in\mathrm{Cl}(\mathfrak{q}')} \left(\frac{\psi_{H_{\mathfrak{q}}}(\mathfrak{q}', C)^{[K(\mathfrak{q}):H_{\mathfrak{q}}K(\mathfrak{q}')]}}{\psi_{H_{\mathfrak{q}}}((1), c(\mathfrak{q}'^{-1}C))^{w[K(\mathfrak{q}):H_{\mathfrak{q}}K(1)]}}\right)^{n_{\mathfrak{q}'}(C)N(\mathfrak{f})/N(\mathfrak{q})}\end{aligned}$$

so finally

$$\text{(1)}\quad \mathbf{N}_{K(\mathfrak{q})/H_{\mathfrak{q}}}\delta_1^{N(\mathfrak{f})/N(\mathfrak{q})} = \prod_{\substack{\mathfrak{q}'|\mathfrak{q}\\ \mathfrak{q}'\neq(1)}} \prod_{C\in\mathrm{Cl}(H_{\mathfrak{q}'})} \left(\frac{\displaystyle\prod_{\substack{C'\in\mathrm{Cl}(H)\\ C'|C}} \psi_H(\mathfrak{f}, C')^{[K(\mathfrak{q}):H_{\mathfrak{q}}K(\mathfrak{q}')]}}{\psi_H((1), c(\mathfrak{q}'^{-1}))^{w[K(\mathfrak{q}):H_{\mathfrak{q}}K(1)]}}\right)^{n_{\mathfrak{q}'}(C)}$$

where for $C \in \mathrm{Cl}(H_{\mathfrak{q}'})$,

$$n_{\mathfrak{q}'}(C) = \sum_{\substack{C''\in\mathrm{Cl}(K(\mathfrak{q}'))\\ C''|C}} n_{\mathfrak{q}'}(C'').$$

Similarly,

$$\text{(2)}\qquad \mathbf{N}_{K(\mathfrak{q})/H_{\mathfrak{q}}}\delta_2^{N(\mathfrak{f})/(\mathfrak{q})} = \left(\frac{\displaystyle\prod_{\substack{C\in\mathrm{Cl}(H)\\ C|c_0 \text{ in } \mathrm{Cl}(H_{(1)})}} \psi_H(\mathfrak{f}, C)}{\psi_H((1), c_0)/\psi_H((1), c(\mathfrak{p})^{-1})}\right)^{wk_{\mathfrak{q}}[K(\mathfrak{q}):HK(1)]}$$

and

$$(3)\quad \mathbf{N}_{K(\mathfrak{q})/H_{\mathfrak{q}}}\delta_3^{N(\mathfrak{f})/N(\mathfrak{q})} = \prod_{c\in \mathrm{Cl}(H_{(1)})}\prod_{\mathfrak{p}_i}\left(\frac{\psi_H((1),c)}{\psi_H((1),cc(\mathfrak{p}_i)^{-1}))^{\mathbf{N}\mathfrak{p}_i}}\right)^{n_{\mathfrak{p}_i}(c)[K(\mathfrak{q}):HK(1)]}$$

where $\{\mathfrak{p}_i\}$ ranges over some finite set of primes, and for $c \in \mathrm{Cl}(H_{(1)})$,

$$n_{\mathfrak{p}_i}(c) = \sum_{\substack{c'\in \mathrm{Cl}(1)\\ c'|c}} n_{\mathfrak{p}_i}(c').$$

We have thus written $\mathbf{N}_{K(\mathfrak{q})/H_{\mathfrak{q}}}\beta^{12hwN(\mathfrak{f})}$ in the form 4.2.0, so it now suffices to prove that these products satisfy 4.2.1 through 4.2.5.

4.2.1 is clear, since β is a unit.

4.2.3 and 4.2.4 follow immediately from the corresponding properties in Theorem 5.3 of Chapter 12.

4.2.2: We use property 5.3.2 from Theorem 5.3 of Chapter 12. Since this congruence held for the δ_1 and δ_2 pieces separately (see the proof of Theorem 5.3, Chapter 12) we may again consider the δ_1 and δ_2 pieces separately. For δ_1 we are given

$$\sum_{\substack{\mathfrak{q}'|\mathfrak{q}\\ \mathfrak{q}'\neq(1)}} \mathbf{N}(\mathfrak{q}/\mathfrak{q}') \sum_{C\in \mathrm{Cl}(\mathfrak{q}')} n_{\mathfrak{q}'}(C)\mathbf{N}\mathfrak{a}(C) \equiv 0 \bmod w_{K(\mathfrak{q})}/w_{K(1)}.$$

Note that

$$w(H_{\mathfrak{q}})/w(H_{(1)}) \quad \text{divides} \quad w(K(\mathfrak{q}))/w(K(1)).$$

If for convenience we write

$$n(\mathfrak{q}') = \sum_{C\in \mathrm{Cl}(\mathfrak{q}')} n_{\mathfrak{q}'}(C)\mathbf{N}\mathfrak{a}(C),$$

then we have

$$\sum_{\substack{\mathfrak{q}'|\mathfrak{q}\\ \mathfrak{q}'\neq(1)}} \mathbf{N}(\mathfrak{q}/\mathfrak{q}')n(\mathfrak{q}') \equiv 0 \bmod w(H_{\mathfrak{q}})/w(H_{(1)}).$$

In terms of the $v(F, F')$ notation of the preceding section, the δ_1-part of the desired congruence reads

$$\sum_{\substack{\mathfrak{q}'|\mathfrak{q}\\ \mathfrak{q}'\neq(1)}} [K(\mathfrak{q}): H_{\mathfrak{q}}K(\mathfrak{q}')] \sum_{C\in \mathrm{Cl}(H_{\mathfrak{q}'})} n_{\mathfrak{q}'}(C)\mathbf{N}\mathfrak{a}(C)v(H, H_{\mathfrak{q}'}) \equiv 0 \bmod w(H)/w(H_{(1)}).$$

Remark. The integer $v(H, H_{\mathfrak{q}})$ is of course defined at first mod $w(H)$, but the congruence on the right is weaker so it makes sense in light of the previous definition.

By the multiplicativity relation

$$v(H, H_{\mathfrak{q}'}) \equiv v(H, H_{\mathfrak{q}})v(H_{\mathfrak{q}}, H_{\mathfrak{q}'}) \bmod w(H),$$

and the fact that

$$n(\mathfrak{q}') \equiv \sum_{C \in \mathrm{Cl}(H_{\mathfrak{q}'})} n_{\mathfrak{q}'}(C)\mathbf{N}\mathfrak{a}(C) \bmod w(H_{\mathfrak{q}})/w(H_{(1)}),$$

the desired congruence may be written more simply as

$$v(H, H_{\mathfrak{q}}) \sum_{\substack{\mathfrak{q}' \mid \mathfrak{q} \\ \mathfrak{q}' \neq (1)}} [K(\mathfrak{q}) : H_{\mathfrak{q}}K(\mathfrak{q}')]v(H_{\mathfrak{q}}, H_{\mathfrak{q}'})n(\mathfrak{q}') \equiv 0 \bmod w(H)/w(H_{(1)}).$$

By Lemma 5.2.1 we know that

$$v(H, H_{\mathfrak{q}}) \equiv 0 \bmod w(H)/w(H_{\mathfrak{q}}),$$

and by Lemma 5.2.2 we have

$$\begin{aligned}\sum_{\substack{\mathfrak{q}' \mid \mathfrak{q} \\ \mathfrak{q}' \neq (1)}} [K(\mathfrak{q}) : H_{\mathfrak{q}}K(\mathfrak{q}')]v(H_{\mathfrak{q}}, H_{\mathfrak{q}'})n(\mathfrak{q}') &\equiv \sum_{\substack{\mathfrak{q}' \mid \mathfrak{q} \\ \mathfrak{q}' \neq (1)}} \mathbf{N}(\mathfrak{q}/\mathfrak{q}')n(\mathfrak{q}') \\ &\equiv 0 \bmod w(H_{\mathfrak{q}})/w(H_{(1)}).\end{aligned}$$

This proves that the δ_1-part of the desired congruence holds.

The δ_2-part trivially satisfies 4.2.2, since the sum in question is just

$$wk_{\mathfrak{q}}[K(\mathfrak{q}) : HK(1)]v(H, H_{(1)}).$$

Finally, we consider 4.2.5. Again, we consider each factor separately. For a factor

$$\frac{\prod\limits_{\substack{C' \in \mathrm{Cl}(H) \\ C' \mid C \text{ in } \mathrm{Cl}(H_{\mathfrak{q}'})}} \psi_H(\mathfrak{f}, C')^{[K(\mathfrak{q}) : H_{\mathfrak{q}}K(\mathfrak{q}')]}}{\psi_H((1), c(\mathfrak{q}'^{-1}C))^{w[K(\mathfrak{q}) : H_{\mathfrak{q}}K(1)]}}$$

arising from δ_1, the left hand side of 4.2.5 is

$$w[K(\mathfrak{q}) : H_{\mathfrak{q}}K(1)] \cdot \mathbf{N}\mathfrak{f} \cdot \mathbf{N}\mathfrak{q}'' \cdot \mathbf{N}\mathfrak{a}(c),$$

where $\mathfrak{q}''$ lies in the class of $\mathfrak{q}'^{-1}$ in $\mathrm{Cl}(\mathrm{H}_{(1)})$. Because of the factor w, we need not bother choosing $\mathfrak{q}''$ or $\mathfrak{a}(c)$ to be $H_{(1)}$-admissible.

The right hand side of 4.2.5 is

$$w[K(\mathfrak{f}):HK(1)][K(\mathfrak{q}):H_{\mathfrak{q}}K(\mathfrak{q}')]v(H,H_{\mathfrak{q}'})\mathbf{N}\mathfrak{a}(C).$$

But by Lemma 5.2.1 we deduce

$$v(H,H_{\mathfrak{q}'}) \equiv [H:H_{\mathfrak{q}'}] \bmod w(H_{(1)}),$$

so it suffices to show that mod $w(H_{(1)})$

$$[K(\mathfrak{q}):H_{\mathfrak{q}}K(1)]\mathbf{N}\mathfrak{f}\cdot\mathbf{N}\mathfrak{q}'' \equiv [K(\mathfrak{f}):HK(1)][K(\mathfrak{q}):H_{\mathfrak{q}}K(\mathfrak{q}')][H:H_{\mathfrak{q}'}].$$

It is an easy exercise to see that

$$\begin{aligned}[K(\mathfrak{f}):HK(1)][K(\mathfrak{q}):H_{\mathfrak{q}}K(\mathfrak{q}')][H:H_{\mathfrak{q}'}] &= [K(\mathfrak{f}):K(\mathfrak{q}')][K(\mathfrak{q}):H_{\mathfrak{q}}K(1)]\\ &= \mathbf{N}(\mathfrak{f}/\mathfrak{q}')[K(\mathfrak{q}):H_{\mathfrak{q}}K(1)].\end{aligned}$$

Since $\mathfrak{q}''$ lies in the class of $\mathfrak{q}'^{-1}$ in $\mathrm{Cl}(H_{(1)})$, we get

$$\mathbf{N}(\mathfrak{f}\mathfrak{q}'') \equiv \mathbf{N}(\mathfrak{f}/\mathfrak{q}') \bmod w(H_{(1)}),$$

so we are done.

The easier cases arising from the δ_2 and δ_3 factors are left to the reader.

Thus we have shown that $E_{\mathrm{mod}}(H)^{12hwN(\mathfrak{f})}$ is contained in the group defined by conditions 4.2.1 through 4.2.5.

It remains to check that the reverse inclusion holds. Suppose that the product 4.2.0 satisfies 4.2.1 through 4.2.5. Pick $\mathfrak{a}(C) \in C$ prime to $w(H)$, and let

$$\sum_{C\in\mathrm{Cl}(H)} n_{\mathfrak{f}}(C)\mathbf{N}\mathfrak{a}(C) = a\cdot w(H)/w(H_{(1)}).$$

Note that a is divisible by w, since $w(H)/w(H_{(1)})$ is prime to $w(H_{(1)})$ and

$$\sum n_{\mathfrak{f}}(C)\mathbf{N}\mathfrak{a}(C) \equiv \sum n_{\mathfrak{f}}(C) = 0 \bmod w.$$

We have already remarked that

$$w(H)/w(H_{(1)}) = \text{g.c.d.}(N(\mathfrak{f}), w(H)),$$

so we may write

$$w(H)/w(H_{(1)}) = b_1 w(H) + b_2 N(\mathfrak{f}) \quad \text{with } b_i \in \mathbf{Z}.$$

Next pick ideals $\mathfrak{a}_i$ in the unit class of $\mathrm{Cl}(H)$ and integers m_i such that

$$\sum_i m_i(\mathbf{N}\mathfrak{a}_i - 1) = w(H).$$

Then let

$$\mathbf{a} = \sum n_{\mathfrak{f}}(C)\mathfrak{a}(C) - ab_1 \sum m_i(\mathfrak{a}_i - \mathfrak{o}).$$

We verify that $\mathbf{a} \in \mathbf{I}_0(\mathfrak{f})$. Indeed, $\mathbf{a}$ has degree 0 because

$$\sum n_{\mathfrak{f}}(C) + \sum n_i(1 - 1) = 0$$

using 4.2.1. For $\mathbf{I}_{\mathfrak{f}}\,\mathbf{2}$, we have

$$\begin{aligned}\sum n_{\mathfrak{f}}(C)\mathbf{N}\mathfrak{a}(C) - ab_1 \sum m_i(\mathbf{N}\mathfrak{a}_i - 1) &= aw(H)/w(H_{(1)}) - ab_1 w(H) \\ &= ab_2 N(\mathfrak{f}),\end{aligned}$$

and a is even, so $\mathbf{I}_{\mathfrak{f}}\,\mathbf{2}$ is satisfied. As for $\mathbf{I}_{\mathfrak{f}}\,\mathbf{3}$, all ideals occurring in $\mathbf{a}$ are odd, and the total degree is 0, so $\mathbf{I}_{\mathfrak{f}}\,\mathbf{3}$ is trivially satisfied. Thus

$$\mathfrak{k}_{\mathfrak{f}}(\mathbf{a}; \mathfrak{o}) \in \mathfrak{R}(\mathfrak{f}).$$

Lemma 6.1. *With the above choice of* $\mathbf{a}$, *there exists* $\beta \in \mathfrak{R}(\mathfrak{o})$ *and integers* $n'_{(1)}(c)$ *for* $c \in \mathrm{Cl}(H_{(1)})$ *such that*

$$\mathfrak{k}_{\mathfrak{f}}(\mathbf{a}; \mathfrak{o})\beta^{-1} \in E_{\mathrm{mod}}(K(\mathfrak{f})),$$

and

$$\mathbf{N}_{K(\mathfrak{f})/H}(\mathfrak{k}_{\mathfrak{f}}(\mathbf{a}; \mathfrak{o})\beta^{-1})^{12hwN(\mathfrak{f})} = \prod_C \psi_H(\mathfrak{f}, C)^{n_{\mathfrak{f}}(C)} \prod_c \psi_H((1), c)^{n'_{(1)}(c)}$$

with the same exponents $n_{\mathfrak{f}}(C)$ *as in* 4.2.0.

Proof. We repeat the calculations of §5, Chapter 12 and of the present section for $\mathfrak{k}_{\mathfrak{f}}(\mathbf{a}; \mathfrak{o})$. For $C \in \mathrm{Cl}(H)$ we let C' denote the class of $\mathfrak{a}(C)$ in $\mathrm{Cl}(\mathfrak{f})$. We have

$$\mathfrak{k}_{\mathfrak{f}}(\mathbf{a}; \mathfrak{o})^{12} = \prod_{C \in \mathrm{Cl}(H)} \left(\frac{g^{12}(1, \mathfrak{f}\mathfrak{a}(C)^{-1})}{\Delta(\mathfrak{f}\mathfrak{a}(C)^{-1})}\right)^{n_{\mathfrak{f}}(C)} \prod_i \left(\frac{g^{12}(1, \mathfrak{f}\mathfrak{a}_i^{-1})/\Delta(\mathfrak{f}\mathfrak{a}_i^{-1})}{g^{12}(1, \mathfrak{f})/\Delta(\mathfrak{f})}\right)^{m_i}$$

so

$$\mathfrak{k}_{\mathfrak{f}}(\mathbf{a}; \mathfrak{o})^{12hwN(\mathfrak{f})} = \prod_{C \in \mathrm{Cl}(H)} \psi_{K(\mathfrak{f})}(\mathfrak{f}, C')^{n_{\mathfrak{f}}(C)} \cdot \delta^{hwN(\mathfrak{f})}$$

where δ is a product of Δ-factors, say

$$\delta = \prod_{\mathfrak{b}} \Delta(\mathfrak{b})^{n(\mathfrak{b})}$$

for convenience. Here

$$\sum n(\mathfrak{b}) = \sum n_{\mathfrak{f}}(C) + \sum m_i(1 - 1) = 0,$$

so we may rewrite

$$\delta^h = \prod_{\mathfrak{b}} g((1), c(\mathfrak{b}))^{n(\mathfrak{b})} \cdot \gamma^{12}$$

where $\gamma \in K$ is a generator for the fractional ideal

$$(\gamma) = \prod \mathfrak{b}^{n(\mathfrak{b})h}.$$

On the other hand, by Corollary 3.12 of Chapter 12, there exists an element $\beta \in \mathfrak{R}(\mathfrak{o})$ such that

$$(\beta) = \prod_{\mathfrak{b}} \mathfrak{b}^{n(\mathfrak{b})},$$

If

$$\beta^{12wh} = \Delta(\mathbf{b}_1) \quad \text{with } \mathbf{b}_1 = \sum m(\mathfrak{b})\mathfrak{b}$$

as in Chapter 12, Theorem 3.9, then

$$\beta^{12wh} = \prod_{\mathfrak{b}} g((1), c(\mathfrak{b}))^{m(\mathfrak{b})} \cdot \gamma_1^{12}$$

with an element $\gamma_1 \in K$ such that

$$(\gamma_1) = \prod_{\mathfrak{b}} \mathfrak{b}^{m(\mathfrak{b})h}.$$

Since the $g((1), c(\mathfrak{b}))$ are units, we have

$$\gamma_1^{12} \sim \beta^{12wh} \sim (\prod \mathfrak{b}^{n(\mathfrak{b})})^{12wh} \sim \gamma^{12w},$$

so $(\gamma_1/\gamma^w)^{12}$ is a unit, and lies in K. Thus $\gamma_1^{12} = \gamma^{12w}$, and

$$(\mathfrak{k}_{\mathfrak{f}}(\mathbf{a}; \mathfrak{o})\beta^{-1})^{12hwN(\mathfrak{f})} = \prod_{C \in \mathrm{Cl}(H)} \psi_{K(\mathfrak{f})}(\mathfrak{f}, C')^{n_{\mathfrak{f}}(C)} \prod_{\mathfrak{b}} \frac{g((1), c(\mathfrak{b}))^{n(\mathfrak{b})wN(\mathfrak{f})}}{g((1), c(\mathfrak{b}))^{m(\mathfrak{b})N(\mathfrak{f})}}.$$

In particular, since $\sum n_{\mathfrak{f}}(C) = 0$, it follows that $\mathfrak{k}_{\mathfrak{f}}(\mathbf{a}; \mathfrak{o})\beta^{-1}$ is a unit, hence lies in $E_{\mathrm{mod}}(K(\mathfrak{f}))$. Since

$$g((1), c(\mathfrak{b}))^{N(\mathfrak{f})} = \psi_{K(\mathfrak{f})}((1), c(\mathfrak{b})),$$

taking norms from $K(\mathfrak{f})$ to H completes the proof of Lemma 6.1.

Let

$$\begin{aligned}\eta &= (4.2.0)/\mathbf{N}_{K(\mathfrak{f})/H}(\mathfrak{k}_{\mathfrak{f}}(\mathbf{a};\mathfrak{o})\beta^{-1})^{12hwN(\mathfrak{f})}\\ &= \prod \psi_H((1),c)^{n_{(1)}(c)-n'_{(1)}(c)}.\end{aligned}$$

By assumption, (4.2.0) satisfies 4.2.1 through 4.2.5. By the first part of the present proof,

$$\mathbf{N}_{K(\mathfrak{f})/H}(\mathfrak{k}_{\mathfrak{f}}(\mathbf{a};\mathfrak{o})\beta^{-1})^{12hwN(\mathfrak{f})}$$

also satisfies 4.2.1 through 4.2.5. Thus η satisfies 4.2.3, 4.2.4, and 4.2.5 with the right hand side of the congruence replaced by 0 since η has no ramified part. By Theorem 3.1,

$$\eta \in E_{\mathrm{mod}}(H_{(1)})^{12hwN(\mathfrak{f})} \subset E_{\mathrm{mod}}(H)^{12hwN(\mathfrak{f})}.$$

Therefore we also get

$$(42.0) = \eta \cdot \mathbf{N}_{K(\mathfrak{f})/H}(\mathfrak{k}_{\mathfrak{f}}(\mathbf{a};\mathfrak{o})\beta^{-1}) \in E_{\mathrm{mod}}(H)^{12hwN(\mathfrak{f})}.$$

This completes the proof of Theorem 4.2.

In the general definition of the Klein group $\mathfrak{K}(\mathfrak{f})$, we needed to take into account intermediate levels. For the present case of prime power $\mathfrak{f} = \mathfrak{p}^n$ with $\mathfrak{p}$ prime to $w(H_{(1)})$, we can limit ourselves to the top level. More precisely:

Theorem 6.2. *The Klein group $\mathfrak{K}(\mathfrak{f})$ is the group of all products*

$$\mathfrak{k}_{\mathfrak{f}}(\mathbf{a}, \mathbf{a}_S, R_S; \mathfrak{c})$$

with $\mathbf{a}$ prime to $\mathfrak{f} = \mathfrak{p}^n$.

Proof. In the proof we have just gone through, the element $\mathbf{a}$ is prime to $\mathfrak{f} = \mathfrak{p}^n$.

Thus the resulting group $E_{\mathrm{mod}}(K(\mathfrak{f}))$ could also have been defined in terms of elements of top level $\mathfrak{f}$. For the same reason, we see that for $\mathfrak{q}\,|\,\mathfrak{f}$,

$$\mathbf{N}_{K(\mathfrak{q})/H_{\mathfrak{q}}} E_{\mathrm{mod}}(K(\mathfrak{q})) \subset \mathbf{N}_{K(\mathfrak{f})/H}(E_{\mathrm{mod}}(K(\mathfrak{f}))\,\mathbf{N}_{K(1)/H_{(1)}}(E_{\mathrm{mod}}(K(1)).$$

Thus in the definition of $E_{\mathrm{mod}}(H)$, we need only take $\mathfrak{q} = \mathfrak{f}$ and $\mathfrak{q} = (1)$. Again, this is true only for a prime power $\mathfrak{f}$ as in the present chapter.

Appendix: The Logarithm of the Siegel Functions

In this appendix, we give the expression found in [Ku 6] for the pure imaginary period of the differential of third kind associated with the logarithm of the Siegel functions. The proof follows that of [Ku 6].

Such periods occurred in a simpler context as the Dedekind sums, for the Dedekind eta function, and were extended to the Siegel functions by Schoeneberg [Sch]. Another way of looking at these was given by Rademacher [Rad] and Kubert [Ku 7]. For purposes of determining certain integral properties and the finer structure of the cuspidal divisor class group, namely the 2-torsion, these previous expressions appeared inadequate, and another one proved more successful.

We shall write matrices σ in $SL_2(\mathbf{Z})$ with the usual notation

$$\sigma = \begin{pmatrix} a & b \\ c & d \end{pmatrix}.$$

We write g_u for the Siegel functions, $u = (u_1, u_2) \in \mathbf{Q}^2$. In fact, we let

$$u \in \frac{1}{N}\mathbf{Z}^2 \quad \text{but } u \notin \mathbf{Z}^2.$$

We let $\sigma \in \Gamma(N)$ so $a \equiv d \equiv 1 \bmod N$ and $b \equiv c \equiv 0 \bmod N$. Then

$$\chi_u(\sigma) = \log g_u(\sigma\tau) - \log g_u(\tau)$$

is equal to $2\pi i$ times a rational number which is independent of which branch of the logarithm of g_u is chosen, and also of τ. To see this, note that there is a single valued branch of $\log g_u(\tau)$ on the upper half plane, since $g_u(\tau)$ has no zeros or poles. Since a power of $g_u(\tau)$ is invariant under $\Gamma(N)$, the difference

$$\log g_u(\sigma\tau) - \log g_u(\tau)$$

must be equal to $2\pi i$ times a rational number. Since the upper half plane is connected, this rational number is independent of τ. We wish to have an explicit expression for $\chi_u(\sigma)$.

If u is changed by adding an integral vector in $\mathbf{Z}^2$, then g_u changes by a root of unity. Then $\log g_u$ changes by a constant. Without loss of generality, we therefore assume that u is **normalized**, that is

$$0 \leqq u_1, u_2 < 1.$$

As usual, we write $\langle t \rangle$ for the normalized representative of a real number mod $\mathbf{Z}$, so we may assume $u_i = \langle u_i \rangle$ for $i = 1, 2$.

Proposition A.1. *If* $c = 0$ *then*

$$\chi_u(\sigma) = 2\pi i b \cdot \operatorname{ord}_{i\infty} g_u(\tau) = 2\pi i b \tfrac{1}{2}\mathbf{B}_2(\langle u_1 \rangle).$$

Proof. We must have $a = 1 = d$, and $b \equiv 0 \bmod N$. The desired result is then an immediate consequence of the residue formula.

For the rest of this section, we assume that $c \neq 0$, and as before,

$$\sigma \equiv id \bmod N.$$

Then

$$\sigma(\tau) = \frac{a\tau + b}{c\tau + d} = \frac{a}{c} - \frac{1}{c^2\tau + cd}.$$

To calculate $\chi_u(\sigma)$ we define for $u \in \mathbf{Q}^2$, $u \notin \mathbf{Z}^2$, u normalized,

$$\log g_u(\tau) = 2\pi i \tfrac{1}{2}\mathbf{B}_2(u_1)\tau + \log(1 - q_z) + \sum_{n=1}^{\infty} (\log(1 - q_\tau^n q_z) + \log(1 - q_\tau^n / q_z)),$$

where

$$z = u_1\tau + u_2 \quad \text{and} \quad q_z = e^{2\pi i z}.$$

The above log is the principal branch on $\mathbf{C}$ from which the negative real axis is deleted. The log chosen is the "natural" one from the q-expansion, except that we omit the term

$$\frac{u_2(u_1 - 1)}{2},$$

which is independent of τ, and for our purposes is an irrelevant constant. Then exp log $g_u(\tau)$ differs from $g_u(\tau)$ by a constant multiple, and we can use this determination of the log to determine $\chi_u(\sigma)$.

Put

$$\tau = -\frac{d}{c} + iy \quad \text{with } y > 0.$$

Then

$$\sigma(\tau) = \frac{a}{c} + \frac{i}{c^2 y}.$$

We calculate $\chi_u(\sigma)$ by letting $y \to 0$. Then $\tau \to -d/c$, and $\sigma(\tau) \to i\infty$. For τ in the upper half plane,

$$\log(1 - q_z) = -\sum_{m=1}^{\infty} \frac{q_z^m}{m},$$

which is valid also for $u_1 = 0$ by the Abel limit formula. Furthermore, using the same standard series for the log under conditions which insure absolute convergence, we get

$$\log g_u(\tau) = 2\pi i \tfrac{1}{2}\mathbf{B}_2(u_1)\tau - Q(z, \tau),$$

where

$$Q(z, \tau) = \sum_{m=1}^{\infty} \frac{1}{m} \frac{q_z^m + (q_\tau/q_z)^m}{1 - q_\tau^m}.$$

Set

$$z_\sigma = u_1\sigma(\tau) + u_2.$$

Then

$$\chi_u(\sigma) = 2\pi i \tfrac{1}{2}\mathbf{B}_2(u_1)\frac{a + d}{c} - Q(z_\sigma, \sigma(\tau)) + Q(z, \tau).$$

Since $\chi_u(\sigma)$ is pure imaginary, we may take the imaginary part of the right hand side to get

$$(1) \qquad \chi_u(\sigma) = 2\pi i \tfrac{1}{2}\mathbf{B}_2(u_1)\frac{a+d}{c} - \operatorname{Im} Q(z_\sigma, \sigma(\tau)) + \operatorname{Im} Q(z, \tau).$$

Since the expression is independent of τ, we may take the limit as τ approaches $-d/c$ and apply the Abel limit formula. To state the resulting limit, we need to introduce a symbol.

Let c, d be integers with $c \equiv 0 \bmod N$, $d \equiv 1 \bmod N$, and $c \neq 0$. Let $t_1, t_2 \in \mathbf{R}$, and let x be an integer such that $c \nmid x$. We use the notation

$$\mathbf{e}(t) = e^{2\pi i t}.$$

We define the **symbol**

$$[x, t_1, t_2]_{d,c} = \frac{\mathbf{e}^x(t_1(1-d)/c + t_2)}{(1-\mathbf{e}(-xd/c))(1-\mathbf{e}(x/c))}.$$

Note at once that

$$[x, t_1, t_2]_{d,c} = [-x, 1-t_1, -t_2]_{d,c}.$$

Proposition A.2. *Normalizing* $0 \leqq u_1 < 1$ *and* $0 \leqq u_2 < 1$, *we have*

$$\chi_u(\sigma) = 2\pi i\left[\tfrac{1}{2}\mathbf{B}_2(u_1)\frac{a+d-2}{c} - \frac{1}{12c}\right] - \frac{2\pi i}{c}\sum_{\substack{x \in \mathbf{Z}/c\mathbf{Z} \\ x \neq 0}} [x, u_1, u_2]_{d,c}.$$

If u_1, u_2 are not normalized as above, then we have to replace u_j by $\langle u_j \rangle$ in the right hand side.

The rest of the section is devoted to the proof.

We shall use constantly the Fourier expansion

$$\sum_{m=1}^{\infty} \frac{1}{m}(\mathbf{e}^m(t) - \mathbf{e}^{-m}(t)) = -2\pi i \mathbf{B}_1(\langle t \rangle),$$

where $\mathbf{B}_1(X) = X - \frac{1}{2}$ is the first Bernoulli polynomial.

We put

$$Q_m(z, \tau) = \frac{q_z^m + (q_\tau/q_z)^m}{1 - q_\tau^m}$$

so that

$$Q(z, \tau) = \sum_{m=1}^{\infty} \frac{1}{m} Q_m(z, \tau).$$

We first consider $\operatorname{Im} Q(z_\sigma, \sigma(\tau))$.

Lemma A.3.

(i) *If* $u_1 = 0$, *then*

$$\lim_{\tau \to -d/c} \operatorname{Im} Q(z_\sigma, \sigma(\tau)) = -\pi i(u_2 - \tfrac{1}{2}).$$

(ii) *If* $u_1 \neq 0$, *then*

$$\lim_{\tau \to -d/c} \operatorname{Im} Q(z_\sigma, \sigma(\tau)) = 0.$$

Proof. Suppose $u_1 = 0$. Then

$$\begin{aligned}
\lim_{\tau \to -d/c} \operatorname{Im} \sum_{m=1}^{\infty} \frac{1}{m} Q_m(z_\sigma, \sigma(\tau)) &= \lim_{\sigma(\tau) \to i\infty} \operatorname{Im} \sum_{m=1}^{\infty} \frac{1}{m} \frac{q^m(u_2) + q^m(\sigma(\tau) - u_2)}{1 - q^m(\sigma(\tau))} \\
&= \sum_{m=1}^{\infty} \frac{e^m(u_2) - e^{-m}(u_2)}{2m} \\
&= -\pi i \mathbf{B}_1(u_2).
\end{aligned}$$

On the other hand, if $u_1 \neq 0$ (and u_1 is normalized), then q_{z_σ} and $q_{\sigma(\tau)}/q_{z_\sigma}$ approach 0, and

$$\lim_{\tau \to -d/c} \operatorname{Im} Q(z_\sigma, \sigma(\tau)) = 0.$$

This proves the lemma.

Next we turn to $Q(z, \tau)$.

With $\tau = -d/c + iy$, we have

$$\begin{aligned}
\lim_{\tau \to -d/c} \operatorname{Im} Q(z, \tau) &= \lim_{\tau \to -d/c} \operatorname{Im} \sum_{c \nmid m} \frac{1}{m} Q_m(z, \tau) + \lim_{\tau \to -d/c} \operatorname{Im} \sum_{c | m} \frac{1}{m} Q_m(z, \tau) \\
&= L' + L'',
\end{aligned}$$

where

$$Q_m(z, \tau) = \frac{q_z^m + (q_\tau/q_z)^m}{1 - q_\tau^m}.$$

The limit L' is that taken for the sum with $c \nmid m$, and L'' is the limit taken for the sum with $c|m$. We let

$$r = e^{-2\pi y}, \qquad M = N|c|.$$

We also consider the following M-th roots of unity:

$$\zeta = \mathbf{e}(-d/c) \quad \text{and} \quad \lambda = \mathbf{e}\left(-\frac{d}{c}u_1 + u_2\right).$$

In determining the limits, we are considering series of type

$$\sum \frac{1}{m} w_m.$$

From summation by parts, to take the limit under the summation sign, it suffices to show that the partial sums of

$$\sum w_m$$

are bounded. For this, we use the Dedekind trick of combining appropriate terms, cf. [L 7], p. 145. We work out one example which is typical. Let

$$r_1 = r^{u_1} \quad \text{with } 0 < u_1, \quad \text{and } 0 < r < 1.$$

Suppose we want to show that the partial sums of the imaginary part

$$\sum\left[\frac{r_1^m\lambda^m}{1 - r^m\zeta^m} - \frac{r_1^m\lambda^{-m}}{1 - r^m\zeta^{-m}}\right]$$

are bounded, uniformly as r tends to 1. We combine the terms with $m = kM + s$ and $m = (k+1)M - s$ for $0 < s < M/2$. [The term with $M/2$ (if it exists) is equal to 0.] This gives rise to four terms. Two of them have the power ζ^s in the denominator. We combine these two, and estimate their sum in absolute value, namely in the case $c \nmid m$:

$$\left|\frac{r_1^{kM+s}\lambda^s}{1 - r^{kM+s}\zeta^s} - \frac{r_1^{(k+1)M-s}\lambda^s}{1 - r^{(k+1)M-s}\zeta^s}\right|.$$

The denominators are bounded away from 0 (for $c \nmid m$). Putting these expressions over a common denominator yields the estimate:

$$\begin{aligned}&\ll r_1^{kM}|r_1^s\lambda^s - r_1^s r^{(k+1)M-s}(\lambda\zeta)^s - r_1^{M-s}\lambda^s + r_1^{M-s}r^{kM+s}(\lambda\zeta)^s|\\ &\leqq r_1^{kM}[|1 - r_1^{M-2s}| + |1 - (r_1/r)^{M-2s}|].\end{aligned}$$

Summing the geometric series for $1 \leqq k \leqq K - 1$ immediately shows that the partial sums are bounded, as desired. The other cases which we meet are treated in the same way and will be left to the reader.

Lemma A.4.

(i) *If* $u_1 = 0$ *then* $L'' = 0$.
(ii) *If* $u_1 = \langle u_1 \rangle \neq 0$, *then* $L'' = -\left(\frac{2\pi i}{c}\right)(u_1 - \frac{1}{2})^2$.

Proof. By definition,

$$L'' = \lim_{\tau \to -d/c} \operatorname{Im} \sum_{c|m} \frac{1}{m} Q_m(z, \tau).$$

We have $q_\tau = r\zeta$. Taking imaginary parts, one sees at once that

$$L'' = \lim_{r \to 1} \sum_{c|m} \frac{r^{u_1 m} - r^{(1-u_1)m}}{1 - r^m} \frac{1}{2m} (\lambda^m - \lambda^{-m}).$$

So using summation by parts and the Dedekind trick we get

$$L'' = \sum_{c|m} (1 - 2u_1) \frac{\lambda^m - \lambda^{-m}}{2m}.$$

If $u_1 = 0$, we note that c divisible by N implies that $L'' = 0$, because λ is an N-th root of unity, and m is divisible by N. Suppose $u_1 \neq 0$. Then changing indices, we have

$$\begin{aligned} L'' &= \sum_{m=1}^{\infty} (1 - 2u_1) \frac{1}{2|c|m} (\lambda^{|c|m} - \lambda^{-|c|m}) \\ &= (1 - 2u_1) \frac{1}{2|c|} \sum_{m=1}^{\infty} [\mathbf{e}^{\varepsilon(c)m}(-u_1) - \mathbf{e}^{-\varepsilon(c)m}(-u_1)] \end{aligned}$$

where $\varepsilon(c) = 1$ if $c > 0$ and -1 if $c < 0$. The factor d has disappeared because $d \equiv 1 \bmod N$, and u_1 has denominator dividing N. Now if t is real and is not an integer, then

$$\sum_{m=1}^{\infty} \frac{1}{m} (\mathbf{e}^m(t) - \mathbf{e}^{-m}(t)) = -2\pi i \mathbf{B}_1(\langle t \rangle),$$

where $\mathbf{B}_1(X) = X - \frac{1}{2}$ is the first Bernoulli polynomial. Thus

$$\begin{aligned} L'' &= -2\pi i(1 - 2u_1)\frac{1}{2|c|}\mathbf{B}_1(\langle -\varepsilon(c)u_1\rangle) \\ &= -\frac{2\pi i}{2c}(u_1 - \tfrac{1}{2})^2. \end{aligned}$$

This proves the lemma.

Finally we turn to the evaluation of the sum L'. We can show:

$$L' = \lim_{\tau \to -d/c} \operatorname{Im} \sum_{c \nmid m} \frac{1}{m} Q_m(z, \tau) = \sum_{c \nmid m} \frac{1}{m} \varphi(m) \tag{2}$$

where

$$\varphi(m) = \frac{\lambda^m + (\zeta/\lambda)^m}{1 - \zeta^m}$$

is obtained by substituting $-d/c$ for τ in $Q_m(z(\tau), \tau)$. Let

$$M = N|c|.$$

Note that

$$\varphi(-m) = -\varphi(m) = \overline{\varphi(m)},$$

so that $\varphi(m)$ is pure imaginary, and is an odd function of m mod M. One verifies that (2) is correct by forming the partial sums of the expressions $Q_m(z, \tau)$, and checking that they are uniformly bounded as $\tau \to -d/c$. We can then apply the Abel summation formula (summation by parts, see [L 7] p. 145). In doing this, we consider each term separately

$$\frac{q_z^m}{1 - q_\tau^m} \quad \text{and} \quad \frac{(q_\tau/q_z)^m}{1 - q_\tau^m}.$$

Subtracting each from its conjugate and using the Dedekind trick yields the boundedness of the partial sums

$$\sum \operatorname{Im} Q_m(z, \tau).$$

For each residue class $x \in \mathbf{Z}/M\mathbf{Z}$, and $2x \notin M\mathbf{Z}$, let

$$f(x) = \sum_{m=1}^{\infty} \frac{a(m, x)}{m}$$

where

$$a(m, x) = \begin{cases} 0 & \text{if } m \not\equiv \pm x \bmod M \\ 1 & \text{if } m \equiv x \quad \bmod M \\ -1 & \text{if } m \equiv -x \bmod M. \end{cases}$$

Then

(3) $$L' = \frac{1}{2} \sum \varphi(x) f(x)$$

where the sum is taken over $x \in \mathbf{Z}/M\mathbf{Z}$, $x \not\equiv 0 \bmod c\mathbf{Z}$, $2x \notin M\mathbf{Z}$.

Lemma A.5. *We have*

$$f(x) = \frac{-\pi i}{M}\left[\frac{1}{1 - \mathbf{e}(x/M)} - \frac{1}{1 - \mathbf{e}(-x/M)}\right].$$

Proof. For a given x, we write down the Fourier expansion of the function $m \mapsto a(m, x)$ on $\mathbf{Z}/M\mathbf{Z}$,

$$a(m, x) = \frac{1}{M} \sum_{y \bmod M} (\mathbf{e}(-xy/M) - \mathbf{e}(xy/M))\mathbf{e}(ym/M).$$

Since $a(m, x)$ is real, the right hand side is equal to its complex conjugate. We add these, and use the Fourier expansion for the first Bernoulli polynomial to obtain

$$f(x) = -\frac{\pi i}{M} \sum_{y=1}^{M-1} (\mathbf{e}(-xy/M) - \mathbf{e}(xy/M))\left(\frac{y}{M} - \frac{1}{2}\right).$$

Let η be an M-th root of unity $\neq 1$. Then

$$\frac{1}{1-\eta} = \frac{-1}{M} \sum_{y=1}^{M-1} y\eta^{y}.$$

This is proved by multiplying the equation by $1 - \eta$ and performing the relevant cancellations on the right hand side. Then in the sum which we obtained for $f(x)$, the term with $\frac{1}{2}$ drops out, and the sum containing y as a factor in each term is then equal to the desired expression, thus proving the lemma.

Let $\omega = e^{2\pi i/N|c|} = \mathbf{e}(1/N|c|)$. We now find:

$$\begin{aligned} L' &= \frac{-\pi i}{2M} \sum_{c \nmid x} \frac{\lambda^x + (\zeta/\lambda)^x}{1 - \zeta^x} \left[\frac{1}{1-\omega^x} - \frac{1}{1-\omega^{-x}} \right], \\ &= \frac{-\pi i}{2M} \sum_{c \nmid x} \left[\frac{\lambda^x}{(1-\zeta^x)(1-\omega^x)} + \frac{(\zeta/\lambda)^x}{(1-\zeta^x)(1-\omega^x)} \right. \\ &\qquad \left. - \frac{\lambda^x}{(1-\zeta^x)(1-\omega^{-x})} - \frac{(\zeta/\lambda)^x}{(1-\zeta^x)(1-\omega^{-x})} \right]. \end{aligned}$$

We change x to $-x$ in the two terms containing ω^{-x}, and we find:

$$\text{(4)} \qquad L' = \frac{-\pi i}{M} \left[\sum_{c \nmid x} \frac{\lambda^x}{(1-\zeta^x)(1-\omega^x)} + \sum_{c \nmid x} \frac{(\zeta/\lambda)^x}{(1-\zeta^x)(1-\omega^x)} \right].$$

This is almost what we want, except that the sum is over $x \bmod M$, with $M = N|c|$, so we have to make one further simplification, as in the next lemma. We shall decompose the sums, and write

$$x = y + k|c|,$$

with

$$0 < y < |c| \quad \text{and} \quad 0 \leqq k \leqq N-1.$$

Lemma A.6. *Let $t = (t_1, t_2) \in (1/N)\mathbf{Z}^2$. Let*

$$\omega = \mathbf{e}(1/Nc), \qquad \zeta = \mathbf{e}(-d/c), \qquad \lambda = \mathbf{e}(-t_1 d/c + t_2).$$

(i) *Suppose $0 \leqq t_1 < 1$. Then*

$$\sum_{k=0}^{N-1} \frac{\lambda^{y+k|c|}}{(1-\zeta^{y+k|c|})(1-\omega^{y+k|c|})} = N[y, t_1, t_2]_{d,c}.$$

(ii) *Suppose $t_1 = 1$. Then that same sum on the left hand side is*

$$= N[y, 1, t_2]_{d,c} + N[y, t_2]_{d,c}$$

where

$$[y, t_2]_{d,c} = \frac{\mathbf{e}^y(-d/c + t_2)}{1 - \mathbf{e}^y(-d/c)}.$$

Proof. Let S be the sum which we are trying to evaluate. Then:

$$\begin{aligned} S &= \frac{\lambda^y}{1 - \zeta^y} \sum_{k=0}^{N-1} \frac{\lambda^{k|c|}}{1 - \omega^{y+k|c|}} \\ &= -\frac{1}{M} \frac{\lambda^y}{1 - \zeta^y} \sum_{r=0}^{M-1} r\omega^{ry} \sum_{k=0}^{N-1} (\lambda\omega^r)^{k|c|}. \end{aligned}$$

The sum on the right over k will be 0 unless $\lambda\omega^r = 1$. Going back to the definitions of the roots of unity in terms of t_1, t_2, d, c, and using the fact that $c \equiv 0 \bmod N$, $d \equiv 1 \bmod N$, we see that $\lambda\omega^r = 1$ if and only if

$$r - t_1 N \equiv 0 \bmod N.$$

Consequently we let $r = t_1 N + sN$ with $0 \leqq s \leqq |c| - 1$, and get

$$S = -\frac{1}{|c|} \frac{\lambda^y}{1 - \zeta^y} \sum_{\substack{r=0 \\ r \equiv t_1 N}}^{M-1} r\omega^{ry}.$$

If $t_1 \neq 1$, then

$$\begin{aligned} S &= -\frac{1}{|c|} \frac{\lambda^y}{1 - \zeta^y} \mathbf{e}^y(t_1/c) \sum_{s=0}^{|c|-1} (t_1 N + sN)\mathbf{e}^y(s/c) \\ &= N \frac{\lambda^y}{1 - \zeta^y} \mathbf{e}^y(t_1/c) \frac{1}{1 - \mathbf{e}^y(1/c)} \\ &= N\mathbf{e}^y\left(t_1 \frac{1-d}{c} + t_2\right) \\ &= N[y, t_1, t_2]_{d,c} \end{aligned}$$

as desired.

On the other hand, for (ii), if $t_1 = 1$, then in the evaluation of S we find one extra term in addition to the previous ones as indicated.

From Lemma A.6 and (4) we find:

If $u_1 \neq 0$ then

$$(5) \qquad L' = -\frac{\pi i}{c}\left[\sum_{\substack{x \in \mathbf{Z}/c\mathbf{Z} \\ x \neq 0}} [x, u_1, u_2]_{d,c} + \sum_{\substack{x \in \mathbf{Z}/c\mathbf{Z} \\ x \neq 0}} [x, 1 - u_1, -u_2]_{d,c}\right].$$

If $u_1 = 0$ then

$$(6) \qquad L' = \frac{-\pi i}{c}\left[\text{same two terms as in (5)} + \sum_{\substack{x \in \mathbf{Z}/c\mathbf{Z} \\ x \neq 0}} [x, -u_2]_{d,c}\right].$$

An easy calculation shows that

$$\sum_{\substack{x \in \mathbf{Z}/c\mathbf{Z} \\ x \neq 0}} [x, -u_2]_{d,c} = \frac{1-c}{2} + u_2 c.$$

Hence if $u_1 = 0$, we find

$$(7) \qquad L' = \frac{-\pi i}{c}[\text{same two terms as in (5)}] - \pi i\left(\mathbf{B}_1(u_2) + \frac{1}{2c}\right).$$

Combining Lemma A.3 and Lemma A.4 together with the evaluation of L' just carried out, we find:

$$\chi_u(\sigma) = 2\pi i \tfrac{1}{2}\mathbf{B}_2(u_1)\frac{a+d}{c} - \frac{2}{c}(u_1 - \tfrac{1}{2})^2$$
$$- \pi i \sum_{\substack{x \in \mathbf{Z}/c\mathbf{Z} \\ x \neq 0}} [x, u_1, u_2]_{d,c} + [x, 1 - u_1, -u_2]_{d,c}.$$

Since we had already noted that

$$[x, t_1, t_2]_{d,c} = [-x, 1 - t_1, -t_2]_{d,c},$$

we can rearrange the terms in the sum arising in $\chi_u(\sigma)$ to get:

Proposition A.2. *With u_1, u_2 normalized, we have:*

$$\chi_u(\sigma) = 2\pi i\left[\tfrac{1}{2}\mathbf{B}_2(u_1)\frac{a+d-2}{c} - \frac{1}{12c}\right] - \frac{2\pi i}{c}\sum_{\substack{x \in \mathbf{Z}/c\mathbf{Z} \\ x \neq 0}} [x, u_1, u_2]_{d,c}.$$

Bibliography

[Ba] A. Baker, Contributions to the theory of diophantine equations, *Phil. Trans. Royal Soc. London*, Series A, Math. and Physical Sciences No. 1139 Vo. **263** (1968) pp. 173–208

[Ba-Co] A. Baker J. Coates, Integer points on curves of genus 1, *Proc. Camb. Phil. Soc.* **67** (1970) pp. 595–602

[Bass] H. Bass, Generators and relations for cyclotomic units, *Nagoya Math. J.* **27** (1966) pp. 401–407

[Ber] R. Bergelson, The index of the Stickelberger ideal of order k on $C^k(\mathrm{N})$, to appear.

[B-SwD] B. J. Birch and P. Swinnerton-Dyer, Notes on elliptic curves II, *J. reine angew. Math.* **218** (1965) pp. 79–108

[Ch 1] C. Chabauty, Sur le théorème fondamental de la théorie des points entiers et pseudo-entiers des courbes algébriques, *C. R. Acad. Sci. Paris* No. 217 (1943) pp. 336–338

[Ch 2] C. Chabauty, Démonstration de quelques lemmes de rehaussement, *C. R. Acad. Sci. Paris* No. 217 (1943) pp. 413–415

[Ch-W] C. Chevalley and A. Weil, Un théoreme d'arithmétique sur les courbes algébriques, *C. R. Acad. Sci. Paris* (1930) pp. 570–572

[Cl] C. H. Clemens, Applications of the Theory of Prym Varieties, Proc. Int. Cong. Math. 1974 pp. 415–421

[Co 1] J. Coates, An effective analogue of a theorem of A. Thue, *Acta Arithmetica*, three papers: I, Vol. **15** (1969) pp. 279–305; II, Vol. **16** (1970), pp. 339–412; III, Vol. **16** (1970) pp. 425–435

[Co 2] J. Coates, On K_2 and some classical conjectures in algebraic number theory, *Ann. of Math.* **95** (1972) pp. 99–116

[Co 3] J. Coates, K-theory and Iwasawa's analogue of the Jacobian, Algebraic K-theory II, Springer *Lecture Notes* **342** (1973) pp. 502–520

[Co-Li] J. Coates and S. Lichtenbaum, On l-adic zeta functions, *Ann. of Math.* **98** (1973) pp. 498–550

[Co-Si 1] J. Coates and W. Sinnott, On p-adic L-functions over real quadratic fields, *Inv. Math.* **25** (1974) pp. 253–279

[Co-Si 2] J. COATES and W. SINNOTT, An analogue of Stickelberger's theorem for higher K-groups, *Invent. Math.* **24** (1974) pp. 149–161

[Co-Si 3] J. COATES and W. SINNOTT, Integrality properties of the values of partial zeta functions, *Proc. London. Math. Soc.* (1977) pp. 365–384

[Co-Wi 1] J. COATES and A. WILES, On the conjecture of Birch-Swinnerton-Dyer, *Invent. Math.* **39** (1977) pp. 223–251

[Co-Wi 2] J. COATES and A. WILES, On P-adic L-functions and elliptic units, *J. Austral. Math. Soc.* **26** (1978) pp. 1–25

[De-Ra] P. DELIGNE and M. RAPOPORT, Les schémas de modules de courbes elliptiques, Modular functions of one variable II, Springer *Lecture Notes* **349**, pp. 143–316

[Dem 1] V. A. DEMJANENKO, Torsion of elliptic curves, *Izv. Akad. Nauk SSSR*, Ser. Math. Tom **35** (1971) No. 2, AMS Translation pp. 289–318

[Dem 2] V. A. DEMJANENKO, On the uniform boundedness of the torsion of elliptic curves over algebraic number fields, *Izv. Akad. Nauk SSSR*, Ser. Math. Tom **36** (1972) AMS translation pp. 477–490

[Deu] M. DEURING, Die Klassenkörper der Komplexen Multiplikation, *Enzyklopädie der Math. Wiss.* Band I, 2. Téil, Heft 10 Teil II

[EGA] A. GROTHENDIECK, Éléments de géométrie algébrique, *Pub. IHES* Chapter IV, 7.8.3, 7.8.6

[Dr] V. G. DRINFELD, Two theorems on modular curves, *Functional Analysis and its applications*, Vol. **7** No. 2, translated from the Russian April–June 1973, pp. 155–156

[Fr] R. FRICKE, *Elliptische Funktionen und ihre Anwendungen*, Vol. **1**, pp. 450–451, Leipzig, Teubner Verlag 1930

[G] A. O. GELFOND, *Transcendental and algebraic numbers*, Moscow (1952); translated, Dover Press, 1960

[Gi-Ro] R. GILLARD and G. ROBERT, Groupes d'unités elliptiques, *Bull. Soc. Math. France* **107** (1979) pp. 305–317

[Gr] R. GREENBERG, On the Iwasawa invariants of totally real number fields, *Am. J. Math.* **98** (1976) pp. 263–284

[Har] M. HARRIS, Kubert-Lang units and elliptic curves without complex multiplication, *Comp. Math.* **41** (1980) pp. 127–136

[Has 1] H. HASSE, Zum Hauptidealsatz der komplexen Multiplikation, *Monat. Math. Physik* **38** (1931) pp. 315–322

[Has 2] H. HASSE, Neue Begrundung der komplexen Multiplikation, I and II, *J. Reine Angew. Math.* **157** (1927) pp. 115–139 and **165** (1931) pp. 64–88

[He 1] Y. HELLEGOUARCH, Une propriété arithmétique des points exceptionnels rationnels d'ordre pair d'une cubique de genre 1, *C. R. Acad. Sci. Paris* **260** (1965) pp. 5989–5992

[He 2] Y. HELLEGOUARCH, Applications d'une propriété arithmétique des points exceptionnels d'ordre pair d'une cubique de genre 1, *C. R. Acad. Sci. Paris* **260** (1965) pp. 6256–6258

[He 3] Y. HELLEGOUARCH, Étude des points d'ordre fini des variétés abéliennes de dimension un définies sur un anneau principal, *J. Reine angew. Math.* **244** (1970) pp. 20–36

[He 4] Y. HELLEGOUARCH, Points d'ordre fini sur les courbes elliptiques, *C. R. Acad. Sci. Paris* Ser. A-B **273** (1971) pp. 540–543

[Iw 1] K. IWASAWA, A class number formula for cyclotomic fields, *Ann. of Math.* **76** (1962) pp. 171–179

[Iw 2] K. IWASAWA, On some modules in the theory of cyclotomic fields, *J.* Math. Soc. Japan **16** (1964) pp. 42–82

[Iw 3] K. IWASAWA, On Γ-extensions of algebraic number fields, *Bull. Amer. Math. Soc.* **65** (1959) pp. 183–226

[Iw 4] K. IWASAWA, On Z_l-extensions of algebraic number fields, *Ann. of Math.* **98** (1973) pp. 246–326

[Iw 5] K. IWASAWA, Lectures on p-adic L-functions, *Annals of Math. Studies* **74**

[Ke 1] D. KERSEY, Modular units inside cyclotomic units, *Ann. Math.* **112** (1980) pp. 361–380

[Ke 2] D. KERSEY, The index of modular units in complex multiplication, to appear

[Kl 1] F. KLEIN, Über die elliptischen Normalkurven der n-ten Ordnung und die zugehörigen Modulfunktionen der n-ten Stufe, *Leipziger Abh. Bd.* **13** (1885) pp. 339

[Kl 2] F. KLEIN, Über die elliptischen Normalkurven der n-ten ordnung, *Abh. math.-phys. Klasse Sächsischen Kgl. Ges. Wiss.* Bd **13**, Nr. IV (1885). *Collected Works* Vol. **3**, Springer Verlag (1923) pp. 198ff.

[K-F] F. KLEIN and R. FRICKE, *Vorlesungen über die Theorie der elliptischen Modulfunktionen*, Vol. **2** Johnson Reprint Corporation, NY and Teubner Verlag, Stuttgart (1966), from the 1890 edition.

[Kli] S. KLIMEK, Thesis, Berkeley, 1975

[Ku 1] D. KUBERT, Universal bounds on the torsion of elliptic curves, *Proceedings London Math. Soc.*, Vol. **XXXIII** (1976) pp. 193–237

[Ku 2] D. KUBERT, Quadratic relations for generators of units in the modular function field, *Math. Ann.* **225** (1977) pp. 1–20

[Ku 3] D. KUBERT, A system of free generators for the universal even ordinary distribution on $\mathbf{Q}^k/\mathbf{Z}^k$, *Math. Ann.* **224** (1976) pp. 21–31

[Ku 4] D. KUBERT, The universal ordinary distribution, *Bull. Soc. Math. France* **107** (1979) pp. 179–202

[Ku 5] D. KUBERT, The $\mathbf{Z}/2\mathbf{Z}$-cohomology of the universal ordinary distribution, *Bull. Soc. Math. France* **107** (1979) pp. 203–224

[Ku 6] D. KUBERT, The square root of the Siegel Group, to appear.

[Ku 7] D. KUBERT, The logarithm of the Siegel functions, *Compositio Math.* **37** (1978) pp. 321–338

[Ku 8] D. KUBERT. Universal bounds on the torsion of elliptic curves, *Composito Math.* **38** (1978) pp. 121–128

[K-L 1] D. KUBERT and S. LANG, Units in the modular function field I, Diophantine applications, *Math. Ann.* **218** (1975) pp. 67–96

[K-L 2] D. KUBERT and S. LANG, idem II, A full set of units, *Math. Ann.* **218** (1975) pp. 175–189

[K-L 3] D. KUBERT and S. LANG, idem III, Distribution Relations, *Math. Ann.* **218** (1975) pp. 273–285

[K-L 4] D. KUBERT and S. LANG, idem IV, The Siegel functions are generators, *Math. Ann.* **227** (1977) pp. 223–242

[K-L 5] D. KUBERT and S. LANG, Distributions on toroidal groups, *Math. Zeit.* **148** (1976) pp. 33–51

[K-L 6] D. KUBERT and S. LANG, The p-primary component of the cuspidal divisor class group on the modular curve $X(p)$, *Math. Ann.* **234** (1978) pp. 25–44

[K-L 7] D. KUBERT and S. LANG, Stickelberger ideals, *Math. Ann.* **237** (1978) pp. 203–212

[K-L 8] D. KUBERT and S. LANG, The index of Stickelberger ideals of order 2 and cuspidal class numbers, *Math. Ann.* **237** (1978) pp. 213–232

[K-L 9] D. KUBERT and S. LANG, Iwasawa theory in the modular tower, *Math. Ann.* **237** (1978) pp. 97–104

[K-L 10] D. KUBERT and S. LANG, Independence of modular units on Tate curves, *Math. Ann.* **240** (1979) pp. 191–201

[K-L 11] D. KUBERT and S. LANG, Cartan-Bernoulli numbers as values of *L*-functions, *Math. Ann.* **240** (1979) pp. 21–26

[K-L 12] D. KUBERT and S. LANG, Modular units inside cyclotomic units, *Bull. Soc. Math. France* **107** (1979) pp. 161–178

[K-L 13] D. KUBERT and S. LANG, Units in the modular function field, Modular Functions of one Variable *V*, Springer *Lecture Notes* **601** (Bonn Conference) 1976 p. 247

[L 1] S. LANG, Integral points on curves, *Pub. IHES* (1960)

[L 2] S. LANG, Division points on curves, *Ann. mat. pura ed applicata IV*, Tomo **LXX** (1965) pp. 229–234

[L 3] S. LANG, Isogenous generic elliptic curves, *Am. J. Math.* **94** No. 3 (1972) pp. 861–874

[L 4] S. LANG, *Algebraic Number Theory*, Addison Wesley, 1970

[L 5] S. LANG, *Elliptic Functions*, Addison Wesley, 1974

[L 6] S. LANG, *Diophantine Geometry*, Interscience (New York) 1962

[L 7] S. LANG, *Introduction to Modular Forms*, Springer Verlag, 1977

[L 8] S. LANG, *Cyclotomic Fields*, Springer Verlag, 1978

[L 9] S. LANG, *Elliptic Curves: Diophantine Analysis*, Springer Verlag, 1978

[Le 1] H. LEOPOLDT, Zur Arithmetik in abelschen Zahlkörpern, *J. reine angew. Math.* **209** (1962) pp. 54–71

[Le 2] H. LEOPOLDT, Eine Verallgemeinung der Bernoullischen Zahlen, *Abh. Math. Sem. Hamburg* (1958) pp. 131–140

[Leu] A. LEUTBECHER, Über Automorphiefaktoren und die Dedekindschen Summen. *Glasgow Math. J.* **11** (1970) pp. 41–57

[Man 1] J. MANIN, Parabolic points and zeta functions of modular curves, *Izv. Akad. Nauk SSSR*, Ser. Mat. Tom **36** (1972) No. 1, AMS translation pp. 19–64

[Man 2] J. MANIN, The *p*-torsion of elliptic curves is uniformly bounded, *Izv. Akad. Nauk SSSR*, Ser. Mat. **33** (1969) No. 3

[Maz] B. MAZUR, Modular curves and the Eisenstein ideal, *Pub. IHES*, Paris (1978)

[M-SwD] B. MAZUR and P. SWINNERTON-DYER, Arithmetic of Weil curves, *Invent. Math.* **25** (1974) pp. 1–61

[Ma-W] B. MAZUR and A. WILES, to appear.

[Me] C. MEYER, *Die Berechnung der Klassenzahl abelscher Zahlkorper uber quadratischen Zahlkorper*, Akademie Verlag, Berlin, 1957

[Mu] D. MUMFORD, A remark on Mordell's conjecture, *Amer. J. Math.* **87** (1965) pp. 1007–1016

[Ne] M. NEWMAN, Construction and application of a class of modular functions, *Proc. London Math. Soc.* (1957) pp. 334–350

[No 1] A. P. NOVIKOV, Sur le nombre de classes des extensions abéliennes d'un corps quadratique imaginaire, *Izv. Akad. Naud SSSR* 3 (1967) pp. 717–726

[No 2] A. P. NOVIKOV, Sur la régularité des idéaux premiers de degré un d'un corps quadratique imaginaire, *Izv. Akad. Nauk SSSR*, Ser. Mat. **33** (1969) pp. 1059–1079 (= *Math. USSR Isv.* **3** (1969) pp. 1001–1018)

[Ogg] A. OGG, Rational points on certain elliptic modular curves, AMS conference, St. Louis, 1972, pp. 211–231

[Rad] H. RADEMACHER, Zur Theorie der Modulfunktionen, *J. Reine angew. Math.* **167** (1932) pp. 312–336

[Ra] K. RAMACHANDRA, Some applications of Kronécker's limit formula, *Ann. of Math.* **80** (1964) pp. 104–148

[Re] S. RECILLAS, A relation between curves of genus three and curves of genus 4, PhD dissertation, Brandeis University 1970 (see especially, pp. 107–108, and Theorem 2 of Chapter IV et sequ.).

[Ri] K. RIBET, A modular construction of unramified extensions of $\mathbf{Q}(\mu_p)$, *Invent. Math.* **34** (1976) pp. 151–162

[Ro 1] G. ROBERT, Unités elliptiques, *Bull. Soc. Math. France Supplément*, Decembre 1973 No. 36

[Ro 2] G. ROBERT, Nombres de Hurwitz et unités elliptiques, *Ann. scient. Ec. Norm. Sup. 4e série* t. **11** (1978) pp. 297–389

[Ro 3] G. ROBERT, Caractères Exceptionnels, *J. Number Theory* **11** (1979) pp. 161–170

[Roh] D. ROHRLICH, Points at infinity on the Fermat curve, *Invent. Math.* **39** (1977) pp. 95–127

[Roth] P. ROTH, Über Beziehungen zwischen algebraischen Gebilden von Geschlechtern drei und vier, *Monatschefte* **22** (1911)

[Sch] B. SCHOENEBERG, *Elliptic Modular Functions*, Springer Verlag, 1974

[Se] J. P. SERRE, Propriétés Galoisiennes des points d'ordre fini des courbes elliptiques, *Invent. Math.* **15** (1972) pp. 259–331

[Sh] G. SHIMURA, *Introduction to the Arithmetic Theory of Automorphic Functions*, Iwanami Shoten and Princeton University Press, 1971

[Sie 1] C. L. SIEGEL, Über einige Anwendungen diophantischer Approximationen, *Abh. Preuss. Akad. Wiss. Phys. Math. Kl.* (1929) pp. 41–69

[Sie 2] C. L. SIEGEL, Lectures on advanced analytic number theory, *Tate Institute Lecture Notes*, 1961 259–331

[Sin] W. SINNOTT, On the Stickelberger ideal and the circular units of a cyclotomic field, *Ann. of Math.* **108** (1978) pp. 107–134

[St 1] H. STARK, Class fields and modular forms of weight one, Modular Functions in one variable V, Springer *Lecture Notes* **601** (1977)

[St 2] H. STARK, Totally Real Fields and Hilbert's twelfth Problem, *Advances in Math.* **22** No. 1 (1976) pp. 64–84

[We] A. WEIL, Arithmétique et géométrie sur les variétés algébriques, *Act. Sci. et Ind.* No. 206 Paris, Hermann, 1935

[Wi] A. WILES, On modular curves and the class group of $\mathbf{Q}(\zeta_p)$, *Invent. Math.* **58** (1980) pp. 1–35

[Wo] K. WOHLFART, Über Dedekindsche Summen und Untergruppen der Modulgruppe, *Hamburg Abh.* **23** (1959) pp. 5–10

[Ya 1] K. YAMAMOTO, The gap group of multiplicative relationships of Gaussian sums, *Symposia Mathematica* No. 15 (1975) pp. 427–440

[Ya 2] K. YAMAMOTO, On a conjecture of Hasse concerning multiplicative relations of Gaussian sums, *J. Combin. Theory* **1** (1966) pp. 476–489

[Yu] J. YU, A cuspidal class number formula for the modular curves $X_1(N)$, *Math. Ann.* **252** (1980) pp. 197–216.

Index

Grundlehren der mathematischen Wissenschaften

A Series of Comprehensive Studies in Mathematics

A Selection

153. Federer: Geometric Measure Theory
154. Singer: Bases in Banach Spaces I
155. Müller: Foundations of the Mathematical Theory of Electromagnetic Waves
156. van der Waerden: Mathematical Statistics
157. Prohorov/Rozanov: Probability Theory. Basic Concepts. Limit Theorems. Random Processes
158. Constantinescu/Cornea: Potential Theory on Harmonic Spaces
159. Köthe: Topological Vector Spaces I
160. Agrest/Maksimov: Theory of Incomplete Cylindrical Functions and their Applications
161. Bhatia/Szegö: Stability of Dynamical Systems
162. Nevanlinna: Analytic Functions
163. Stoer/Witzgall: Convexity and Optimization in Finite Dimensions I
164. Sario/Nakai: Classification Theory of Riemann Surfaces
165. Mitrinovic/Vasic: Analytic Inequalities
166. Grothendieck/Dieudonné: Eléments de Géometrie Algébrique I
167. Chandrasekharan: Arithmetical Functions
168. Palamodov: Linear Differential Operators with Constant Coefficients
169. Rademacher: Topics in Analytic Number Theory
170. Lions: Optimal Control of Systems Governed by Partial Differential Equations
171. Singer: Best Approximation in Normed Linear Spaces by Elements of Linear Subspaces
172. Bühlmann: Mathematical Methods in Risk Theory
173. Maeda/Maeda: Theory of Symmetric Lattices
174. Stiefel/Scheifele: Linear and Regular Celestial Mechanic. Perturbed Two-body Motion–Numerical Methods–Canonical Theory
175. Larsen: An Introduction to the Theory of Multipliers
176. Grauert/Remmert: Analytische Stellenalgebren
177. Flügge: Practical Quantum Mechanics I
178. Flügge: Practical Quantum Mechanics II
179. Giraud: Cohomologie non abélienne
180. Landkof: Foundations of Modern Potential Theory
181. Lions/Magenes: Non-Homogeneous Boundary Value Problems and Applications I
182. Lions/Magenes: Non-Homogeneous Boundary Value Problems and Applications II
183. Lions/Magenes: Non-Homogeneous Boundary Value Problems and Applications III
184. Rosenblatt: Markov Processes. Structure and Asymptotic Behavior
185. Rubinowicz: Sommerfeldsche Polynommethode
186. Handbook for Automatic Computation. Vol. 2. Wilkinson/Reinsch: Linear Algebra
187. Siegel/Moser: Lectures on Celestial Mechanics
188. Warner: Harmonic Analysis on Semi-Simple Lie Groups I
189. Warner: Harmonic Analysis on Semi-Simple Lie Groups II
190. Faith: Algebra: Rings, Modules, and Categories I
191. Faith: Algebra II, Ring Theory
192. Mallcev: Algebraic Systems
193. Pólya/Szegö: Problems and Theorems in Analysis I
194. Igusa: Theta Functions
195. Berberian: Baer*-Rings
196. Athreya/Ney: Branching Processes
197. Benz: Vorlesungen über Geometric der Algebren

198. Gaal: Linear Analysis and Representation Theory
199. Nitsche: Vorlesungen über Minimalflächen
200. Dold: Lectures on Algebraic Topology
201. Beck: Continuous Flows in the Plane
202. Schmetterer: Introduction to Mathematical Statistics
203. Schoeneberg: Elliptic Modular Functions
204. Popov: Hyperstability of Control Systems
205. Nikollskii: Approximation of Functions of Several Variables and Imbedding Theorems
206. André: Homologie des Algébres Commutatives
207. Donoghue: Monotone Matrix Functions and Analytic Continuation
208. Lacey: The Isometric Theory of Classical Banach Spaces
209. Ringel: Map Color Theorem
210. Gihman/Skorohod: The Theory of Stochastic Processes I
211. Comfort/Negrepontis: The Theory of Ultrafilters
212. Switzer: Algebraic Topology—Homotopy and Homology
213. Shafarevich: Basic Algebraic Geometry
214. van der Waerden: Group Theory and Quantum Mechanics
215. Schaefer: Banach Lattices and Positive Operators
216. Pólya/Szegö: Problems and Theorems in Analysis II
217. Stenström: Rings of Quotients
218. Gihman/Skorohod: The Theory of Stochastic Processes II
219. Duvant/Lions: Inequalities in Mechanics and Physics
220. Kirillov: Elements of the Theory of Representations
221. Mumford: Algebraic Geometry I: Complex Projective Varieties
222. Lang: Introduction to Modular Forms
223. Bergh/Löfström: Interpolation Spaces. An Introduction
224. Gilbarg/Trudinger: Elliptic Partial Differential Equations of Second Order
225. Schütte: Proof Theory
226. Karoubi: K-Theory. An Introduction
227. Grauert/Remmert: Theorie der Steinschen Räume
228. Segal/Kunze: Integrals and Operators
229. Hasse: Number Theory
230. Klingenberg: Lectures on Closed Geodesics
231. Lang: Elliptic Curves: Diophantine Analysis
232. Gihman/Skorohod: The Theory of Stochastic Processes III
233. Stroock/Varadhan: Multi-dimensional Diffusion Processes
234. Aigner: Combinatorial Theory
235. Dynkin/Yushkevich: Markov Control Processes and Their Applications
236. Grauert/Remmert: Theory of Stein Spaces
237. Köthe: Topological Vector-Spaces II
238. Graham/McGehee: Essays in Commutative Harmonic Analysis
239. Elliott: Probabilistic Number Theory I
240. Elliott: Probabilistic Number Theory II
241. Rudin: Function Theory in the Unit Ball of $\mathbb{C}^n$
242. Blackburn/Huppert: Finite Groups I
243. Blackburn/Huppert: Finite Groups II